Theory of Physical Chemistry Processes at a Gas-Solid Interface

Advances
in
Science
and
Technology
in
the USSR

Chemistry Series

ADVANCES IN
SCIENCE AND TECHNOLOGY
IN THE USSR

Theory of Physical Chemistry Processes at a Gas-Solid Interface

YU. K. TOVBIN

Translated from the Russian by G. Leib

Mir Publishers
Moscow

CRC Press
Boca Raton   Ann Arbor   Boston   London

Library of Congress Cataloging-in-Publication Data

Tovbin, IU. K. (IUriĭ Konstantinovich)
[Teoriia fiziko-khimicheskikh protsessov na granitse gaz-tverdoe telo. English]
Theory of physical chemistry processes at a gas-solid interface / Yu.K. Tovbin ; translated from the Russian by G. Leib.
p cm. — (Advances in science and technology in the USSR)
Translation of: Teoriia fiziko-khimicheskikh protsessov na granitse gaz-tverdoe telo.
Includes bibliographical references and index.
ISBN 0-8493-7133-3
1. Chemical reaciton, rate of. 2. Surface chemistry. I. Title.
II. Series.
QD502.T6813 1991
541.3′3—dc20

91-30674
CIP

International Standard Book Number 0-8493-7133-3

Library of Congress Card Number 91-30674
Printed in the United States of America 1 2 3 4 5 6 7 8 9 0

Preface

The surface processes at a gas-solid interface can be either an independent object of investigation or a constituent part in the investigations of many practical applications. They include the adsorption of gas-phase molecules, migration of adsorbed particles, reactions between adsorbed particles and between adsorbed particles and atoms of the solid, the dissolution of adsorbed particles, surface segregation of components of the solid, processes causing motion of the phase interface, and so on. The enumerated processes have a general nature and are realized on the surfaces of all solids such as metals, alloys, and variable composition compounds. The occurrence of a surface process may change the state not only of the surface and subsurface region, but also of the entire bulk of the solid.

The task of the theory of physicochemical processes is to give a quantitative description of the state of a system for known external conditions and a characteristic time scale. A gas-solid system is a condensed one, hence when constructing a theory of physicochemical processes the methods of the theory of a condensed state must be used. Condensed systems are featured by the determining role of the interactions between particles. The calculations of these interactions are based on the quantum theory, but its practical application encounters many difficulties. In practice, more approximate models of particle interactions are employed, namely, model or empirical potentials whose parameters are determined by fitting the predicted characteristics of a system to the experimental ones.

All the above processes occurring at a gas-solid interface can be described from a single standpoint corresponding to the atomic-molecular level of the statistical theory (or simply to the molecular level, although atoms, ions, radicals, or molecules can be the particles in question) using model interaction potentials. The present book is devoted to such a single approach for constructing mathematical models of the physicochemical processes in a gas-solid system with a view to its chief real properties—the interaction between all the components and the inhomogeneity of the system. Two problems had to be solved to take these real factors into account: it was necessary to construct a theory of chemical kinetics considering the interaction between all the components of a reaction system and a theory describing the distribution of the interacting particles by the sites of inhomogeneous lattices. These problems were solved by employing the cluster approach. It is set out from simple cases to more involved ones so that the reader will be able to master the method and use it independently for constructing equations describing processes occurring on the surface and in the subsurface region of a solid. The usual knowledge acquired at higher schools is sufficient for mastering the cluster approach. The intermediate transformations are given in detail to curtail technical difficulties.

The book is intended for a broad circle of specialists in physical chemistry dealing with surface phenomena, adsorption, heterogeneous catalysis, gas chromatography, and the growth of crystals from the gas phase. It is also intended for post-graduates and senior students specializing in physical chemistry and related subjects. Experimenters will obtain notions on the physical fundamentals of molecular-kinetic models of surface processes, on the field of application of the traditionally employed models, and on the possibilities placed at their disposal by modern statistical physics, while specialists in mathematical modelling can use the material for constructing mathematical models for a broad range of processes being studied.

The first chapter introduces the basic notions of a gas-solid interface, the fundamental concepts of lattice systems and multiparticle distribution functions that are used later on. Chapters 2-7 can be divided conditionally into two parts. In Chapters 2-5, the cluster approach is employed for the simplest interaction potential—the interaction of the nearest neighbors. This simplifies acquaintance with the new method and enables one to reveal the general laws of models of nonideal systems. These laws are usually hidden by the specific nature of the particle interactions in the bulk of a solid, on its surface, and between the adsorbed particles. In Chapters 6 and 7, the cluster approach is used for more complicated potentials—the two-body potential with an arbitrary finite radius of interaction and multiparticle potentials.

The paramount task of the molecular-kinetic theory of processes in condensed phases consists in describing the distribution of particles, a knowledge of which allows one to calculate any thermodynamic characteristics of a system and the flows of particles. First the equilibrium distributions of particles on homogeneous (Chap. 2) and inhomogeneous (Chap. 3) lattices are considered, and then the kinetics of surface processes with fast (Chap. 4) and slow (Chap. 5) mobility of the particles. The distribution of particles with more involved interaction potentials is set out in the same sequence. The discussed models cover the entire time interval from the instant of contact of the gas phase with the solid, when the state of the latter does not have time to change under the influence of the gas molecules (a nonreorganizing surface), to the instant when this process terminates (a reorganized surface or bulk of a solid, or both).

The book deals with the adsorption of molecules of the gas phase, the diffusion of adsorbed particles over the surface and their dissolution in the bulk of the solid, catalytic processes, the motion of the phase interface when a crystal grows from the gas phase, the surface segregation of the components of alloys in vacuum and when they contact the gas phase. A detailed treatment of each of these topics, as well as topics such as the statistical theories of the condensed state, the theories of elemetary processes in a condensed phase, and their applications is impossible within the small volume of the present book.

Contents

1 Introduction

1.1. Gas-Solid Interface

Let us consider a volume of space containing a part of the bulk phase of a solid, its surface, and part of the volume of a gas phase. This volume can be divided into three regions (Fig. 1.1). The singled out region of the solid has the same composition and distribution of its components as the remaining part of the solid. Similarly, the singled out region of the gas phase contains components in the same state as in the remaining part of this phase. The second region is one in which the composition and distribution of the particles differs from the other two phases. The molecular interpretation of a phase interface differs from the singling out of the dividing surface between two phases adopted in thermodynamics [1, 2]. It has a finite width which in the general case can change in time.

Let us discuss the main concepts and initial propositions used in the following.

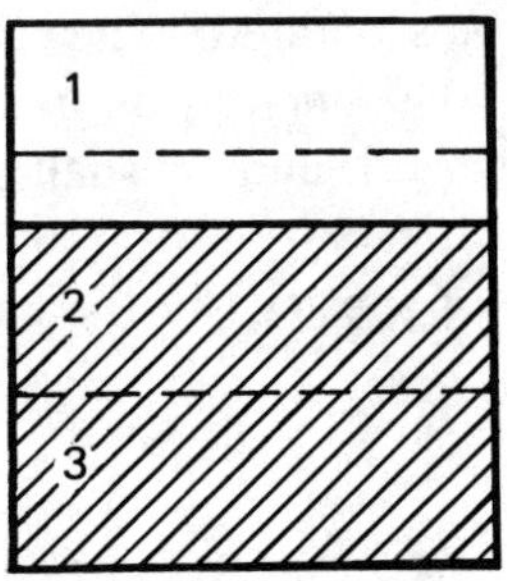

Fig. 1.1. Gas-solid interface—region *2* (between the dashed lines):
1—region of gas phase; *3*—region of solid

Gas Phase. We shall limit ourselves to very simple notions of a gas phase as an ideal mixture of molecules [3, 4]. The total pressure P in a gas phase is determined as $P = \sum_i P_i$, where P_i is the partial pressure of the i-th component of the gas mixture; $1 \leqslant i \leqslant s$, s is the number of components in the mixture. The partial pressure of a component is related to its concentration c_i by the expression

$$P_i = \beta^{-1} c_i \tag{1.1}$$

where

$$\beta = (N_A k_B T)^{-1} = (RT)^{-1} \tag{1.1a}$$

$$c_i = \frac{N_i}{V} \tag{1.1b}$$

Here k_B is the Boltzmann constant, N_A is the Avogadro constant, R is the molar gas constant, T is the absolute temperature, N_i is the number of molecules of species i, and V is the volume of the system.

The molecules of a gas phase are considered to be particles that do not interact with one another, and each of them is a statistically independent system. The statistical sum of a molecule of species i is

$$F_i^0(T) = \sum_n g_{in} \exp(-\beta E_{in}) \tag{1.2}$$

where E_{in} is the energy of a molecule of species i in the state n, and g_{in} is the degree of degeneracy of the given state. The energy of a molecule is usually represented as the sum of the relevant contributions of translational, rotational, and vibrational motions (if they can be separated). The calculation of the rotational and vibrational multipliers of the function of states of gas phase molecules is a quantum-statistical problem [5], and we shall assume them to be known.

The chemical potential of a molecule of species i in the gas phase can be written as follows:

$$\mu_{ig} = \mu_i^0 + \beta^{-1} \ln \frac{\beta P_i}{F_i^0} \tag{1.3}$$

where μ_i^0 determines the reference point. For a free particle in a gas phase, $\mu_i^0 = 0$, while if the adsorbed state is the reference one, we have $\mu_i^0 = \varepsilon_i$, where ε_i is the binding energy.

The influence of the nonideal nature of the gas phase is not considered in this book, but we shall note that the lattice model of the condensed state which is used can also be employed for these purpose [6-8].

Solid. We shall consider that the bulk of a solid is a homogeneous substitutional, interstitial, or mixed-type solution [9, 10]. We shall treat only the simplest Bravais lattices [11] in which depending on the temperature and concentration of the components their ordered arrangement is possible [12-14].

The nature of the distribution of the atoms in a solid solution depends on the interaction of its components. It is the simplest to consider the spectrum of possible distributions of the atoms for a binary alloy. We shall use the concepts of short-range and long-range order for their description. Short-range order characterizes the trend of local distribution of neighboring atoms of an alloy. The atoms of an alloy are said to tend to become ordered if the neighboring location of atoms of different species is more preferable than their chaotic arrangement, and they are said to tend to stratify if the neighboring location of atoms of one species is preferable.

In an equilibrium state, lowering of the temperature in the first case is attended by an ordered arrangement of the alloy atoms, and regular sublattices mainly filled with atoms of one species can be separated from the crystal lattice. The degree of deviation from this arrangement of the atoms characterizes long-range order. In the second case, lowering of the temperature is attended by the separation of a new phase, and the initially homogeneous phase becomes macroinhomogeneous.

In more complicated cases, different combinations of phase transitions of stratification and ordering are possible. For example, a system may pass over to a disordered and ordered phases or to two different ordered phases. The type of the phase transitions and the regions of their realization are determined by the concentrations of the components, the temperature, and the potentials of particle interaction. Similar transitions also occur in multicomponent solutions. An increase in the number of components increases the number of different combinations of the phase transitions [15].

Subsurface Region of a Solid. The nature of particle distribution near a surface generally differs from that in the bulk [16-20]. This difference is associated with the absence of a solid phase in the other half-plane. This disturbs the periodicity of the crystalline structure along a normal to the surface and results in a difference in the forces acting on an atom from the side of the other atoms of the alloy surrounding it on the surface and in the bulk (an atom in the bulk has more neighbors than one on the surface). As a result, a different distribution of the atoms ensuring a minimum of the free energy sets in in equilibrium.

The perturbation produced by a surface affects a subsurface region of a solid. Its width depends on the temperature, concentrations of the solid solution components, and on the potentials of particle interaction.

A very important characteristic of the subsurface region is the surface concentration of the atoms. For alloys, it is customary to speak of the surface segregation of the component whose surface concentration exceeds the bulk one. With an increase in the distance from the surface, the local concentrations of the particles tend to their bulk values. This also relates to the other characteristics of particle distribution determining short- and long-range orders, which in the subsurface region can have a great anisotropy.

How the atoms of a solid are distributed in the subsurface region has a determining influence on all the surface processes: the surface composition determines the properties of the surface on which the gas-phase molecules are adsorbed and the surface reactions proceed, while how the atoms are distributed in the subsurface region governs the dissolution of the adsorbed particles and the rearrangement of the region itself.

A much more complicated situation is encountered for a "frozen" highly nonequilibrium state of a surface. Under these conditions, the ways of forming the surface are exceedingly important, namely, by splitting off part of a crystal in vacuum, sputtering onto a support, precipitation from a solution, sintering of a polycrystalline powder, etc. [21, 22], because the procedure employed to prepare a surface and the treatment of a specimen prior to running an experiment affect the state of the surface and the subsurface region first of all.

Adsorption. The surface atoms of a solid have a definite ability of forming bonds with molecules of the gas phase, which increases the concentration of these molecules near the surface [23-29]. The nature of the adsorption bond depends on the nature of the solid and the gas-phase molecules. It is customary practice to distinguish physical and chemical adsorption. In the former case, no collectivization of the electrons of the adsorbent and adsorbate occurs, while in the latter case a conventional chemical bond forms with redistribution of the electrons.

At the atomic level, the surface of a monatomic crystal is not homogeneous. It is a distributed system of atoms. The different positions of an adsorbed molecule over the adsorbent atoms are not equivalent energetically. The change in the binding energy when a molecule travels along a surface is shown schematically in Fig. 1.2. In a number of positions, its energy is minimal. Such positions and their environment form regions of localization of adsorbed molecules called adsorption centers of the surface. If the difference $E_1 - E_2$ is large and is much larger than the thermal energy of the adsorbed particles,

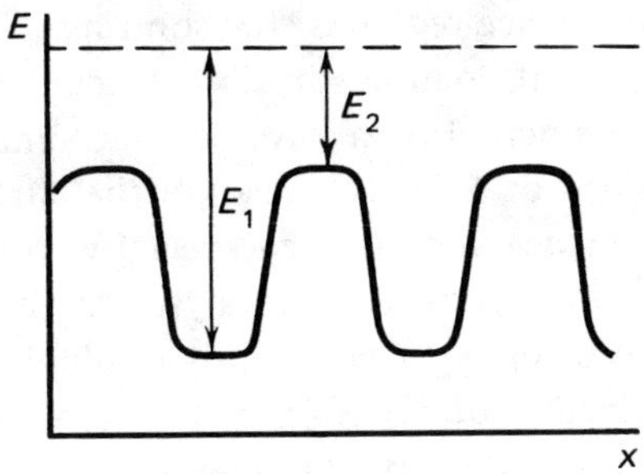

Fig. 1.2. Potential relief of a homogeneous surface.
The dashed line depicts the energy of a free particle in a gas;
E = activation energy of desorption;
$E_1 - E_2$ = activation energy of surface migration

the latter are predominately in the localization regions, and we have to do with localized adsorption. It generally occurs in chemical adsorption.

If the values of the energies E_1 and E_2 are close and have the order of magnitude of the thermal energy of the adsorbed particles, we have to do with nonlocalized adsorption. It occurs most frequently in physical adsorption. Here too lowering of the temperature may be attended by a transition from nonlocalized to localized adsorption (with disregarding of the tunneling effects possible for light particles [30, 31]). The nature of adsorption also depends on the state of the adsorbed particle. With a slow exchange of energy between a particle and the atoms of a solid, the particle may be in excited states [26, 32], which increases the degree of its delocalization.

Most solids have regions of localization of adsorbed particles with different binding energies on their surface. This is due to the different local chemical composition of the surface and a different geometric arrangement of the solid's atoms. Both these factors usually manifest themselves simultaneously. Such surfaces are said to be inhomogeneous. The scale of the inhomogeneities varies within broad limits from point imperfections and impurities to various macroregions [33, 34].

Effect of Adsorbed Particles on the State of a Surface. If the binding energy of an adsorbed particle to a surface is commensurable with the binding energy between the atoms of the solid, adsorption of molecules may cause substantial changes in the state of the surface. This is largely associated with the adsorbed particles compensating the absence of neighbors at the surface atoms. As a result, the state of the entire subsurface region of the solid may change. This, in turn, will change the number of different adsorption centers on the surface.

As a whole, considerable experimental data have been accumulated in the literature concerning the influence of the gas phase on the surface state of adsorbents and catalysts. The surface state is used specially to preliminarily produce the required surface composition of a catalyst. At the same time, quantitative characteristics of these changes are mainly available only for binary alloys contacting with molecules of H_2, CO, and O_2 [35-39]. Investigations conducted in recent years reveal that the influence of adsorbed particles on the state of a solid surface is apparently more significant than is customarily considered at present. The structural transformations in the surface layers (an example is the rearrangement of the surface layer of platinum in the adsorption of CO and O_2 molecules [40, 41]) and the processes of formation of new phases in them, which are similar to three-dimensional topochemical processes [42-44], may be of a major significance.

Absorption. Many solids absorb molecules of the gas phase (H_2, O_2, N_2, etc.) in considerable amounts [45, 46]. The dissolved atoms occupy the interstices of the crystal lattices and form interstitial alloys [47, 48]. They also greatly affect the state of the bulk phase of the solid, changing the short- and long-range orders in alloys. With great saturations of the bulk by dissolved atoms, rearrangement of the bulk phase and transitions to multiphase systems are possible. Similar changes are possible in the subsurface region of a solid that alter the state of the surface and affect the surface processes.

Particle Mobility. A change in the distributions of the condensed phase particles (the bulk of a solid, its subsurface region, and the adsorbed particles) in the absence of chemical transformations occurs as a result of their migration. The most widespread mechanisms of particle migration in solids are the vacancy mechanism and the mechanism of migration along the interstices [49-52]. Similar mechanisms are most probable for the migration of particles in the subsurface region and for adsorbed particles. But on a surface, jumps of particles over various lengths are possible [53, 54]. Their fraction depends greatly on the nature of interaction of a particle with the surface atoms and on the temperature. With an increasing degree of delocalization of adsorption, the fraction of the long jumps grows. On the other hand, an increase in the coverage of the surface reduces the average length of a jump of an adsorbed particle, hence in the following we shall limit ourselves to a treatment of migration to neighboring free centers.

The components of a phase interface have a wide range of mobilities. The mobility of adsorbed particles exceeds that of the atoms in the bulk of a solid.

It is usually considered that the mobility of the adsorbed particles is greater than that of the surface atoms of the solid, and when the state of the solid's atoms changes, the adsorbed particles at each instant are distributed practically in equilibrium. The same relation between the mobilities is presumed for absorbed particles and the atoms of a solid in the bulk. This is apparently not always true. In the general case, each component has its own ratio of the migration activation energies, and cases are possible when the mobilities of the surface atoms of a solid and of the adsorbed particles are commensurable. The mobilities of the atoms of a solid and of absorbed particles may also be commensurable.

When chemical transformations occur, the state of a system changes at the expense of the migration of particles and their participation in reactions. The relations between the reaction rates and the mobilities of the particles may vary within very broad limits. Cases are possible when chemical transformation is the limiting step (the traditional region of chemical kinetics [55, 56]), and when the migration of particles is the limiting factor (diffusion-controlled reactions [57, 58]). As a whole, this leads to various conditions under which processes proceed on a surface and in the subsurface region and determines the diversity of models used to describe them.

1.2. Tasks of the Theory of Physicochemical Processes

The existence of a solid itself, the phenomena of adsorption and absorption of gases are due to the interactions between different components of a system. The nature of the interaction between the particles of a gas-solid system is quite diverse. It depends on the nature of the solid's atoms and the gas-phase molecules. The theory of particle interactions is studied by quantum chemistry [59, 60]. To date, we can consider that the prospective trends in the development of this theory for metals and semiconductors [61-63] and alloys [64] have been formulated. They enable one to describe the thermodynamic characteristics of solids, particularly of phase equilibria, the conditions of stability of systems, and the nature of phase transitions [65-67]. Lately, methods of calculating the interactions of adsorbed particles with a surface and between adsorbed particles have been developing intensively [68-71]. But the practical use of quantum-chemical methods for describing physicochemical processes is hampered by mathematical difficulties. This makes one employ rougher models of particle interaction—model or empirical potentials. Their choice depends on the problems being considered, therefore we shall discuss the central problems that must be solved first of all.

We shall describe a gas-solid interface with the aid of a lattice model. The volume of a crystalline solid is characterized by its natural lattice. With a good approximation, we can consider that the subsurface region of a solid is characterized by the same crystal lattice. When the structure of the surface layers changes in the subsurface region (as occurs in the system CO/Pt(100) [72]) or the lattice parameters of the surface layers change in the subsurface region [73, 74], the latter can be divided into a number of subregions with different lattice structures.

Consider that the adsorption centers form a surface structure. This assumption is justified even for nonlocalized adsorption because at low surface coverages the results for a model of localized and nonlocalized adsorption virtually coincide [75]. When the surface coverage grows, the adsorbed particles behave like a two-dimensional liquid which a lattice model can be used for [76-78] (see also [6-8]).

Hence, the lattice model is a general one for the condensed phase of a gas-solid system. That the surface and subsurface region structures may differ from the structure of the bulk of the solid is of no significance. This situation fits naturally into the concept of inhomogeneous lattices that is needed at least to take account of the well known fact of the inhomogeneity of surfaces. In the following, we shall deal only with lattice models.

The present state of the theories of atomic and molecular processes in condensed phases is characterized by great nonuniformity of its development. The achievements made in describing individual equilibrium and kinetic characteristics of surface processes do not generally enable one to describe the processes as a whole. The equilibrium theory of solid solutions has been developed the most [12-15, 48, 79, 80]. The equilibrium theory of phase interfaces has been developed to a smaller extent [81-83]. Matters are much worse in the theory of the kinetics of processes at a molecular level—the first steps are being made in using models that take into consideration particle interactions (the transfer of a substance in the bulk [84] and on the surface of solids, and also the kinetics of the simplest heterogeneous reactions on homogeneous faces of crystals—for greater detail on this matter see Appendix 2). The kinetics of surface processes mainly employs models taking no account of the interaction of the adsorbed particles (the law of mass or surface action) [85-90]. This does not reflect the real properties of a gas-solid interface.

There is also a diversity of models when considering the interaction of the particles because various approximations are used (equilibrium is described with a view to the correlation effects, while kinetics ignores them [12, 48, 91]). The problem of approximations is of a fundamental significance in the theory of condensed systems. Interaction between the particles causes all the particles to be bound to one another—a change in the state of a particle affects its neighbors, the latter, in turn, change their influence on their neighbors, and so on. This substantially hinders a description of the state of a system and the solution of the problem on the distribution of the particles.

Exact solutions of the problems for many interacting particles are possible in a few cases [92-94]. They include one-dimensional systems and a number of individual cases for two-dimensional systems. In practice, we have to do with approximate methods that make it possible to describe systems within broad regions of a change in the concentrations of particles and the temperature. An important characteristic of an approximation in the appraisal of the extent to which the effects of correlation of the mutual distribution of the particles are taken into account. More accurate approximations take the correlation effects into account more accurately, but the growth in the accuracy is attended by a rapid increase in the intricacy of the model (which makes one find a reasonable compromise).

The parameters of the models, which include the equilibrium constants, the rate constants of elementary processes, and the interaction potentials, differ when the same experimental data are described by using different approximations. This is why the description of complicated processes requires that all the subsystems be considered with one level of accuracy, i.e. with the employment of the same approximation. For this purpose, it is essential that the physical and chemical fundamentals introduced into a mathematical model correspond to the real conditions within a broad range of variations of the external conditions. Now the parameters of models found from an analysis of individual subsystems can also be used when describing a process as a whole.

The task of theory consists in constructing mathematical models that would satisfy different limiting states of a system and the joint influence of real factors with an identical accuracy of their consideration. This task in a certain sense is opposite to the traditional situation when a single factor is taken as the basis, and the contributions of all other factors are ignored. To take into account the real properties of surface systems, the competition of the contributions of various factors must be considered.

This necessity also follows from the requirement of the simultaneous description of various experimental characteristics. In recent years, complex procedures of measuring various characteristics are growing in popularity [95-97]. Illustrative in this respect is the studying of the adsorption of nitrogen molecules on the face (110) of nickel [98]—to establish how matters stand, seven different procedures were compared. The use of simplified models for describing individual characteristics does not guarantee the description of other characteristics of a system with their aid. For instance, the adsorption isotherms are often described by the Langmuir isotherm, although the heat of adsorption changes with the coverage.

Hence, the central task of theory is a self-consistent description of the main real factors of a gas-solid system with the aid of a single set of parameters of models for different time intervals including a transition to the equilibrium state if it exists under given external conditions. The problem of the self-consistent description of the evolution of a system is of major significance in the theory of processes in condensed phases, especially when we have in view the simultaneous description of various elementary processes, e.g. the stages of transfer and the stages of chemical transformation. To date, this problem had a positive solution only for models taking no account of particle interaction. Even for processes in nonideal gases, the fact of the self-consistent description of equilibrium and kinetics must be proved in each specific model. This problem is also of an important practical significance because part of the parameters of a model can be determined independently from the data on the equilibria of individual subsystems and from the kinetics of simpler processes.

We shall single out the theory describing the distribution of interacting particles among the sites of inhomogeneous lattices and the theory of chemical kinetics with account taken of the interaction between all the components of the reaction system as auxiliary problems. Their solution will make it possible to approach the solution of the central problem, but they also have an independent significance for surface processes.

The traditional apparatus of statistical physics employed to construct models of physicochemical processes is the method of calculating the partition function [4, 12, 99, 100]. The alternative method of correlation functions or distribution functions [101] is more flexible. It is now the main method in the theory of the condensed state both for solid [102-104] and liquid [105-107] phases. This method has also found an application for lattice systems [108-113]. It enables one to obtain new approximations that are difficult to reproduce via the partition function and yields a clear physical interpretation of the employed approximation and the ways of refining it.

A new variant of the method of correlation functions—the cluster approach—is treated in the book. This method was developed for the solution of two problems mentioned above. The cluster approach provides a procedure for the self-consistent calculation of the complete set of probabilities of particle configurations on a cluster being considered. This makes it possible to take account of the local inhomogeneities of a lattice in the equilibrium and nonequilibrium states of a system of interacting particles. The novel approach is considerably simpler than other analogous methods, and the usual knowledge acquired by physicochemists is adequate for mastering it. Questions of the theory of the condensed state that are not traditional for them are given attention to the extent needed for general understanding of the subject. The technical aspect is based on the simplest concepts of the theory of probability and is given in detail with indication of all the intermediate operations. The local configurations of particles, which are treated in the following section, are the main concepts of the cluster approach.

We shall first consider the simplest interaction potential—the interaction of nearest neighbors (Chaps. 2-5), and then more involved potentials (Chaps. 6, 7), namely, two-body interaction for any finite number of neighbors and multiparticle interactions.

1.3. Local Configurations of Particles

In a lattice model, the particles are localized at the lattice sites. The discrete arrangement of the particles simplifies the way of describing their position—it is sufficient to set the number of the relevant site. Let us first discuss the distribution of particles on a homogeneous lattice whose sites are equivalent to one another, and then on an inhomogeneous lattice.

Consider a regular periodic lattice whose constant in its order of magnitude equals the diameter of the particles. For unambiguity, we shall presume that we have to do with a homogeneous solid solution of the substitutional type. Let the number of solution components be s. Let the subscript f signify the number of the lattice site, $1 \leqslant f \leqslant N$, where N is the number of sites. Every lattice site has z nearest neighbors forming its first coordination sphere. For a one-dimensional lattice, we have $z = 2$, for a two-dimensional one we have $z = 3$, 4, or 6, and for a three-dimensional one we have $z = 4$, 6, 8, or 12 (Fig. 1.3).

The occupancy of each lattice site is determined by the species of the particle at it. Since at any lattice site there can be a particle of any species, then each specific state of occupancy of site f is a random variable determined [114-116] as follows:

$$\gamma_f^i = \begin{cases} 1 & \text{if a particle of species } i \text{ is at site } f \\ 0 & \text{if a particle of species } j \text{ is at site } j,\ j \neq i \end{cases} \tag{1.4}$$

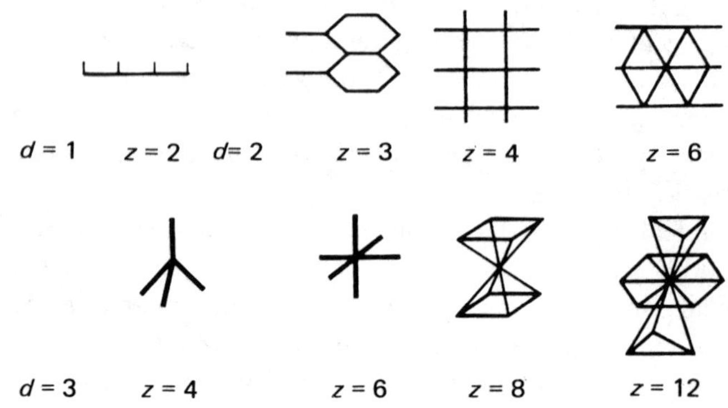

Fig. 1.3. Nearest neighbors of one- ($d = 1$), two- ($d = 2$), and three-dimensional ($d = 3$) lattices

This definition shows that (1) at any site there must be a particle of one species or another:

$$\sum_{i=1}^{s} \gamma_f^i = 1 \tag{1.5}$$

and (2) any site is occupied by a particle of only one species:

$$\gamma_f^i \gamma_f^j = \gamma_f^i \Delta_{ij}, \quad \Delta_{ii} = 1, \quad \Delta_{ij} = 0, \quad i \neq j \tag{1.6}$$

where Δ_{ij} is Kronecker's symbol. The following identity is observed for the random variables (1.4):

$$\exp(b\gamma_f^i) = 1 + \gamma_f^i(\exp b - 1), \quad b = \text{const} \tag{1.7}$$

The state of the entire lattice system is given by the complete set of random variables $\{\gamma_f^i\} = \gamma_1^i \ldots \gamma_N^i$. They provide detailed information on a particle of which species is at each site of a lattice. The probability of finding a lattice system at the instant τ in a state given by the set $\{\gamma_f^i\}$ will be designated by $P(\{\gamma_f^i\}, \tau)$. In the limiting case when $\tau \to \infty$ corresponding to a steady, including an equilibrium, state of the system, we shall omit the argument τ.

The number of different states of a system is s^N, and it grows sharply with both an increase in s and in N. In observed macroscopic characteristics, the effect of a large number of particles tells, e.g. for surface systems N is at least

10^9 particles. In this connection, it is impossible to use a detailed description of systems with the aid of complete sets of $\{\gamma_f^i\}$ and it is necessary to go over to their abridged description. (A real detailed description is impossible already at quite small values of s and N, for instance for $s = 3$ and $N = 12$.)

The description of a system is abridged by employing multiparticle correlation functions (or simply correlators) determined as follows:

$$\theta_{f_1 \ldots f_m}^{i_1, \ldots i_m}(\tau) = \langle \gamma_{f_1}^{i_1} \ldots \gamma_{f_m}^{i_m} \rangle_\tau = \sum_{i_1=1}^{s} \ldots \sum_{i_N=1}^{s} \prod_{n=1}^{m} \gamma_{f_n}^{i_n} P(\{\gamma_f^i\}, \tau) \tag{1.8}$$

where the sums are taken over all states of occupancy of all the lattice sites, i_f is the species of the particle at site f, m is the dimension or order of the correlator showing the probability of realizing the following configuration of the particles at the instant τ: the particle i_1 is at site f_1, the particle i_2 is at site f_2, and so on up to the m-th particle, inclusively. We are not interested in the state of occupancy of the remaining $(N - m)$ sites of the lattice, and averaging will be performed for them.

To operate with the introduced correlators, it is sufficient to know the following two main properties of the probabilistic functions [117-119].

1. The probability $P(AB)$ of a set of events AB (a complex event) can be represented in the form of the product $P(AB) = P(A)P(B|A)$, where $P(A)$ is the probability of event A, and $P(B|A)$ is the probability of event B provided that event A occurs (the events A and B themselves, in turn, may be complex events).

2. The probabilities of the complex events $P(A_1 \ldots A_m)$ and $P(A_1 \ldots A_{m-1})$ of the dimension m and $m - 1$, respectively, are related by the expression $P(A_1 \ldots A_{m-1}) = \sum_{A_m} P(A_1 \ldots A_{m-1} A_m)$, where the sum is taken over all the realizations of the event A_m, $m = 2, 3, \ldots$. The condition of normalization of all the probabilities is set by the relation $\sum_{A_1} P(A_1) = 1$. These properties relate to any instant of time.

In lattice systems, the elementary events are described by the quantities (1.4), while their different products are complex events. The probabilities of these complex events characterize the probabilities of the relevant local configurations of the particles. At $m = N$, correlator (1.8) is a complete N-particle distribution function in terms of which any correlator is expressed as

$$\theta_{f_1 \ldots f_m}^{i_1 \ldots i_m}(\tau) = \sum_{i_{m+1}=1}^{s} \ldots \sum_{i_N=1}^{s} \theta_{f_1 \ldots f_N}^{i_1 \ldots i_N}(\tau) \tag{1.9}$$

This enables us to relate the correlators of dimensions m and $m-1$ as follows:

$$\theta_{f_1 \ldots f_{m-1}}^{i_1 \ldots i_{m-1}}(\tau) = \sum_{i_m=1}^{s} \theta_{f_1 \ldots f_m}^{i_1 \ldots i_m}(\tau) \tag{1.10}$$

Consider the most important correlators of a lattice system. The (unary) correlator

$$\theta_f^i(\tau) = \langle \gamma_f^i \rangle_\tau, \quad \sum_{i=1}^{s} \theta_f^i(\tau) = 1 \tag{1.11}$$

is the simplest one. It shows the probability of finding the particle i at the instant τ at f, i.e. the local density of component i at f at the instant τ expressed in mole fractions.

The term *local density* is generally used when considering distributed systems, e.g. in problems on the transfer of particles in the kinetic theory of gases [120-122] or in condensed phases [123-125]. Traditionally, a physically small volume is singled out that contains a macroscopic number of particles. It is exactly the average density of the particles in such a small volume that is the local density. In expression (1.11), an elementary volume of space has a size of the order of magnitude of that of a site accommodating only one particle. The need of such great localization of an elementary volume follows from the great inhomogeneity of the surface systems (inhomogeneity of the adsorption centers, of the subsurface region of a solid, etc.) considered below. These systems have a typical size of the order of magnitude of that of a particle.

It should also be noted that the transition to local densities (1.11) enables one to observe the evolution of a system during small time intervals beginning with ones typical of one jump of a particle. This is why the average value of the local density (1.11), as in the general case of expressions (1.8), should be understood as the mean value taken at a given instant τ over a macroscopic ensemble of equivalent lattices on which the process being studied occurs. For an equilibrium distribution of the particles, by the ergodic hypothesis [1, 77, 126], the average value over an ensemble is equivalent to the average value over time, and we can consider a single lattice on which we shall count the frequency of appearance of various particle configurations during a prolonged period ($\tau \to \infty$).

Two-body correlators are second correlators as regards their dimension,

$$\theta_{fg}^{i_1 i_2}(\tau) = \langle \gamma_f^i \gamma_g^j \rangle_\tau \tag{1.12}$$

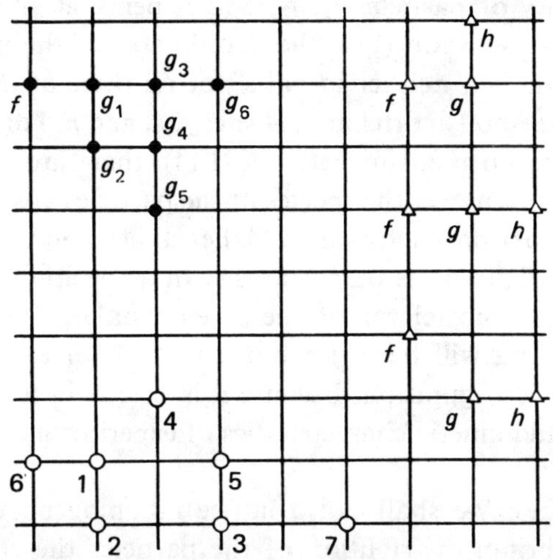

Fig. 1.4. Correlators on a square lattice.
Two-body correlators fg_n, where $n = 1$ to 6 is the number of the coordination sphere of site f; three-body correlators fgh, and a correlator of seven particles 1-7

They show the probability of particle i_1 being at site f and i_2 at site g at the instant τ. Owing to condition (1.6), we have $f \neq g$. Figure 1.4 shows two-body correlators at different distances r_{fg} from each other. The discrete nature of the lattice causes the values of r_{fg} to change discretely. A multitude of sites at the same distance from the site adopted as the central one form its coordination sphere. Let their number be $z(r_{fg})$. For lattice systems, we must distinguish the geometric distance between sites f and g and the number of jumps during which a particle from site f can reach site g provided that each jump is performed only between the nearest sites. For instance, for a square lattice, a particle from site f can be at site g_2 in two jumps ($r_{fg_2} = a\sqrt{2}$; a is the lattice constant), or at site $g_3(r_{fg_3} = 2a)$.

The conditions of normalization of two-body correlators have the form

$$\sum_{i_1=1}^{s} \theta_{fg}^{i_1 i_2}(\tau) = \theta_g^{i_2}(\tau), \qquad \sum_{i_2=1}^{s} \theta_{fg}^{i_1 i_2}(\tau) = \theta_f^{i_1}(\tau) \tag{1.13}$$

Three-body correlators are the following ones as regards their dimension:

$$\theta_{fgh}^{i_1 i_2 i_3}(\tau) = \langle \gamma_f^{i_1} \gamma_g^{i_2} \gamma_h^{i_3} \rangle_\tau, \qquad \sum_{i_1=1}^{s} \theta_{fgh}^{i_1 i_2 i_3}(\tau) = \theta_{gh}^{i_2 i_3}(\tau) \tag{1.14}$$

They show the probability of particles i_1, i_2, and i_3 being at sites f, g, and h, respectively. Owing to condition (1.6), the coordinates of the particles do not coincide with one another, neither in pairs nor all three at once. Figure 1.4 shows examples of three-body correlators at sites f, g, and h. For three-body correlators, in addition to normalizing relation (1.14), there are similar relations in which the sum is taken over the species of the particles at sites g and h.

Similarly, we can consider correlators of a higher dimension that show the probabilities of local configurations of particles with a specific arrangement thereof. Figure 1.4 shows a correlator of the seventh order.

Our goal in the following will be to set out a method of calculating the probabilities of the local configurations because it is exactly in terms of them that all the equilibrium and kinetic characteristics of experimental systems are expressed.

Inhomogeneous Lattices. We shall call a lattice inhomogeneous if under conditions of the equilibrium distribution of the particles the states of occupancy of different sites vary. Such a definition combines two different factors leading to various states of occupancy of different sites by particles of one species. The first factor is inhomogeneity due to different interactions with the surrounding particles that do not change their state at the characteristic times of the process. Examples are various adsorption centers of a surface and various absorption centers in the bulk of a solid that differ in the binding energies with gas-phase molecules. The second factor is inhomogeneity associated with the nature of distribution of the particles at the expense of interaction with one another. Examples are two- and multiphase systems and interfaces of stratifying phases including the subsurface region of a gas-solid interface, and also ordering of the particles on a homogeneous surface or in the bulk of a solid. In the second case, inhomogeneity is due to the cooperative properties of the system leading to the realization of phase transitions of stratification or ordering, or both.

It should be noted that the simultaneous "effect" of various factors causing inhomogeneity of a system also results in its inhomogeneity. We shall indicate as an example the case of ordering of the adsorbed particles (the second factor) on a stepped surface (the first factor).

Inhomogeneous lattices reflect a broad range of real situations, therefore in the following we shall formulate a general approach disregarding the physical nature of the inhomogeneity (Chap. 3).

Let us introduce the quantities η_f^q showing the type q of the lattice site numbered f, here $1 \leqslant f \leqslant N$ and $1 \leqslant q \leqslant t$, where t is the number of site types [127]. The quantities η_f^q are analogues of the quantities γ_f^i. In other words,

$\eta_f^q = 1$ if site f is one of type q, and $\eta_f^q = 0$ otherwise. Similar relations hold for them:

$$\sum_{q=1}^{t} \eta_f^q = 1, \quad \eta_f^q \eta_f^p = \eta_f^q \Delta_{qp} \tag{1.15}$$

A multitude of values of $\{\eta_f^q\} = \eta_1^q \ldots \eta_N^q$ unambiguously determine the inhomogeneity of lattice sites. But there is a fundamental distinction between the quantities γ_f^i and η_f^q, namely, the set $\{\eta_f^q\}$ is considered to be preset and constant, while averaging is conducted over the set $\{\gamma_f^i\}$. As for the quantities γ_f^i, it is impossible to work with a complete set of values of η_f^q, and it is necessary to go over to the distribution functions of inhomogeneous sites. Their dimension, as will be seen below, depends on the interaction potential, on the approximation used in taking account of the interaction, on the type of elementary process, and on the dimension of the particle configurations.

The state of occupancy of the sites of an inhomogeneous lattice, as above, is characterized by the random variables γ_f^i, but a site itself is additionally characterized by its types with the aid of the preset quantity η_f^q. In a general form, the correlators on an inhomogeneous lattice have a form similar to (1.8):

$$\theta_{f_1 \ldots f_m}^{i_1 \ldots i_m}(\tau | q_1 \ldots q_m) = \langle \gamma_{f_1}^{i_1} \ldots \gamma_{f_m}^{i_m} | \eta_{f_1}^{q_1} \ldots \eta_{f_m}^{q_m} \rangle_\tau \tag{1.16}$$

Formula (1.16) additionally indicates the type of sites at which the correlator of dimension m has been determined. The same addition is present in expressions (1.9)-(1.14), hence there is no need to rewrite them for inhomogeneous lattices.

The condition of statistical independence is written for the correlators as follows:

$$\theta_{f_1 \ldots f_m}^{i_1 \ldots i_m}(\tau | q_1 \ldots q_m) = \prod_{n=1}^{m} \theta_{f_n}^{i_n}(\tau | q_n) \tag{1.17}$$

It signifies that the occupation of each site does not depend on the state of occupancy of the other sites, i.e. correlation effects are absent. Strictly speaking, this is possible only when there is no interaction between the particles. A particular case of employing condition (1.17) is the use of the law of mass (surface) action for calculating the rate of chemical transformations in the kinetic mode.

When we know the local correlators, it is not difficult to construct macroscopic characteristics of particle distribution by lattice sites. The nature of occupation of sites of each type is of major significance in equilibrium on inhomogeneous lattices (surfaces). For this purpose, an indication of the types of sites is present in expressions (1.16). The quantities η_f^q play a more "active" role in the expressions for the energy of lattice systems.

For equilibrium particle distribution, the normalized distribution function will be written as follows [77]:

$$P(\{\gamma_f^i\}) = \frac{1}{Q}\exp\left[-\beta H(\{\gamma_f^i\})\right] \tag{1.18}$$

where Q is a normalizing factor generally called the partition function, and $H(\{\gamma_f^i\})$ is the effective Hamiltonian function (or Hamiltonian) of the lattice system:

$$H(\{\gamma_f^i\}) = \tilde{H}(\{\gamma_f^i\}) - \sum_{f,\, i} \mu_i \gamma_f^i,$$

$$\tilde{H}(\{\gamma_f^i\}) = H_{\mathrm{K}}(\{\gamma_f^i\}) + H_{\mathrm{P}}(\{\gamma_f^i\}) \tag{1.19}$$

In formula (1.19), $\tilde{H}(\{\gamma_f^i\})$ is the Hamiltonian that characterizes the total energy of a system, i.e. the sum of the kinetic $H_{\mathrm{K}}(\{\gamma_f^i\})$ and potential $H_{\mathrm{P}}(\{\gamma_f^i\})$ energies, and μ_i is the chemical potential of particles i. [Distribution (1.18) has been written for a large canonical ensemble.] The expression for the partition function of a lattice system is:

$$Q = \sum_{i_1 = 1}^{s} \dots \sum_{i_N = 1}^{s} \exp\left[-\beta H(\{\gamma_f^i\})\right] \tag{1.20}$$

The specific form of writing the Hamiltonian $\tilde{H}$ is unambiguously related to the physical model. A change in the latter requires the construction of a new Hamiltonian. By changing the interaction potential, we shall alter the form of $H_{\mathrm{P}}(\{\gamma_f^i\})$.

In lattice models, the particles are considered to be fixed at the lattice sites. Their kinetic energy contains no translational degrees of freedom, but consists of vibrational ones and, if a particle has an intricate structure, of rotational degrees of freedom. Let us consider that the motion of each particle is independent of that of the other ones—an analogue of Einstein's model of oscillators [11]. The expressions for $H_{\mathrm{K}}(\{\gamma_f^i\})$ will be determined in Sec. 1.5 for ideal models, and they do not change when particle interaction is taken into account.

1.4. The Cluster Approach

Normalizing relations (1.10) provide a relation between correlators of a various dimension. If we knew the correlators of dimension m, it would be easy to find the correlators of dimension $m - 1$ with their aid. But such a situation is not realized. On the other hand, the number of relations (1.10) is not sufficient to calculate the correlators of dimension m from the known correlators of dimension $m - 1$. It is therefore necessary to construct additional equations relating the correlators of the same dimension to one another. Liouville's equations for time distribution functions are employed for this purpose, while differentiation with respect to the coordinates of a system's particles is employed for the equilibrium distribution functions [101, 103-107, 126, 128-130].

A typical feature of the equations obtained for the distribution functions of dimension m is the appearance in them of an unknown distribution function of the order of magnitude $m + 1$. In this way, coupling Bogolyubov systems of equations form for time distribution functions, and Bogolyubov-Born-Green-Kirkwood-Ivone systems of equations form for equilibrium distribution functions. To close such a chain of equations, one must use approximate ways of expressing the unknown distribution functions of a higher order in terms of distribution function of a lower order. In practice, it is difficult to advance in this way farther than equations for two-body distribution functions even for homogeneous equilibrium systems, while for inhomogeneous equilibrium systems and especially for dynamic systems it is very often necessary to limit oneself to equations for unary distribution functions [103, 104, 131]. Here a major part of the correlation effects is absent, which can seriously distort the nature of the particle distributions.

When analyzing inhomogeneous systems, one must know the specific distribution of neighboring particles relative to a selected lattice site because a knowledge of the mean number of interacting pairs of particles is insufficient here. The same problem exists for describing the dynamics of processes in the condensed phase—all the surrounding particles affect the probability of an elementary process taking place. This requires procedures for the self-consistent calculations of a large number of probabilities of various multiparticle configurations, and not only for the calculation of the mean number of different pairs. Such a possibility is provided by a new method of calculating the probabilities of different configurations of particles—the cluster approach [132]. (We shall understand the term cluster to signify a finite number of lattice sites bound to one another.)

The essence of the cluster approach is as follows: (1) we make an image of the initial lattice system on a set of clusters (Fig. 1.5), "remembering" the se-

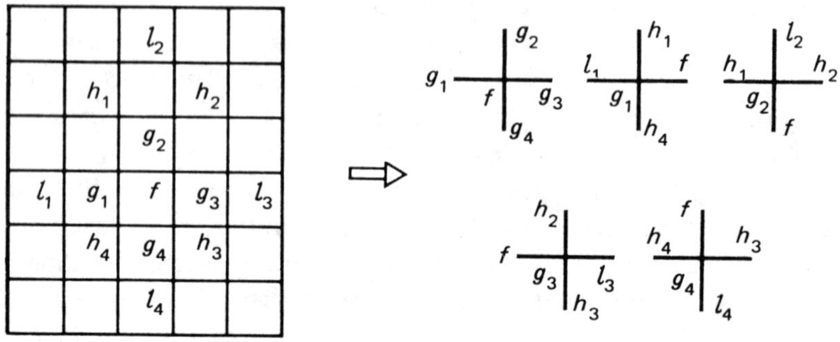

Fig. 1.5. An image of a lattice fragment on a set of clusters consisting of a central site and the sites of its first coordination sphere

quence of their arrangement on the initial lattice; (2) for each cluster we construct an exact system of equations with respect to the probabilities of the various particle configurations that can exist on it (equations for the cluster correlators); and (3) the obtained systems of equations are finite, but, as indicated above, are not closed, therefore we shall introduce a single way of closing them for all the systems of equations with account taken of the real sequence of cluster arrangement on the initial lattice. The result is a system of equations for the entire lattice as a whole.

Figure 1.5 shows an example of depicting a flat square lattice on a set of clusters consisting of five sites. Each lattice site is a central site of one of the clusters and is included in the sites of the coordination sphere when its neighbors are the central sites of other clusters.

In general, the procedure of forming an image consists in singling out three lattice regions, namely, 1—the central sites, 2—the sites surrounding the central one, and 3—the remaining sites of the lattice. The cluster itself contains sites of the first and second regions. We can choose any number of sites as the central ones. We shall confine ourselves to discussing cases with one and two central sites, as follows from the condition of the simultaneous participation of particles in an elementary process (mono- and bimolecular processes). The sites of the second region should separate the central sites from those of the third region so that a change in the state of occupancy of the third region sites would not directly affect the state of occupancy of the central sites (and vice versa). In other words, the size of the second region must not be smaller than the radius R of interaction between particles. The number of sites in the second region depends on the dimension of the lattice and on

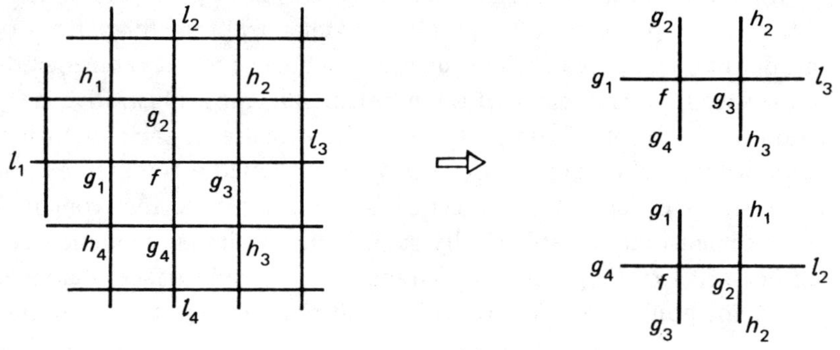

Fig. 1.6. An image of a lattice fragment on a set of clusters consisting of two central sites and their nearest neighbors

the radius of interaction. The image of a square lattice (see Fig. 1.5) on a set of clusters with one central site at $R = 2$ yields clusters consisting of nine sites (sites with the subscripts f, g, and h), while when $R = 3$ it yields clusters consisting of 13 sites (in which there are additionally included sites with the subscripts l).

Figure 1.6 shows an image of a square lattice on clusters with two central sites when $R = 1$.

The cluster correlators characterize the probabilities of the configurations of particles at the sites of the first and second regions. Their dimension is determined by the number of central sites, the dimension of the lattice, and the radius of the interaction potential. Two types of correlators can be introduced for each set of species of the central particles. The types differ in the ways of presetting the sets of species of the particles in the second region.

The first type contains an exact predetermination of the number of particles of various species at a specific distance from the central particles. The second type of correlators is obtained from those of the first type by averaging over the states of occupancy of a part of the sites of the second region. The correlators of the first type determine the probabilities of a configuration with a known value of the energy of the central particles, while the correlators of the second type are used to express the thermodynamic characteristics of the system and construct the closure of the system of equations. The construction of the system of equations for the correlators and the ways of their closure form the technical part of the cluster approach. This material will be set out on following pages.

Any process at a molecular level in the cluster approach is treated as a set of elementary processes of migration and chemical transformations occurring on a definite sequence of lattice sites. The sequence of sites considered depends on the specific formulation of the problem. The most illustrative is the propagation of the front of the particles at the expense of their diffusion from a region with a higher local concentration to a rarefied region. Here the sequence of the sites can represent a direction at right angles to the front if the latter is one-dimensional or spherically symmetric. A similar sequence of the sites can be used for the processes of transfer in the subsurface region of a solid or near the boundaries of various faces, at right angles to steps on the surface, etc.

For the equilibrium distribution of adsorbed particles on an inhomogeneous surface, of importance are how the sites of one type are surrounded by sites of another type and how the nature of this environment changes on different macroscopic sections of the surface. The real nature of the "linking" of the sites of various types to one another is reflected when constructing the relevant distribution functions.

The application of the cluster approach to various systems is dealt with in the main part of the book. We shall complete this chapter by discussing ideal models of lattice systems. It will be useful to set out the traditional material in the language of the methods of the condensed state theory.

1.5. Ideal Models

We shall limit ourselves to the equilibrium distribution of the particles of a lattice system (1.18). In the absence of interaction between the particles, their potential energy is characterized by the energy binding them to the lattice. (For adsorption or absorption processes, this is the binding energy to atoms of the solid.) The energy states of the internal degrees of freedom of a particle are counted from the binding energy. The total energy of a system for a homogeneous lattice is written as follows:

$$\tilde{H} = H_k + H_p, \quad H_k = -\beta^{-1} \sum_{f=1}^{N} \sum_{i=1}^{s} \ln (F_i)\gamma_f^i$$

$$H_p = H_v = -\sum_{f=1}^{N} \sum_{i=1}^{s} \frac{\varepsilon_i}{m_i} \gamma_f^i \tag{1.21}$$

where F_i is the partition function of particle i with the assumption that the internal degrees of freedom of each particle are independent of the presence of other particles [in this case the partition function of the particles has a form similar to Eq. (1.2)], m_i is the degree of dissociation of a gas-phase molecule; for undissociating molecules $\varepsilon_i = \tilde{\varepsilon}_i (\tilde{\varepsilon}_i$ is the binding energy of particle i to the lattice), for dissociating molecules $\varepsilon_i = m_i \tilde{\varepsilon}_i - D_i$ (D_i is the energy of dissociation of a molecule in the gas phase). We shall consider the vacant sites to be particles v for which the internal degrees of freedom, the binding energy to the lattice, and the chemical potential yield zero contributions (these zero values will be taken into account only in the final expressions).

When enumerating the system components i from 1 to s, we shall assume that the species s corresponds to vacant sites. The sums over f in formulas (1.21) are taken over all the lattice sites.

Inserting expression (1.21) into formula (1.19), we obtain the effective Hamiltonian for an ideal lattice system:

$$H = \sum_{f=1}^{N} \sum_{i=1}^{s} \nu_i \gamma_f^i, \quad \nu_i = -(\beta^{-1} \ln F_i + \mu_i + \varepsilon_i/m_i) \tag{1.22}$$

With a view to the determination of the chemical potential of a molecule M_m in the gas phase by Eq. (1.3) and its relation to the chemical potential of a particle M of the lattice system in (1.19) $\mu_M = (1/m)\mu_{M_m}$, the expression for the parameter ν_i can be rewritten as follows:

$$\nu_i = -\beta^{-1} \ln y_i, \quad y_i = (a_i P_i)^{1/m_i}, \quad a_i = (F_i)^{m_i} \beta \exp \beta\varepsilon_i / F_i^0 \tag{1.23}$$

where a_i is the adsorption or absorption coefficient, and p_i is the partial pressure of the molecules i. The parameter ν_i shows the contribution to the effective Hamiltonian of the system introduced by one particle of species i (the one-particle contribution).

We obtain equations of adsorption or absorption equilibrium. In the absence of interaction, condition (1.17) is observed, hence it is sufficient to obtain equations for θ_f^i. For this purpose, we insert (1.18) into definition (1.8) for θ_f^i with formula (1.22) in which we single out a site g from the sum over f: $H = G_g + h_g$, where $h_g = \sum_{j=1}^{s} \nu_j \gamma_g^j$, $G_g = \sum_{f=1}^{N} \sum_{j=1}^{s}{}' \nu_j \gamma_f^j$ and the prime in the sum over f signifies the absence of a site numbered g. Hence

$$\begin{aligned}\theta_g^i &= \frac{1}{Q} \sum_{i_1=1}^{s} \cdots \sum_{i_N=1}^{s} \gamma_g^i \exp[-\beta(h_g + G_g)] \\ &= \frac{1}{Q} \sum_{i_1=1}^{s} \cdots \sum_{i_{g-1}=1}^{s} \sum_{i_{g+1}=1}^{s} \cdots \sum_{i_N=1}^{s} \exp(-\beta G_g) \sum_{i_g=1}^{s} \gamma_g^i \exp(-\beta h_g) \\ &= \Lambda \sum_{i_g=1}^{s} \gamma_g^i \exp\left(-\beta \sum_{j=1}^{s} \nu_j \gamma_g^j\right)\end{aligned} \tag{1.24}$$

The coefficient Λ includes contributions from all the lattice sites except g, and the partition function of the system. It should be noted that its magnitude does not depend on the state of occupancy of site g.

To continue our calculation of the quantity θ_g^i, we shall first use expression (1.7) [for transformation of the exponential multiplier (1.24)], and then expressions (1.5) and (1.6):

$$\begin{aligned}\exp\left(-\beta \sum_{j=1}^{s} \nu_j \gamma_g^j\right) &= \prod_{j=1}^{s} \exp(-\beta \nu_j \gamma_g^j) \\ &= \prod_{j=1}^{s} [1 + \gamma_g^j(\exp(-\beta\nu_j) - 1)] \\ &= \sum_{j=1}^{s} \gamma_g^j \exp(-\beta\nu_j); \sum_{i_g=1}^{s} \gamma_g^i \sum_{j=1}^{s} \gamma_g^j \exp(-\beta\nu_j) \\ &= \sum_{i_g=1}^{s} \gamma_g^i \exp(-\beta\nu_i) = \exp(-\beta\nu_i)\end{aligned} \tag{1.25}$$

The second relations (1.25) lead to the final result: $\theta_g^i = \Lambda \exp(-\beta\nu_i)$. Here we again use relation (1.6) and definition (1.4) according to which an event when particle i is at g corresponds to the correlator θ_g^i. This is why the contribution only from a particle of this species will be other than zero in the last sum. The final result can be obtained in a simpler way directly on the basis of definition (1.4) without using the formal transformations (1.6) and (1.7) in expressions (1.25): if we assume that $\gamma_g^i = 1$, which corresponds to the authentic location of a particle of species i at a site numbered g, then by (1.4) and (1.5) both sums in (1.25) with respect to i_g and j "vanish", and a single addend remains other than zero.

The chief factor in the considered determination of a local correlator is that it is expressed in terms of the contributions introduced by the singled out particle and the remaining part of the system. This feature is also retained in nonideal systems.

It is easy to obtain equations of interphase equilibrium on the basis of the determination for the local correlator. Let us consider the ratio of functions (1.24) and (1.25) for different species of particles i and j:

$$\theta_g^i = \theta_g^j \exp \beta(\nu_j - \nu_i) \tag{1.26}$$

The unknown quantities (the contributions of the remaining lattice sites and the partition function) cancel out, and expressions remain relating the probabilities of occupation of site g by particles of various species. The number of relations (1.26) is $(s - 1)$. By supplementing them with normalizing relation (1.11), we obtain the sought expressions:

$$\theta_g^i = \exp(-\beta\nu_i) \Big/ \sum_{j=1}^{s} \exp(-\beta\nu_j) \tag{1.27}$$

that determine the probability of occupation of adsorption or absorption centers depending on the partial pressures of the molecules in the gas phase. For homogeneous lattices, owing to the equivalence of all the sites, the same result is obtained when going over to macroscopic means:

$$\theta_i = \frac{1}{N} \sum_{f=1}^{N} \theta_f^i = y_i \Big/ \left(1 + \sum_{j=1}^{s-1} y_j\right) \tag{1.28}$$

where θ_i is the probability of any lattice site being occupied by particle i (the mole fraction of component i). The properties of the vacant sites are taken into account in the explicit form in formula (1.28): $y_s = 1$.

Inhomogeneous Lattice. Depending on the site type, particles have different internal degrees of freedom and binding energies to atoms of the solid. The total energy of a system consists of the same addends as for a homogeneous lattice, but the molecular parameters become dependent on the type of lattice site. Formula (1.22) is generalized as follows:

$$H = \sum_{f=1}^{N} \sum_{i=1}^{s} \nu_q^i \gamma_f^i \eta_f^q, \quad \nu_q^i = -\beta^{-1} \ln y_q^i$$
$$y_q^i = (a_q^i P_i)^{1/m_i}, \quad a_q^i = (F_q^i)^{m_i} \beta \exp \beta\varepsilon_q^i / F_i^0 \tag{1.29}$$

where a_q^i is the coefficient of adsorption or absorption of a molecule i for a center of type q, $1 \leqslant q \leqslant t$, F_q^i and $\tilde{\varepsilon}_q^i$ are the partition function and the binding energy of a particle i for the center q (in the absence of dissociation of gas phase molecules $\varepsilon_q^i = \tilde{\varepsilon}_q^i$, while when they dissociate, $\varepsilon_q^i = \tilde{\varepsilon}_q^i m_i - D_i$).

The equilibrium relations between the local probabilities of site occupancies and the pressure of the molecules in the gas phase are obtained similarly. For an inhomogeneous lattice, formulas (1.24) and (1.25) are written as follows:

$$\theta_g^i(q) = \sum_{i_1=1}^{s} \cdots \sum_{i_N=1}^{s} \gamma_g^i \exp\left[-\beta\left\{\sum_{j=1}^{s} \nu_q^j \gamma_g^j \eta_g^q + \sum_{f=1}^{N}{}' \sum_{j=1}^{s} \nu_q^j \gamma_f^j \eta_f^q\right\}\right] \Big/ Q = \Lambda e^{-\beta\nu_q^i} \tag{1.30}$$

As earlier, an event when a particle i is at a site corresponds to correlator (1.30), while the contribution of the particle depends on the type of the site set by the quantity η_g^q. The unknown coefficient Λ depends on the state of the entire lattice system except for site g. It is excluded if any pairs of relations between components of the system are considered:

$$\theta_g^i(q) = \theta_g^j(q) \exp \beta(\nu_q^j - \nu_q^i) \tag{1.31}$$

By supplementing the system of equations (1.31) with normalizing expression (1.11), we obtain the final expression:

$$\theta_g^i(q) = y_q^i \Big/ \left(1 + \sum_{j=1}^{s-1} y_q^j\right) \tag{1.32}$$

Owing to the equivalence of sites of one type, the macroscopic value of the probability of finding particles i at sites of type q corresponds to the local value of (1.32):

$$\theta_q^i = \frac{1}{N_q} \sum_{h=1}^{N_q} \theta_h^i(q) = \theta_g^i(q), \quad N_q = \sum_{f=1}^{N} \eta_f^q \tag{1.33}$$

where h runs through all the lattice sites of type q, while g is one of the numbers of the sites of type q, and N_q is the number of sites of type q on the lattice. The macroscopic probability of finding particles of species i on an inhomogeneous lattice is determined as follows:

$$\theta_i = \frac{1}{N} \sum_{f=1}^{N} \theta_f^i(q)\eta_f^q = \sum_{q=1}^{t} f_q \theta_q^i, \quad f_q = N_q/N \tag{1.34}$$

Here f runs through all the lattice sites, f_q is the fraction of sites of type q on an inhomogeneous lattice. If we take relation (1.33) into consideration, then insertion of (1.32) into (1.34) yields a well known expression for the adsorption or absorption isotherm of a mixture of gases [33, 85-88, 116].

Rates of Surface Reactions. To calculate the rate of elementary processes in an ideal lattice system, we shall employ the model of the transition state proposed by Eyring [133, 134]. The foundations and propositions of this model are well known, and there is no need to treat them in detail. Within the scope of the model of the transition state, we consider that (1) an activated complex is a particle with modified properties—one of the vibrational degrees of freedom has been replaced with translational motion along the reaction coordinate; (2) an activated complex is in equilibrium with the reaction subsystem which it is formed from; (3) the number of activated complexes is small;

and (4) the volume of an activated complex does not differ a lot from that of the reactants.

Homogeneous Surface. Let us first consider the monomolecular reaction $i \to j$. The condition that the number of activated complexes is small allows us to limit ourselves to the consideration of one complex. Assume that an activated complex forms at site n from the particle i at it. In its absence, the total energy of the system is described by expression (1.22). The appearance of the complex changes the energy. The total energy of a system with one activated complex occupying one lattice site is written in the form

$$H_1 = \sum_{f=1}^{N}{}' \sum_{l=1}^{s} \nu_l \gamma_f^l + \overset{*}{\nu}_i \gamma_n^i,$$

$$\nu_i^* = -(\beta^{-1} \ln \tilde{F}_i^* + \varepsilon_i^* + \mu_i^*) \tag{1.35}$$

The sum over f is taken over all the sites except the site n occupied by the activated complex. Since it forms only from the particle i, there is no summation over the species of particles at site n. The characteristics of the activated complex having the same meaning as for particles in the ground state are indicated by an asterisk.

Formula (1.35) can be written as follows by using expression (1.22):

$$H_1 = H + \delta h_n^i, \quad \delta h_n^i = (\nu_i^* - \nu_i)\gamma_n^i \tag{1.36}$$

where δh_n^i is the change in the energy of the system when the activated complex forms.

The rate of a reaction in the theory of absolute reaction rates is a quantity proportional to the equilibrium concentration of the activated complexes:

$$U_i = \varkappa \omega \theta_i^* / \delta \tag{1.37}$$

where $\varkappa$ is a transmission coefficient, ω is the mean velocity of passing a barrier with a length of δ, $\omega = \{2\pi M^* \beta\}^{1/2}$ (here M^* is the reduced mass of a complex), and U_i is the rate per surface site.

The quantity θ_i^* can be expressed as the probability of a complex event: (1) at site n there should be a particle i [the equality $\theta_n^i = \theta_i$ for a homogeneous surface owing to the equivalence of the sites (1.28)], and (2) the energy of the system changes by the quantity δh_n^i:

$$\theta_i^* = \langle \gamma_n^i \exp(-\beta \delta h_n^i) \rangle \tag{1.38}$$

The technique employed for determining the equilibrium concentration of an activated complex is identical to the conventional one in terms of the equilibrium constant between a complex and the reactants.

The mean value on the right-hand side of expression (1.38) is calculated formally on the basis of identity (1.7) and equation (1.6):

$$\langle \gamma_j^i e^{b\gamma_f^i} \rangle = \langle \gamma_f^i (1 + \gamma_f^i (e^b - 1)) \rangle = \langle \gamma_f^i \rangle e^b, \quad b = \nu_i - \nu_i^* \tag{1.39}$$

The same result can be obtained in a simplified way using expression (1.4): the energy changes by the amount $\nu_i - \nu_i^*$ only when there is a particle i at the site $f = n$. As a result:

$$\theta_i^* = \theta_i \exp \beta(\nu_i - \nu_i^*) \tag{1.40}$$

By singling out the translational degree of freedom along the reaction coordinate from the partition function of a complex (h is Planck's constant)

$$\tilde{F}_i^* = F_i^* \frac{\delta}{h} (2\pi M^* \beta)^{1/2} \tag{1.41}$$

we obtain an expression for the rate of an elementary process:

$$\begin{aligned} U_i &= K_i \theta_i, \quad K_i = \hat{K}_i \exp(-\beta E_i) \\ \hat{K}_i &= \frac{F_i^* \varkappa}{F_i h \beta} \exp \beta(\mu_i^* - \mu_i), \quad E_i = \varepsilon_i - \varepsilon_i^* \end{aligned} \tag{1.42}$$

where K_i is the rate constant, $\hat{K}_i$ is its preexponential factor, and E_i is the energy of activation of an elementary process. To ensure a single form of writing the expressions, let us agree to include the partial pressures in the preexponential factor of the rate constant, and concentrate our main attention on the species of the particles of the condensed phase. The specific expressions for the rate constant K_i depend on the reaction subsystem the activated complex is in equilibrium with.

(a) Consider the desorption process $i\mathrm{Z} \rightarrow \mathrm{vZ} + i$, where iZ is a site occupied by particle i, vZ is a vacant site, and i is a molecule in the gas phase. If we count the energy from that of a free particle in the gas, we have $\varepsilon_i^* < 0$. The activated complex is in equilibrium with the adsorbed particle i, $\mu_i = \mu_i^*$:

$$K_i = \frac{F_i^* \varkappa}{F_i h \beta} \exp(-\beta E_i) \tag{1.43}$$

(b) Consider the adsorption process $i + vZ \rightarrow iZ$. An activated complex is in equilibrium with molecules i in the gas phase and with vacant sites, $\mu_v^* = \mu_v + \mu_{i,g}$, where $\mu_{i,g}$ has been determined by Eq. (1.3). Having in view that $\mu_v = 0$ and $F_v = 1$, we obtain:

$$K_v = \frac{F_v^* \varkappa}{F_i^0 h} P_i \exp(-\beta E_v), \quad E_v = E_i - \varepsilon_i, \; F_i^* = F_v^* \tag{1.44}$$

(c) Consider the reaction of gas phase molecules by an impact mechanism with an adsorbed particle $A + BZ \rightarrow CZ + D$. An activated complex is in equilibrium with an A molecule and an adsorbed B particle, $\mu_B^* = \mu_B + \mu_{Ag}$:

$$K_B = \frac{F_B^* \varkappa}{F_B F_A^0 h} P_A \exp(-\beta E_B) \tag{1.45}$$

The kinetic schemes $A_1 + \ldots + A_m + BZ$ are formal generalizations of the considered reaction. This leads to a new meaning of the multipliers of the reaction rate constant:

$$P_A = \beta^{m-1} \prod_{i=1}^{m} P_{A_i}, \quad F_A^0 = \prod_{i=1}^{m} F_{A_i}^0 \tag{1.46}$$

For the reverse process, the subscripts A and B are altered respectively to D and C. In a similar way, monomolecular processes of the activated surmounting of barriers with the transition of particles to excited states, isometric forms, and various space orientations are considered.

Let us now go over to the bimolecular reaction $i + j \rightarrow$ products. Let an activated complex form from the reactants i and j at neighboring sites n and k. In the absence of a complex, the total energy of the system has the form of expression (1.22). In the presence of an activated complex that occupies two lattice sites, the total energy will be

$$H_2 = \sum_{f=1}^{N}{}' \sum_{i=1}^{s} \nu_i \gamma_f^i + \nu_{ij}^* \gamma_n^i \gamma_k^j, \quad \nu_{ij}^* = -(\beta^{-1} \ln \tilde{F}_{ij}^* + \mu_{ij}^* + \varepsilon_{ij}^*) \tag{1.47}$$

The sum over f is taken over all the sites except n and k. Summation over the species of the particles at the sites occupied by a complex is absent because the latter forms only from particles i and j. The characteristics $\tilde{F}_{ij}^*$, ε_{ij}^*, and μ_{ij}^* of the activated complex have the meaning indicated earlier.

Expression (1.47) can be rewritten as follows:

$$H_2 = H + \delta h_{nk}^{ij}, \quad \delta h_{nk}^{ij} = \nu_{ij}^{*}\gamma_n^i\gamma_k^j - \nu_i\gamma_n^i - \nu_j\gamma_k^j \tag{1.48}$$

The probability of finding an activated complex formed from particles i and j at two neighboring sites can be written as the probability of a complex event: (1) particle i is at site n and particle j is at site k, and (2) the energy of the system changes by the amount δh_{nk}^{ij}:

$$\theta_{ij}^{*} = \langle \gamma_n^i\gamma_k^j \exp(-\beta\delta h_{nk}^{ij})\rangle \tag{1.49}$$

Expression (1.49) is calculated in exactly the same way as indicated for a monomolecular reaction with the use of formulas (1.4), (1.6), and (1.7). By inserting the obtained expression into a formula similar to (1.37) and separating the translational degree of freedom along the reaction coordinate from the partition function of the complex, by analogy with Eq. (1.41) we obtain the following expression for the rate of a bimolecular reaction on a pair of neighboring sites:

$$U_{ij} = K_{ij}\theta_i\theta_j, \quad K_{ij} = \hat{K}_{ij}\exp(-\beta E_{ij}),$$
$$\hat{K}_{ij} = \frac{F_{ij}^{*}\varkappa}{F_iF_jh\beta}\exp\beta(\mu_{ij}^{*} - \mu_i - \mu_j), \; E_{ij} = \varepsilon_i + \varepsilon_j - \varepsilon_{ij}^{*} \tag{1.50}$$

where K_{ij} is the rate constant, $\hat{K}_{ij}$ is the preexponential factor, and E_{ij} is the activation energy of the reaction. The form of the rate constant of the reaction depends on which reaction subsystem the activated complex is in equilibrium with.

(a) Desorption with association: $2iZ \to 2vZ + i_2$. The activated complex of desorption is in equilibrium with two adsorbed particles, $\mu_{ii}^{*} = 2\mu_i$:

$$K_{ii} = \frac{F_{ii}^{*}\varkappa}{(F_i)^2h\beta}\exp(-\beta E_{ii}) \tag{1.51}$$

(b) Dissociative adsorption: $2vZ + i_2 \to 2iZ$. The activated complex is in equilibrium with two neighboring vacant sites and molecule i_2 in the gas phase, $\mu_{vv}^{*} = 2\mu_v + \mu_{i_2g}$. With a view to the properties of vacant sites indicated earlier, we obtain:

$$K_{vv} = \frac{F_{vv}^{*}\varkappa}{F_{i_2}^{0}h}P_{i_2}\exp(-\beta E_{vv}), \quad F_{vv}^{*} = F_{ii}^{*} \tag{1.52}$$

(c) Exchange reaction: AZ + BZ → CZ + DZ. The activated complex is in equilibrium with particles of A and B, $\mu_{AB}^{*} = \mu_A + \mu_B$, and the preexponential factor in expression (1.50) is simplified; this is also done in formula (1.51). Particular cases of exchange reactions are processes of particle migration by a vacancy mechanism iZ + vZ ⇄ vZ + iZ and a reaction by an adsorption mechanism between adsorbed particles AZ + BZ → 2Z + AB.

(d) Heterogeneous reactions: AZ + BZ + E, where $E = E_1 \ldots E_m$.

$$K_{AB} = \frac{F_{AB}^{*} \varkappa \beta^{m-1}}{F_A F_B F_E^0 h} P_E \exp(-\beta E_{AB}),$$

$$P_E = \prod_{i=1}^{m} P_{E_i}, \quad F_E^0 = \prod_{i=1}^{m} F_{E_i}^0 \tag{1.53}$$

Inhomogeneous Surface. Let us consider the monomolecular reaction $i \to j$. It results in the formation of particle j from particle i. For inhomogeneous surface, the rate of the reaction depends on the species of the site where it occurs. The total energy of the system when there is one activated complex of species q at a site n will be written as

$$H_1 = \sum_{f=1}^{N}{}' \sum_{l=1}^{s} \nu_q^l \gamma_f^l \eta_f^q + \nu_q^{i*} \gamma_n^i \eta_n^q,$$

$$\nu_q^{i*} = -(\beta^{-1} \ln \tilde{F}_q^{i*} + \varepsilon_q^{i*} + \mu_i^*) \tag{1.54}$$

where $\tilde{F}_q^{i*}$ and ε_q^{i*} are the partition function and the binding energy of the activated complex formed from particle i for a center of type q. The sum over f excludes site n occupied by the complex. With a view to expression (1.29), formula (1.54) will be rewritten as follows:

$$H_1 = H + \delta h_n^i(q), \quad \delta h_n^i(q) = (\nu_q^{i*} - \nu_q^i)\gamma_n^i \eta_n^q \tag{1.55}$$

The probability of formation of an activated complex can be expressed as

$$\theta_n^{i*}(q) = \langle \gamma_n^i \exp(-\beta \delta h_n^i(q)) | \eta_n^q \rangle = \theta_n^i(q) \exp \beta(\nu_q^i - \nu_q^{i*}) \tag{1.56}$$

The second equality is obtained when all the operations indicated above for monomolecular reactions on a homogeneous surface are repeated.

The rate of a reaction at sites of type q is determined by analogy with formula (1.37):

$$U_q^i = \varkappa_q \omega_q \theta_q^{i*}/\delta_q = K_q^i \theta_q^i, \quad K_q^i = \hat{K}_q^i \exp(-\beta E_q^i)$$
$$\hat{K}_q^i = \frac{F_q^{i*} \varkappa_q}{F_q^i h\beta} \exp \beta(\mu_i^* - \mu_i), \quad E_q^i = \varepsilon_q^i - \varepsilon_q^{i*} \tag{1.57}$$

where K_q^i is the rate constant, $\hat{K}_q^i$ is its preexponential factor, and E_q^i is the activation energy of the reaction $i \to j$ at a site of type q. In the first formula, $\varkappa_q$, ω_q, and δ_q have their usual meaning (1.37). For a site q, a translational degree of freedom along the coordinate of the reaction has been singled out in the second equality. The specific forms of formula (1.57) depend on the kind of reaction subsystem in which the activated complex forms. In the general form, a one-site process can be written as A + BZ, and for it we have

$$K_q^{\mathrm{B}} = \frac{F_q^{\mathrm{B}*} \varkappa_q}{F_q^{\mathrm{B}} F_{\mathrm{A}}^0 h} P_{\mathrm{A}} \exp(-\beta E_q^{\mathrm{B}}) \tag{1.58}$$

The transition from formula (1.58) to the rate constants of the adsorption and desorption of undissociating molecules and for the major part of the gas-phase molecules participating in an elementary event is the same as has been shown on an earlier page for homogeneous surfaces.

Now let us consider the bimolecular reaction $i + j \to$ products. On an inhomogeneous surface, the rate of a reaction depends on the type of the two neighboring sites at which it occurs. Repeating what has been said above word by word, the probability of formation of an activated complex from particles i and j at sites n and k of types q and p will be written respectively as follows:

$$\theta_{nk}^{ij*}(qp) = \langle \gamma_n^i \gamma_k^j \exp(-\beta \delta h_{nk}^{ij}(qp)) | \eta_n^q \eta_k^p \rangle$$
$$\delta h_{nk}^{ij}(qp) = \nu_{qp}^{ij*} \gamma_n^i \gamma_k^j \eta_n^q \eta_k^p - \nu_q^i \gamma_n^i \eta_n^q - \nu_p^j \gamma_k^j \eta_k^p \tag{1.59}$$
$$\nu_{qp}^{ij*} = -(\beta^{-1} \ln \tilde{F}_{qp}^{ij*} + \varepsilon_{qp}^{ij*} + \mu_{ij}^*)$$

where $\tilde{F}_{qp}^{ij*}$ and ε_{qp}^{ij*} are the partition function and the binding energy of the complex of the reaction of $i + j$ at a pair of sites of type qp.

The rate of a bimolecular reaction at a pair of sites of type qp is

$$U_{qp}^{ij} = \varkappa_{qp} \omega_{qp} \theta_{qp}^{ij*}/\delta_{qp} = K_{qp}^{ij} \theta_q^i \theta_p^j$$
$$K_{qp}^{ij} = \hat{K}_{qp}^{ij} \exp(-\beta E_{qp}^{ij}), \tag{1.60}$$
$$\hat{K}_{qp}^{ij} = \frac{F_{qp}^{ij*} \varkappa_{qp}}{F_q^i F_p^j h\beta} \exp \beta(\mu_{ij}^* - \mu_i - \mu_j)$$

where $E_{qp}^{ij} = \varepsilon_q^i + \varepsilon_p^j - \varepsilon_{qp}^{ij*}$ is the activation energy of the reaction of i and j at sites of types q and p, K_{qp}^{ij} is the rate constant of this reaction, and $\hat{K}_{qp}^{ij}$ is its preexponential factor. The specific form of the rate constant depends on the reaction subsystem containing the reactants i and j. As for a homogeneous surface, we can consider different reactions. Let us limit ourselves to the reaction AZ + BZ + E:

$$K_{qp}^{AB} = \frac{F_{qp}^{AB*} \varkappa_{qp}}{F_q^A F_p^B h} P_E \exp(-\beta E_{qp}^{AB}) \tag{1.61}$$

More strictly, the rate constants (1.57) [similar to (1.42)] should be called rate constants of one-site processes, and the rate constants (1.61) [similar to 1.50]—rate constants of two-site processes. A distinction consists in that if a gas-phase molecule participates in the reaction, the molecularity of the latter grows by unity. For instance, the adsorption processes (1.44) and (1.52) are bi- and trimolecular ones. Moreover, the rate constants, if at least one vacant site v participates in a reaction, contain the partial pressures of the gas-phase molecules as multipliers. The introduction of the rate constants of one- and two-site processes enables us to write the expressions for the rates of a reaction irregardless of the kind of reaction subsystem.

From the obtained expressions for the rates of one- and two-site processes for sites q and pairs of sites qp when constructing the kinetic equations determining the rate of change in the total number of particles of A in various reactions, it is necessary to go over to the surface as a whole. Let Q_A and Q_{AC} be the number of steps in which a particle of A participates in one-site ($A \rightleftarrows B_\alpha$) and two-site ($A + C_\alpha \rightleftarrows B_\alpha + D_\alpha$) processes, and α be the ordinal number of the step. When going over to the entire surface, the contribution of the one-site process is multiplied by the fraction of sites of type q (1.34), while the contribution of the two-site process is multiplied by f_{qp}—the fraction of pairs of sites of types qp, and by z_q—the number of neighbors of sites q accommodating a particle of C reacting with a particle of A. As a result, the fractional coverage of the surface by particles of A changes in time in accordance with the following kinetic equation:

$$\frac{d\theta_A}{d\tau} = \sum_{\alpha=1}^{Q_A} [U_B(\alpha) - U_A(\alpha)] + \sum_{\alpha=1}^{Q_{AC}} [U_{BD}(\alpha) - U_{AC}(\alpha)] \tag{1.62}$$

$$U_i(\alpha) = \sum_{q=1}^{t} f_q K_q^i(\alpha) V_q^i(\alpha), \quad U_{ij}(\alpha) = \sum_{q=1}^{t} f_q z_q \sum_{p=1}^{t} d_{qp} K_{qp}^{ij}(\alpha) V_{qp}^{ij}(\alpha) \tag{1.63}$$

$$z_q f_{qp} = z_p f_{pq}, \; d_{qp} = f_{qp}/f_q,$$

$$f_{qp} = \frac{1}{N} \sum_{f=1}^{N} \sum_{g} \frac{1}{z_q} \eta_f^q \eta_g^p, \; \sum_{p=1}^{t} d_{qp} = 1 \tag{1.64}$$

where $V_q^i(\alpha) = \theta_q^i$ and $V_{qp}^{ij}(\alpha) = \theta_q^i \theta_p^j$ are the concentration components of the rates of one- and two-site processes; for ideal models they are expressed in terms of the partial coverages of sites of various species [Eq. (1.33)]. In the following, our task will be to find these components for nonideal reaction systems. In formula (1.64), d_{qp} is the conditional probability of finding a site of type p next to one of type q. This function characterizes the structure of an inhomogeneous surface. Examples of constructing these functions for point and linear imperfections have been given by the author in other publications [116, 135]. In the theory of surface processes, it was mainly assumed that sites of various types are arranged in patches, $d_{qp} = \Delta_{qp}$, or are arranged chaotically, $d_{qp} = f_p$ [33, 85-88]. The functions d_{qp} allow one to describe the correlated distribution of adsorption centers of various types. The author has shown [116] in model calculations of the rate of a reaction by an adsorption mechamism and of the rate of dissociative adsorption that the influence of the surface structure may be commensurable with the influence of the change in the surface composition of an adsorbent within broad limits.

In expression (1.63), the number z_q is generally assumed to be constant [88, 136] as for a homogeneous surface [137], and is included as a multiplier in the rate constant of a reaction.

Hence, the application of the methods of the condensed state theory even for ideal models made it possible to obtain new results, namely, take into consideration the differences in the numbers of neighbors for sites of various types due to the structural imperfections of surfaces, and also take into consideration the correlated arrangement of pairs of sites of various types. The assumptions actively employed in models of the kinetics of heterogeneous processes [85-87, 138, 139] have been analyzed by the author in [140].

2 The Equilibrium State of Homogeneous Systems

2.1. Homogeneous Systems

Consider a homogeneous lattice system, any site of which can be occupied by a particle of species i, $1 \leqslant i \leqslant s$, where s is the number of components. Such a model corresponds to three situations [77, 109].

1. If a lattice is two-dimensional and one of the states of site occupancy relates to vacant sites, this is a model of adsorption of $(s - 1)$ components of the gas phase on a homogeneous surface. The lattice sites are regions of localization of the adsorbed particles formed as a result of their interaction with atoms of the solid.

2. If a lattice is three-dimensional and one of the states of site occupancy relates to vacant sites, this is a model of the absorption of $(s - 1)$ components of the gas phase by a homogeneous absorbent. The lattice sites are interstices of the lattice of the solid's atoms.

3. The initial formulation corresponds to an s-component solid solution of the substitutional type in a disordered state. The lattice sites are the positions of the particle centers, while the lattice itself exists at the expense of interaction between the atoms of the solid solution.

We shall take account of the interaction between particles at neighboring lattice sites. For adsorption, this relates to particles in adjacent localization regions (at adjacent centers), while for absorption — in adjacent interstices. We shall assume the interactions of particles with vacant sites to be zero. These zero values must be taken into account only in the final expressions of the adsorption and absorption theories. In the present chapter, we shall be interested only in single-phase disordered systems when the probability of finding a particle of any species does not depend on the number of the lattice site.

We shall consider that the internal degrees of freedom of the particles do not depend on the change in the energy of particle interaction of neighboring particles when particles of one species are replaced by particles of another one. This allows us to use the one-particle contributions determined in (1.21) and (1.22) for an ideal lattice system in expression (1.18) for the distribution function of the particles of a lattice system. If we denote by ε_{ij} the parameters of two-body interaction of neighboring particles of species i and j, the potential energy of the system will be composed of all possible two-body contributions of the interacting particles, and the effective Hamiltonian (1.19) will be written as follows [141]:

$$H = \sum_{f,i} \nu_i \gamma_f^i - \frac{1}{2} \sum_{f,g} \sum_{i,j} \varepsilon_{ij} \gamma_f^i \gamma_g^j \tag{2.1}$$

where the sum over f is taken over all the lattice sites, and the sum over g is taken over all the z neighbors of site f. The sums over i and j are taken over all the states of occupancy of the lattice sites. The factor 1/2 takes into account that each pair participates twice in the summation.

In the cluster approach, the initial lattice system is depicted on a set of clusters. Let us consider only clusters with one central site. The initial lattice will be depicted on clusters consisting of $z + 1$ sites, i.e. the central site and its z neighbors (see Fig. 1.5). For each cluster, we can introduce the quantities h_j showing the contribution of the central particles to the total energy of the systems (the cluster Hamiltonians). When account is taken of the interaction of the nearest neighbors, the cluster Hamiltonian for a cluser with one central site has the form

$$h_f = \sum_{i=1}^{s} h_f^i, \; h_f^i = (\nu_i - \sum_g \sum_{j=1}^{s} \varepsilon_{ij} \gamma_g^j) \gamma_f^i \tag{2.2}$$

Let us determine two types of cluster correlators. Those of the first type, by definitions (1.8) and (1.18), can be represented as follows:

$$\theta_{fg_1 \ldots g_z}^{ij_1 \ldots j_z} = \frac{1}{Q} \sum_{i_1=1}^{s} \cdots \sum_{i_N=1}^{s} \gamma_f^i \prod_g \gamma_g^j \exp(-\beta H)$$

$$= \frac{1}{Q} \sum_{i_1=1}^{s} \cdots \sum_{i_N=1}^{s} \gamma_f^i \prod_{j=1}^{s} (\gamma_g^j)^{n_j} \exp[-\beta(h_f + G_f)] \equiv \theta_{\{[n]\}}^i \tag{2.3}$$

where $G_f = H - h_f$. The second equality determines the transition from an indication of the species of a particle at site g to the number of energy bonds of central particle i with neighboring particles j in the coordination sphere of a cluster. On a homogeneous lattice, the energy of a central particle does not depend on the numbers of the sites at which the particle and its neighbors are, therefore for our purposes the subscripts f, $g_1 \ldots g_z$ have been omitted. The square brackets in the functions $\theta^i_{\{[n]\}}$, where $\{[n]\} = [n_1 \ldots n_s]$, signify that the coordination sphere of a cluster contains exactly n_1 particles of species 1, exactly n_2 particles of species 2, etc. For unambiguously setting the state of occupancy of the sites of a coordination sphere, it is sufficient to set $(s - 1)$ values of n_j because

$$n_s = z - \sum_{j=1}^{s-1} n_j, \quad 0 \leqslant n_j \leqslant z \tag{2.4}$$

i.e. the particles of species s can be considered as an addition for the complete occupancy of the sites of the coordination sphere of a cluster. The braces signify a complete set of the values of n_j.

The correlations of the first type in (2.3) show the probability of finding a particle i at the center of a cluster, and n_j particles of species j, $1 \leqslant j \leqslant s$, at the sites of its coordination sphere.

The expression for G_f in (2.3) does not depend on the state of occupancy of site f, but does depend on the states of occupancy of the sites of its coordination sphere, consequently for any specific set of particles $\{[n]\}$ formula (2.3) can be rewritten as

$$\theta^i_{\{[n]\}} = \Lambda \exp \beta\Big(-\nu_i + \sum_{j=1}^{s} \varepsilon_{ij} n_j\Big), \quad \Lambda = \mathrm{const}\,(\{[n]\})/Q \tag{2.5}$$

where the unknown constant quantity const $(\{[n]\})$ takes into account the energy contributions from all the lattice sites except for the central site of the cluster accommodating a particle i.

The correlators of the second type are obtained from the correlators in (2.3) with averaging over the states of occupancy of part of the sites of the coordination sphere. Using the numbers of the energy bonds of the central particle with its neighbors, these correlators will be written in the form

$$\theta^i_{\{n\}} = \langle \gamma^i_f \prod_{j=1}^{s-1} (\gamma^j_g)^{n_j} \rangle, \quad 0 \leqslant n_j \leqslant z \tag{2.6}$$

Functions (2.6) characterize the probability of there being a particle i at the center, and at least n_1 particles of species 1, n_2 particles of species 2, etc., and at least n_{s-1} particles of species $s - 1$ in the coordination sphere of the cluster. In the given case, Eq. (2.4) is not observed, while in designation (2.6) the square brackets have been omitted. All the thermodynamic characteristics of a system are expressed in terms of the correlators (2.6). For instance, by (1.11), $\theta^i_{\{0\}} = \theta^i_{0\ldots0} = \langle\gamma^i_f\rangle = \theta_i$ is the probability of finding a particle i at any site of a lattice; fixing of the value of θ_i in the equations of the sorption (adsorption or absorption) isotherms determines the external pressure of the gas phase molecules, or vice versa. The functions $\theta^i_{0\ldots1_j\ldots0} = \langle\gamma^i_f\gamma^j_g\rangle$ (with the j-th subscript equal to unity) are the probabilities of finding particles i and j at neighboring sites of a lattice. Knowing these mean values, we can find the heats of sorption or mixing, the heat capacities, and so on.

The introduction of two types of correlators makes it possible to reduce the entire procedure of constructing systems of equations for correlators only to work with probability relations between them. We shall consider the stages of this work in detail for a binary system. The obtained equations relate to solid solutions and to the adsorption and absorption of molecules of the gas phase. After this, we shall go over to multicomponent systems.

2.2. Distribution of the Particles in a Binary Solution

Let us construct equations describing the local distribution of particles of A and B in a binary solution in the equilibrium state. For visualization, we shall take into account only the direct correlations between the central particle and its neighbors and disregard the mutual influence between the particles in the coordination sphere of a cluster (i.e. indirect correlations, their consideration will be treated later). To "justify" this assumption, we shall use the fact that the particles in the coordination sphere of a cluster can be at a larger distance from one another than their distance to the center of a cluster; the latter takes place for two-dimensional lattices with z = 2, 3, 4 and three-dimensional ones with z = 4, 6, or 8 (see Fig. 1.3).

In the first stage, we shall use expression (2.5) and consider the relation of the correlators of the first type differing in the species of the central particle, but having the same state of occupancy of the coordination sphere sites:

$$\theta^{\mathrm{B}}_{[n]} = \theta^{\mathrm{A}}_{[n]} M_n, \quad \mathrm{M}_n = \exp\left[-\beta(\nu + n\omega)\right]$$
$$\nu = \nu_{\mathrm{B}} - \nu_{\mathrm{A}} + z(\varepsilon_{\mathrm{AB}} - \varepsilon_{\mathrm{BB}}), \quad \omega = \varepsilon_{\mathrm{AA}} + \varepsilon_{\mathrm{BB}} - 2\varepsilon_{\mathrm{AB}} \tag{2.7}$$

where n is the number of particles of A in the coordination sphere of a cluster. The ratio of functions (2.5) cancels out the unknown quantities Q and const $(\{[n]\})$.

In the second stage, we shall consider the relation between the correlators of the first and second types. By the definition of correlators of the second type containing at least n particles of A in the coordination sphere of a cluster, we can write

$$\theta_n^{\mathrm{A}} = \sum_{k=0}^{z-n} b_n^k \theta_{[n+k]}^{\mathrm{A}}, \quad 0 \leqslant n \leqslant z \tag{2.8}$$

The coefficients of relation (2.8) are easy to find by the following reasoning. By adding another particle of A to n particles, we go over to a configuration containing $n + 1$ particles of A and characterized by the correlator $\theta_{[n+1]}^{\mathrm{A}}$. Such a transition is possible in $(z - n)$ ways. A similar transition to $(n + 2)$ particles is possible when two particles of A are added in $(z - n)(z - n - 1)/2$ ways, and so on. Consequently,

$$\theta_n^{\mathrm{A}} = \sum_{k=0}^{z-n} C_{z-n}^k \theta_{[n+k]}^{\mathrm{A}}, \quad C_m^k = m!/(k!(m - k)!) \tag{2.9}$$

Transforming linear system of equations (2.9) relative to correlations of the first type, we find

$$\theta_{[n]}^{\mathrm{A}} = \sum_{k=0}^{z-n} (-1)^k C_{z-n}^k \theta_{n+k}^{A} \tag{2.10}$$

Similar expressions exist for the central particle of B.

In the third stage, we insert expressions (2.10) for the central particles of A and B into relations (2.7):

$$\sum_{k=0}^{z-n} C_{z-n}^k (-1)^k \theta_{n+k}^{\mathrm{B}} = M_n \sum_{k=0}^{z-n} (-1)^k C_{z-n}^k \theta_{n+k}^{\mathrm{A}} \tag{2.11}$$

The linear system of equations between θ_n^{A} and θ_n^{B} is solved consecutively beginning with $n = z$. Its solution has the form

$$\theta_n^{\mathrm{B}} = \sum_{k=0}^{z-n} C_{z-n}^k M_{n+k} \sum_{r=0}^{z-n-k} (-1)^r C_{z-n-k}^r \theta_{n+k+r}^{\mathrm{A}}$$

$$= \exp\left[-\beta(\nu + n\omega)\right] \sum_{k=0}^{z-n} C_{z-n}^k x^k \theta_{n+k}^{\mathrm{A}},$$

$$x = \exp(-\beta\omega) - 1 \tag{2.12}$$

In the second equality, the addends have been regrouped so that the first sum changes the dimension of the correlators, while the energy factors are in the inner sum. The explicit form of M_n in (2.7) has also been taken into consideration.

The system of equations (2.12) is the initial one for calculating the cluster correlators. With the relevant consecutive consideration of all the lattice sites,

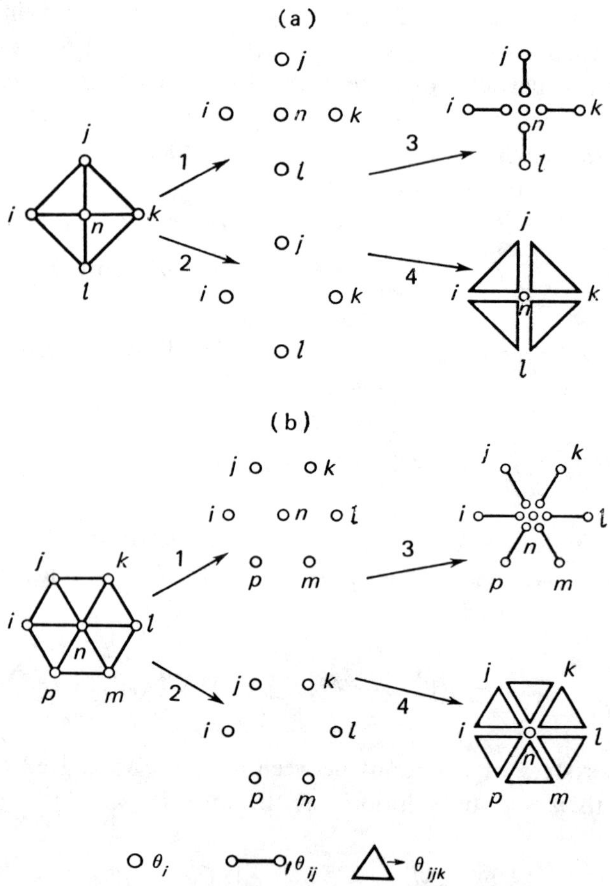

Fig. 2.1. Approximations of the probabilities of multiparticle configurations in two-dimensional lattices with $z = 4$ (*a*) and 6 (*b*): *1*—molecular-field approximation (MFA) and chaotic approximation (CA); *2*—polynomial approximation (PA); *3*—quasichemical approximation (QCA); *4*—account taken of indirect correlations in QCA

system (2.12) enables one in principle to obtain an exact solution. This has been done for a one-dimensional lattice [108] when employing an equivalent system of equations (Ap. 1.3).

For two- and three-dimensional lattices, the number of equations of system (2.12) is less than the number of unknowns θ_n^A and θ_n^B [with account of the normalizing condition (1.11) and the condition of the invariance of occupying the central and its neighboring sites of a cluster], consequently it must be closed. Depending on how system of equations (2.12) is closed by approximating the higher correlators in terms of correlators with a smaller dimension, we obtain various approximations differing in the accuracy of taking the correlation effects into account. The structure of the obtained equations is such that it is necessary to formulate a general rule for the approximation of all the higher correlators in terms of the lower ones that determines the functional dependence of the closed system for the correlators that are not uncoupled. This, in turn, enables us, after finding the correlators remaining uncoupled, to self-consistently calculate all the higher correlators of both the first and second types in the approximation being considered.

Let us treat the simplest ways of approximating system of equations (2.12) that result in quasichemical, chaotic, molecular-field, and polynomial approximations (see Appendix 1) (Fig. 2.1).

Quasichemical Approximation [142-144]. Leaving the two-body correlator $\theta_{AA} = \langle \gamma_f^A \gamma_g^A \rangle$ coupled, let us write $\theta_n^A = \theta_A t_{AA}^n$, where $t_{AA} = \theta_{AA}/\theta_A$ is the conditional probability of finding an A particle next to another A particle. Now system (2.12) can be rewritten as

$$\theta_n^B = \exp(-\beta\nu)\theta_A(1 + t_{AA}x)^{z-n}(t_{AA}\exp(-\beta\omega))^n \tag{2.13}$$

where the unknown θ_A and t_{AA} are determined from expressions (2.13) with $n = 0$ and $n = 1$:

$$\begin{aligned} \theta_B &= \theta_A \exp(-\beta\nu)(1 + xt_{AA})^z, \quad \theta_0^B = \theta_B = 1 - \theta_A \\ t_{AA} &= 1 - 2\theta_B/(1 + \delta), \quad \delta = \{1 + 4\theta_A\theta_B(\exp\beta\omega - 1)\}^{1/2} \end{aligned} \tag{2.14}$$

Chaotic Approximation [145]. We postulate the uncoupling (1.17) $\theta_n^A = \theta_A^{n+1}$; hence from (2.12) we obtain

$$\theta_n^B = \theta_A^{n+1}\exp[-\beta(\nu + \omega n)](1 + x\theta_A)^{z-n} \tag{2.15}$$

where θ_A is determined by the equation (for $n = 0$)

$$1 - \theta_A = \theta_A \exp(-\beta\nu)(1 + x\theta_A)^z \tag{2.16}$$

Molecular-Field Approximation [146]. This approximation is also based on uncoupling (1.17), but with additional account of the limiting process $\beta\omega \to 0$: $1 + x\theta_A \approx \exp(-\beta\omega\theta_A)$ with the subsequent extrapolation of the obtained relation for θ_A to the terminal values of $\beta\omega$:

$$1 - \theta_A = \theta_A \exp[-\beta(\nu + z\omega\theta_A) \quad (2.17)$$

Having obtained self-consistent expressions for correlators of the second type by formula (2.10), let us determine the correlators of the first type:
for the quasichemical approximation (i = A, B)

$$\theta^i_{[n]} = \theta_i(t_{iA})^n(t_{iB})^{z-n}, \quad t_{AB} = \frac{\theta_A(1 - t_{AA})}{1 - \theta_A}, \quad t_{BB} = \frac{1 - 2\theta_A + \theta_A t_{AA}}{1 - \theta_A} \quad (2.18)$$

for the chaotic and molecular-field approximations:

$$\theta^i_{[n]} = \theta_i(\theta_A)^n(\theta_B)^{z-n} \quad (2.19)$$

In the quasichemical and polynomial approximations, account is taken of the effects of correlation of the mutual distribution of the particles, while in the chaotic and molecular-field approximations, the distribution of the particles is considered to be just as chaotic as for particles that do not interact. It has been shown by the author [141] that the last condition can be considered together with system of equations (2.12) only when $\beta\omega \to 0$. Therefore, the failure to take into consideration the correlation effects does not enable us to self-consistently calculate the complete set of correlators for large values of $\beta\omega$. In particular, in the chaotic approximation, this leads to the disturbance of symmetry relative to the choice of the particles of A and B from their mixture (here B or A are considered as a supplement) [141].

It will be shown below that the correlation effects play an important role in describing the equilibrium and kinetic characteristics of processes in condensed states. The final expressions of the mathematical models depend not only on the initial physical and physicochemical prerequisites of the models, but also on the way of closing the systems of equations describing the distribution of the particles. The accuracy of describing a process being considered depends on the accuracy of taking into account the correlation effects.

Approximations Taking Indirect Correlations into Account. A more accurate description of particle distribution is associated with the consideration of what are called indirect correlations between the particles in the coordination sphere of a cluster (in contrast to the direct correlations between the cen-

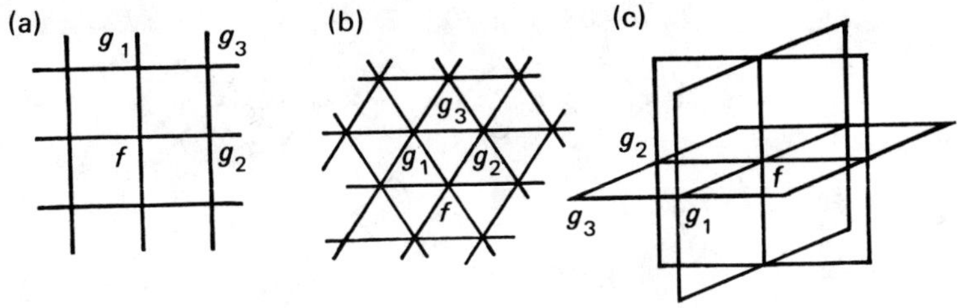

Fig. 2.2. Nature of the indirect correlations between sites g_1 and g_2

tral particle of a cluster and its neighbors in the coordination sphere). In Fig. 2.2, the particles at sites g_1 and g_2 are in the coordination sphere of site f. The states of occupancy of particles g_1 and g_2 are related to each other at the expense of their interaction with the particle at site g_3 in the same way as with the particle at site f. It must therefore be taken into account that the states of occupancy of sites g_1 and g_2 are not independent, as was assumed above. [Figure 2.1 shows schematically the approximations of multiparticle correlations used for closing the system of equations (2.12).] This signifies that if the indirect correlations are taken into account, the multitude of configurations of particles with an identical number n will break up into a set of configurations differing in the probabilities of their realization. We shall characterize these sets by the index $\sigma(n)$. Figure 2.3 presents a complete set of configurations of coordination sphere particles for two-dimensional lattices with $z = 4$ and 6.

As previously, we shall introduce correlators of two types [132]. The correlators of the first type $\theta^i_{[n\sigma]}$ show the probability of finding exactly n particles of A around the central particle i; they form the configuration $\sigma(n)$. The correlators of the second type $\theta^i_{n\sigma}$ indicate the probability of finding around a particle i at least n particles of A in various permissible configurations $\sigma(n + k)$ that are realized with all possible ways of adding the next k particles to the initial configuration $\sigma(n)$. By definition, we have

$$\theta^i_{n\sigma} = \sum_{k=0}^{z-n} \sum_{\sigma(n+k)} b^{\sigma(n)}_{\sigma(n+k)} \theta^i_{[n+k,\sigma]}, \quad \sum_{\sigma(n+k)} b^{\sigma(n)}_{\sigma(n+k)} = C^k_{z-n} \qquad (2.20)$$

where $b^{\sigma(n)}_{\sigma(n+k)}$ is the number of ways of going over from the configurations $\sigma(n)$ to the configuration $\sigma(n + k)$. The sum over $\sigma(n + k)$ is taken over all the possible kinds of configurations with the given number $(n + k)$ of A parti-

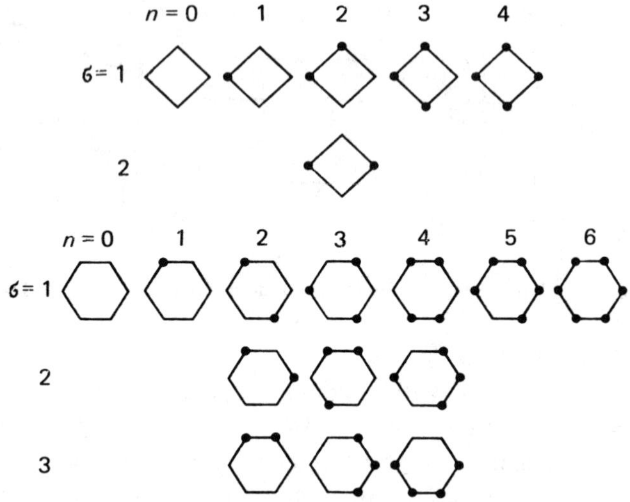

Fig. 2.3. Configuration of particles in the first coordination sphere of central sites of two-dimensional lattices with $z = 4$ and 6

cles that are realized with their consecutive addition. The quantities $b^{\sigma(n)}_{\sigma(n+k)}$ are calculated elementarily by the kind of configurations.

Transformation of system (2.20) allows us to write

$$\theta^{i}_{[n\sigma]} = \sum_{k=0}^{z-n} (-1)^k \sum_{\sigma(n+k)} b^{\sigma(n)}_{\sigma(n+k)} \theta^{i}_{n+k,\sigma} \tag{2.21}$$

Let us insert expressions (2.21) into a system of equations of the type of (2.7) differing in the presence of the subscript σ for correlators of the first type and solve it with respect to the correlators of the second type. After regrouping of the addends and having in view that the interaction of the central particle with its neighbors does not depend on their configuration, we obtain a generalization of Eqs. (2.12):

$$\theta^{\mathrm{B}}_{n\sigma} = \exp[-\beta(\nu + n\omega)] \sum_{k=0}^{z-n} x^k \sum_{\sigma(n+k)} b^{\sigma(n)}_{\sigma(n+k)} \theta^{\mathrm{A}}_{n+k,\sigma} \tag{2.22}$$

Comparing expressions (2.12) and (2.22), we can conclude that the procedure of obtaining equations taking the indirect correlators into account can be simplified if in the initial stage we consider only the number of particles in a coordination sphere, and then go over from $C^{k}_{z-n}\theta^{\mathrm{A}}_{n+k}$ to

$\sum_{\sigma(n+k)} b^{\sigma(n)}_{\sigma(n+k)} \theta^{A}_{n+k,\sigma}$ and take into account the indirect correlators in the stage of closing the system of equations.

The consideration of the indirect correlations increases the dimension of the lower correlators in terms of which the higher ones are uncoupled and, accordingly, increases the dimension of the closed system of equations. For example, for a two-dimensional lattice $z = 6$, the higher correlators are uncoupled in terms of the correlators θ_A, θ_{AA}, and θ^A_{23} (see Figs. 2.1 and 2.3):

$$\theta^A_{21} = \theta^A_{22} = \theta^2_{AA}/\theta_A, \ \theta^A_{33} = (\theta^A_{23})^2/\theta_{AA}, \ \theta^A_{32} = \theta^A_{23}\theta_{AA}/\theta_A$$
$$\theta^A_{31} = \theta^3_{AA}/\theta^2_A, \ \theta^A_{43} = (\theta^A_{23})^3/\theta^2_{AA}, \ \theta^A_{41} = \theta^A_{42} = (\theta^A_{23})^2/\theta_A$$
$$\theta^A_{51} = (\theta^A_{23})^4/\theta^3_{AA}, \ \theta^A_{61} = \theta_A(\theta^A_{23}/\theta_{AA})^6$$

By inserting these expressions into the right-hand sides of Eqs. (2.22), we shall have closed equations relative to θ_A, θ_{AA}, and θ^A_{23}.

The equations taking into account the indirect correlators in a polynomial approximation are constructed similarly (Appendix 1, Sec. 1).

2.3. Isotherm and Heat of Adsorption

A binary system in which component A signifies adsorbed particles and component B — vacant sites on the surface with the parameters

$$\nu_v = 0, \ \varepsilon_{i,v} = 0, \ i = A, v \tag{2.23}$$

is a model of an adsorption process.

The one-particle contribution of component A is determined by expression (1.23); we shall omit the subscript A in θ_A, t_{AA}, P_A, and in the parameters of the model ε_{AA}, y_A, and m_A in the present section.

Equations (2.14), (2.16), and (2.17) obtained in Sec. 2.2, and also Eqs. (Ap. 1.5) and (Ap. 1.6) relating the fraction θ of occupied surface sites θ and the pressure P of the A molecules in the gas phase are equations of adsorption isotherms in various approximations. With a view to Eqs. (1.23) and (2.23), they can be rewritten as follows:

$$y = \frac{\theta}{1-\theta} S^z, \ S = \begin{cases} 1 + xt & \text{QCA} \\ 1 + x\theta & \text{CA} \\ \exp(-\beta\varepsilon\theta) & \text{MFA} \end{cases} \tag{2.24}$$

where x and t are determined in Eqs. (2.13) and (2.14) for $\omega = \varepsilon$. The factor S takes into consideration the nonideal nature of the adsorption system in the quasichemical (QCA), chaotic (CA), and molecular-field (MFA) approximations. The equation of the isotherm in the quasichemical approximation differs from the traditional one [147], but both shapes are equivalent to each other. In the absence of interaction, all the isotherms transform into the Langmuir isotherm [23], and in the region of exceedingly small coverages — into the Henry isotherm, namely, $y = \theta$ for any interactions.

In the region of large surface coverages, the adsorption isotherms in various approximations can be written in the form

$$y \sim \frac{\exp(-\beta z\varepsilon)}{1-\theta} \tag{2.25}$$

Repulsion between particles ($\varepsilon < 0$) causes an increase, and attraction ($\varepsilon > 0$), a decrease in the value of the pressure at which large surface coverages are achieved in comparison with the absence of interaction.

The interaction between particles changes the concentration dependences of the thermodynamic characteristics of adsorption systems. The contribution of this interaction is proportional to the average number of pairs of neighboring particles, therefore we shall briefly consider the concentration dependence of $\theta_{AA}(\theta)$. By the formulas of Sec. 2.2, this dependence has the form

$$\theta_{AA} = \begin{cases} \theta^2 & \text{MFA, CA} \\ \theta t & \text{QCA} \end{cases} \tag{2.26}$$

where t is determined in formula (2.14).

In the region of low surface coverages for quasichemical and polynomial (Appendix 1) approximations, the average fraction of pairs of particles is appraised as

$$\theta_{AA} = \theta^2 \exp \beta\varepsilon \tag{2.27}$$

A growth in the repulsion of particles lowers their probability of being at neighboring sites in comparison with a chaotic distribution, while a growth in the attraction increases this probability. Figure 2.4 shows the concentration dependences of θ_{AA} for repelling and attracting particles in various approximations. In the quasichemical approximation, the correlation effects are overstated [77, 148], therefore the region of coverages for which relation (2.27) is observed is larger than in the polynomial approximation.

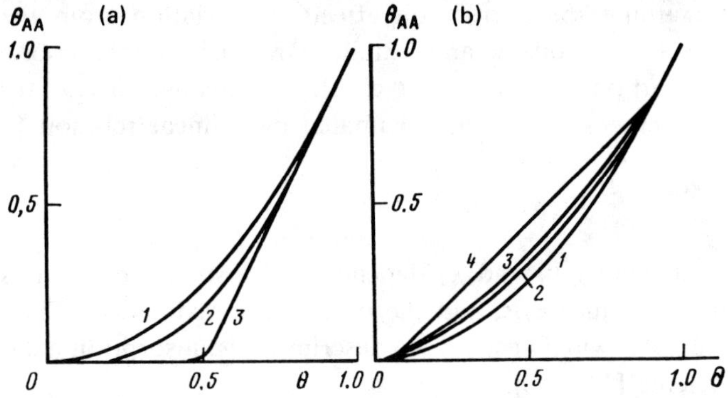

Fig. 2.4. Concentration dependences of θ_{AA} for repelling (*a*) and attracting (*b*) particles:

1—no correlation effects; *2*—polynomial approximation; *3*—quasichemical approximation; *4*—account taken of two-dimensional condensation in the QCA

Adsorption isotherms in various approximations are depicted in Fig. 2.5. These curves are typical for chemisorbed systems ($\varepsilon < 0$). In the regions of small coverages $\theta < 0.2$, the adsorption isotherms in the approximations taking into account the correlation effects differ insignificantly from the Langmuir isotherm, while at $\theta \geqslant 0.2$ owing to repulsion of the particles, a sharp increase in the pressure is needed to achieve this coverage. In the approxima-

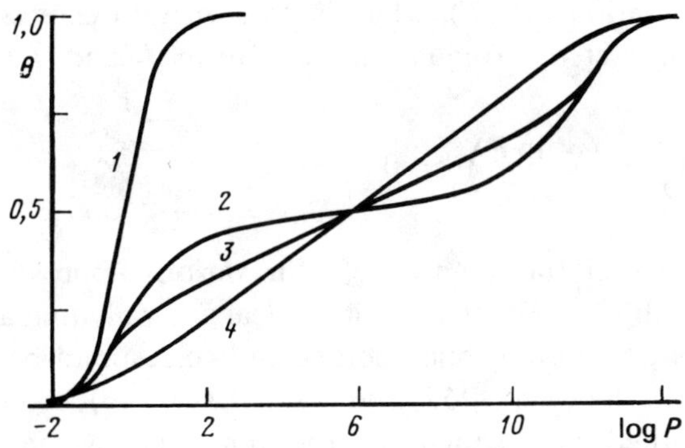

Fig. 2.5. Chemisorption isotherms:

1—no interaction; *2*—quasichemical approximation; *3*—polynomial approximation; *4*—molecular-field approximation

tions disregarding the correlation effects, a deviation from the Langmuir isotherm occurs at considerably smaller coverages. In the coordinates θ vs. $\ln y$, the adsorption isotherms in all the approximations in the region of moderate coverages can be approximated by a linear relation [149]:

$$\theta = \frac{1}{k_1} \ln y + k_2, \quad 0.3 < \theta < 0.7 \tag{2.28}$$

where the constants k_1 and k_2 depend on the structure of the surface, the dimensionless quantity $\beta\varepsilon$, and the type of approximation. The logarithmic isotherm (2.28) is widely used when describing chemisorption data in the most diverse systems [33, 150].

The attraction between adsorbed particles causes a more rapid growth in the surface coverage with an increase in the pressure than when there is no interaction. Lowering of the temperature may be attended by the phenomenon of two-dimensional condensation (stratification), as a result of which the adsorption system becomes a two-phase one. This results in the appearance on the isotherm of portions with a negative compressibility (a loop of the van der Waals type). The two-dimensional pressure π of the particles is expressed in terms of the adsorption isotherm as follows [151]

$$\pi = \beta^{-1} \int_0^P \theta(P) d \ln P \tag{2.29}$$

The critical conditions for the density of the particles θ_c and the temperature T_c below which stratification of the phases occurs is determined by the point of inflection of the relation $\pi(\theta, T)$, which with a view to expression (2.29) leads to the following system of equations relative to θ_c and T_c:

$$\left(\frac{\partial \ln P}{\partial \theta}\right)_T = \left(\frac{\partial^2 \ln P}{\partial \theta^2}\right)_T = 0 \tag{2.30}$$

The numerical values of the quantity $\beta_c\varepsilon$ for various approximations at $\theta_c = 0.5$ are given in Table 2.1. For the quasichemical and polynomial approximations, these values are close to one another and correspond better to the exact values of $\beta_c\varepsilon = 1.76$ ($z = 4$) [152] and $\beta_c\varepsilon = 1.1$ ($z = 6$) [153] for $d = 2$ than the molecular-field approximation (the remaining values given in Table 2.1 are considered below).

It is known from the theory of phase transitions [154] that the cluster methods give only a qualitative description in the critical regions. All the

Table 2.1. Critical Temperature ($\beta_c\varepsilon$) for a 50% Lattice Coverage

Lattice	MFA	PA	QCA	QCA1
$z = 3$, $d = 2$	1.3(3)	1.92	2.2	—
$z = 4$, $d = 2$	1.0	1.30	1.39	1.433
$z = 6$, $d = 2$	0.6(6)	0.79(0.86)	0.81(0.85)	0.850
$z = 6$, $d = 3$	0.6(6)	0.79	0.81	0.845

Note. The results in parentheses for $z = 6$ and $d = 2$ correspond to account being taken of indirect correlations with a three-body correlator. QCA1 signifies that account has been taken of the indirect correlations in a two-body approximation.

considered approximations, differing in the accuracy of taking into account the effects of correlation of short-range order in the vicinity of the critical points, do not reflect large-scale fluctuations, and scaling approximations should be used instead of them. The region of critical temperatures is appraised by the relation $|1 - T/T_c| \leqslant 1/z$. Outside this region, the cluster methods may be employed for a quantitative description of the characteristics of a system.

When stratification occurs, the adsorption isotherm breaks up into three portions (Fig. 2.6). At low densities, the curve grows monotonically up to a value of θ^- equal to the density of the rarefied coexisting phase. On the portion $\theta^- < \theta < \theta^+$, where θ^+ is the density of the dense coexisting phase, a

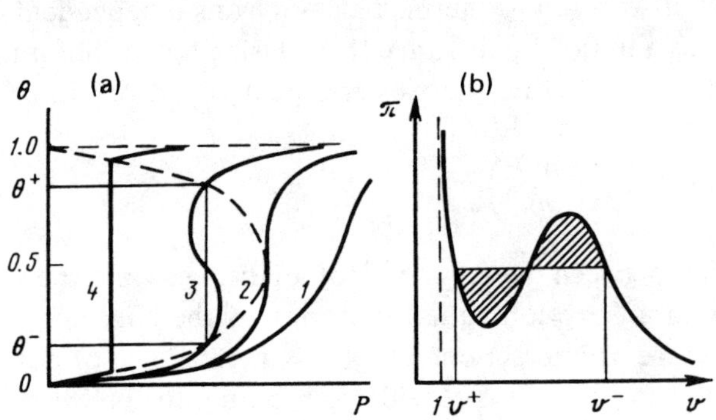

Fig. 2.6. Adsorption isotherms for attracted particles (*a*) and "Maxwell's rule" in the coordinates π vs. v (equality of the hatched areas) (*b*):

1—$T > T_c$; 2—$T = T_c$; 3, 4—$T < T_c$

jump in the density occurs at a constant pressure of the gas phase. At coverages exceeding θ^+, the fractional coverage grows monotonically with an increasing pressure. The position of the density jump can be determined with the aid of Maxwell's rule [92] enabling one to exclude a section with a negative compressibility on isotherms. In lattice systems, this rule is employed for the equation of state $\pi(v)$, where $v = 1/\theta$ is the volume per particle, because the number of lattice sites is constant, and the number of particles changes instead of the traditional change in the volume of the gas with a constant number of particles. The dashed curve indicates the region of coexistence of the two phases. For illustrative purposes, the curves in Fig. 2.6 have been plotted for a constant value of the adsorption coefficient (for a number of substances having different interaction parameters at a certain temperature). When the temperature lowers for a substance, the adsorption coefficient increases, and the pressure at which stratification of the adsoprtion system occurs sharply drops.

Curve *4* in Fig. 2.4 has been calculated with a view to two-dimensional condensation:

$$\theta_{AA}(\theta) = \lambda\theta_{AA}(\theta^+) + (1 - \lambda)\theta_{AA}(\theta^-), \ \lambda = \frac{\theta - \theta^-}{\theta^+ - \theta^-} \tag{2.26a}$$

where λ is the fraction of the particles in the dense phase.

A very important characteristic of adsoprtion systems is the heat of adsorption. It reflects the energetics of an adsorption process considerably more accurately than an adsorption isotherm. This is why its independent calorimetric determination has a major significance [155]. Using a relation of the Clausius-Clapeyron type, let us define the isosteric heat of adsorption:

$$Q(\theta) = -\left(\frac{\partial \ln P}{\partial \beta}\right)_\theta \tag{2.31}$$

where the relation between the pressure P of an ideal gas and the surface coverage is given by an adsorption equation in one of the considered approximations. In the absence of interaction, i.e. $Q(\theta) = \varepsilon_A$, the heat of adsoprtion does not depend on the coverage. It is considered here that the preexponential factor of the adsorption coefficient depends only slightly on the temperature.

In accordance with definition (2.31) and the equations of the adsoprtion isotherms indicated above, the concentration dependence of the heat of adsorption can be expressed as follows:

$$Q = \varepsilon_A + z\varepsilon\varphi(\theta)m, \tag{2.32}$$

$$\varphi(\theta) = \begin{cases} [1 - (1 - 2\theta)/\delta]/2 & \text{QCA} \\ \theta \exp(-\beta\varepsilon)/(1 + x\theta) & \text{CA} \\ \theta & \text{MFA} \end{cases}$$

The nature of the change in the heat of adsorption with increasing coverage is determined by the sign of the interaction parameter: repulsion diminishes, and attraction increases the heat of adsorption.

In the region of low coverages in the quasichemical and polynominal (see Appendix 1) approximations, we have

$$\varphi \approx \theta \exp \beta\varepsilon \tag{2.33}$$

For the repulsion of particles, this suppresses the contribution of particle interaction because the particles tend to take up positions at greater distances than their nearest neighbors, and the heat of adsorption remains practically constant. For the attraction of particles, the latter tend to come together, and the heat of adsorption grows. In the molecular-field approximation, regardless of the magnitude of the interaction parameter, a linear change of $\varphi(\theta)$ occurs. In a chaotic approximation, where symmetry with respect to the choice of the species of particles is disturbed, we have $\varphi(\theta) = \theta \exp(-\beta\varepsilon)$. This sharply diminishes the heat in the repulsion of particles and slightly increases it in their attraction (the region of low coverages).

Figure 2.7 shows the concentration φ against the coverage θ in various approximations. All the curves except for the one calculated in the chaotic approximation pass through the point $[\theta = 0.5;\ \varphi(0.5) = 0.5]$ and are antisymmetric relative to it. In the quasichemical approximation, the region of the virtual constancy of the heat of adsorption with increasing repulsion tends to $\theta \approx 0.5$, while in the polynomial approximation it reaches $\theta \approx 0.2\text{-}0.25$. In the polynomial approximation, $\varphi(\theta)$ has typical steps associated with the consecutive filling of the units of the coordination sphere of a cluster. When repulsion grows, the steps become sharper, but at $\beta z\varepsilon \leqslant 20\text{-}24$ they become blurred and in the region of moderate coverages they can be approximated by a linear relation [156, 157]. As a result, in the region of moderate coverages, each approximation has its own slope of φ against θ.

The total drop in the heat of adsorption is expressed as follows:

$$\Delta Q = Q(\theta = 0) - Q(\theta = 1) = -z\varepsilon m \tag{2.34}$$

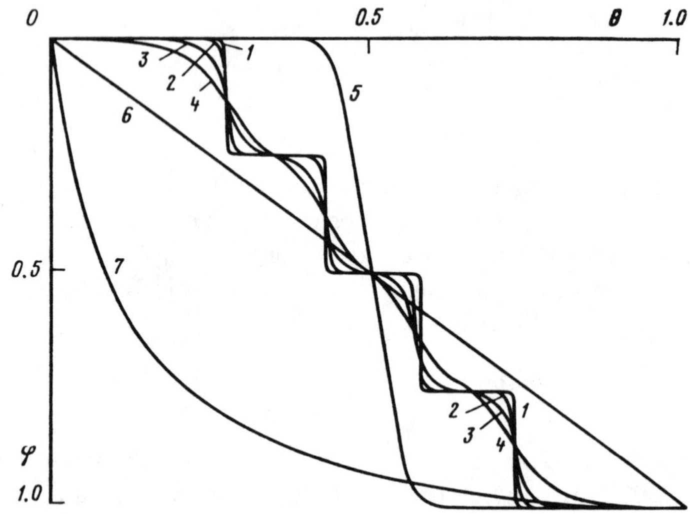

Fig. 2.7. Heats of chemisorption in PA (*1-4*), QCA (*5*), MFA (*6*), and CA (*7*) at $z = 4$:

1—$\beta\varepsilon = -10.8$; *2, 5, 7*—$\beta\varepsilon = -7.8$; *3*—$\beta\varepsilon = -6.0$; *4*—$\beta\varepsilon = -5.0$

This expression makes it possible to unambigously relate the parameter ε of theory to the experimental characteristic of the adsorption process and use it to verify whether the model being considered ensures a self-consistent description of various thermodynamic characteristics of a system, i.e. whether a model on the basis of measurements of one of the characteristics of a system from which its parameters are determined can correctly predict the behavior of the other characteristics. This was verified for the chemisorption systems H_2-Fe, Mo, W [149]. The interaction parameter was determined from the complete drop in the heat of adsorption according to calorimetric data, after which the adsorption isotherms being calculated were compared with the experimental ones. The taking into consideration of the correlation effects in the polynomial approximation was found to ensure the self-consistence of the description of the equilibrium adsorption characteristics. Overstating or understating of the correlation effects leads to a discrepancy exceeding the error of the experimental measurements. One must regard the obtained result critically because the experiments were run with polycrystalline specimens. But this does not lower the value of the very fact of self-consistence because all the approximations "are" in identical conditions. Similar conclusions on the

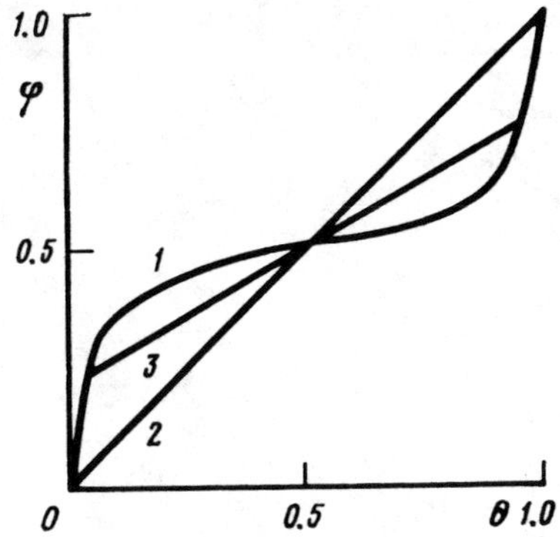

Fig. 2.8. Heats of adsorption of attracted particles ($\beta\varepsilon = 1.65$):
1—QCA; *2*—MFA; *3*—QCA (two-phase system)

need to take account of the correlation effects for the self-consistent description of the system's characteristics were obtained [158] for the chemisorption of potassium and barium on tungsten.

The heats of adsorption of attracted particles (physical adsorption) are shown in Fig. 2.8. With increasing coverage, the heat of adsorption grows. The nature of how the factor φ [see formula (2.32)] depends on the coverage is influenced by account being taken of the correlation effects and by the stratification of the adsorption dependence. Curve *3* consists of three regions: with small and large coverages, relations *1* and *3* coincide, while when $\theta^- < \theta < \theta^+$, the heat of adsorption changes linearly with an increasing coverage.

Consideration of the indirect correlations changes the concentration dependences of the adsorption characteristics. Attention was first given to this in works on the theory of solutions for three-dimensional lattices $z = 12$ [159] and on the theory of adsorption for a two-dimensional lattice $z = 6$ [160]. Indirect correlations change the relation between Q and θ. With repulsion of the particles, two jumps in this relation are observed at $\theta \approx 1/3$ and $2/3$ instead of one at $\theta \approx 1/2$ in the quasichemical approximation. Consideration of indirect correlations in a polynomial approximation results in quantitative changes, leaving the nature of the relation between Q and θ unchanged [161]. Similar quantitative changes are observed for the adsorption isotherm, the contribution of the indirect correlations to the change in the values of the

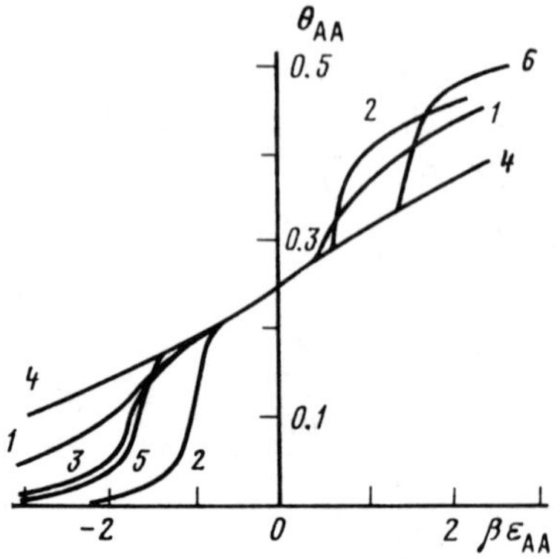

Fig. 2.9. Dependence of θ_{AA} on $\beta\varepsilon_{AA}$ at $\theta = 0.5$ for a square lattice (where $z = 6$ and $d = 3$ only for curve *2*):

1, 2—account of indirect correlations; *3*—Onsager's exact solution; *4*—quasichemical approximation; *5*—account of ordering in QCA; *6*—account of two-dimensional condensation in QCA

parameters k_1 and k_2 [see Eq. (2.28)] being greater than their difference for different lattices $z = 4$ and 6, and for the relation between θ_{AA} and θ. This is why it is good to consider the contributions of the indirect correlations for a self-consistent description of various thermodynamic characteristics of adsorption systems.

For the attraction of particles, the indirect correlations improve the critical values of the quantity $\beta_c\varepsilon$ somewhat (see Table 2.1). The table also shows the results with two-body approximation of the three-body correlator θ_{23}^{A} in terms of θ_{AA}, namely, $\theta_{23}^{A} = \theta_A^3(\xi_{AA})^3$, where $\xi_{AA} = \theta_{AA}/\theta_A^2$. This enables us to reduce by one the dimension of the closed system of equations (2.22). The results of calculations ($\varepsilon_{AA} \gtreqless 0$) with an uncoupled three-body correlator and its two-body approximation also retaining the effects of indirect correlations point to their virtually complete coincidence.

In the critical region, all the approximations yield only a qualitative description. Outside of it, consideration of the indirect correlations improves the description of the equilibrium characteristics of sorption. This can be seen quite clearly from the relation between θ_{AA} and $\beta\varepsilon_{AA}$ with a constant lattice coverage (Fig. 2.9). The consideration of the indirect correlations is much closer to Onsager's exact solution [162] than the quasichemical approximation, although it is also a rather rough approximation at $\beta\varepsilon < -2.0$ (in the region of order of the particles). For attracted particles at $\beta\varepsilon > 0.7$, curves *1* and *2* begin to differ appreciably from curve *4* as they approach the point of a first-order phase transition. Their behavior shows that the consideration of the indirect correlations yields a correct description of particle distribution in the region of condensation. Curve *6* reflects the behavior of θ_{AA} in a two-phase adsorption system (calculations in the quasichemical approximation with the use of Maxwell's rule).

2.4. Particle Distribution in a Multicomponent Solution

The construction of systems of equations for the local correlators of a multicomponent solution [141], as for a binary one, consists of three stages. We shall again, as previously, consider clusters with one central site and limit ourselves to taking only the direct correlations into account. To simplify the notation, let $s = 3$. We shall characterize the state of occupancy of the coordination sphere sites by the indices $\{[n]\} = [n_1 n_2]$, where n_1 is the number of particles of the first species, n_2 — of the second, while the particles of the third species are considered a supplement to the complete occupation of the coordination sphere (2.4), we shall denote them by s.

The coupled relations between the correlators of the first type (2.5) yield the following expressions for $i \neq j$:

$$\theta^i_{[n_1 n_2]} = \theta^j_{[n_1 n_2]} M^{ij}_{n_1 n_2}, \quad M^{ij}_{n_1 n_2} = y^i_{n_1 n_2} / y^j_{n_1 n_2}$$
$$y^i_{n_1 n_2} = \exp \beta[-\nu_i + z\varepsilon_{is} + \sum_{j=1}^{2} n_j(\varepsilon_{ij} - \varepsilon_{is})] \qquad (2.35)$$

According to the definition of the correlators of the second type $\theta^i_{n_1 n_2}$ in (2.6) containing at least n_1 particles of the first species and at least n_2 particles of the second species, they can be represented as the sum of all the correlators of the first type for which the subscripts of the particles of the first and second species are greater than or equal to the subscripts n_1 and n_2, respectively. As for a binary solution, it is simple to show that the number of ways of going over from a configuration containing n_1 and n_2 particles of the first and second species (its probability is $\theta^i_{[n_1 n_2]}$) to a configuration containing $(n_1 + k_1)$ and $(n_2 + k_2)$ particles of the first and second species (its probability is $\theta^i_{[n_1+k_1, n_2+k_2]}$) equals the number of combinations of $\tilde{z} = (z - n_1 - n_2)$ sites of the coordination sphere of a cluster taken k_1, k_2 particles at a time:

$$\theta^i_{n_1n_2} = \sum_{k_1=0}^{\tilde{z}} \sum_{k_2=0}^{\tilde{z}-k_1} C_{\tilde{z}}^{k_1k_2}\theta^i_{[n_1+k_1,n_2+k_2]}, \quad C_{\tilde{z}}^{k_1k_2} = \frac{\tilde{z}!}{k_1!k_2!(\tilde{z}-k_1-k_2)!} \tag{2.36}$$

System of equations (2.36) can be inverted by expressing the correlators of the first type in terms of those of the second type:

$$\theta^i_{[n_1n_2]} = \sum_{k_1=0}^{\tilde{z}} \sum_{k_2=0}^{\tilde{z}-k_1} (-1)^{k_1+k_2} C_{\tilde{z}}^{k_1k_2}\theta^i_{n_1+k_1,n_2+k_2} \tag{2.37}$$

Inserting expressions (2.37) into relations (2.35), we obtain a linear system of equations relating the correlators of the second type to various central particles. Its form is similar to that of system (2.11). The system is solved consecutively beginning with $n_1 = z$ and $n_2 = 0$. Next the relevant expressions are written for the correlators with $n_1 = z - l$, $n_2 = l$, $0 \leqslant l \leqslant z$, and then for the correlators with $n_1 = z - 1 - l$, $n_2 = l$, $0 \leqslant l \leqslant z - 1$, and so on, which results in a system of equations of the type of (2.12). The right-hand sides of the obtained system can be regrouped by taking into account the explicit form of $M^{ij}_{n_1n_2}$ and collecting the addends with identical functions $\theta^j_{n_1n_2}$. As a result, the system of equations for cluster correlators has the form

$$\begin{gathered}
\theta^i_{n_1n_2} = \exp\beta\left(\nu_{ij} - \sum_{m=1}^{2}\omega_{ij_m}n_m\right)\sum_{k_1=0}^{\tilde{z}}\sum_{k_2=0}^{\tilde{z}-k_1} C_{\tilde{z}}^{k_1k_2}x_{ij1}^{k_1}\,x_{ij2}^{k_2}\,\theta^j_{n_1+k_1,n_2+k_2} \\
\nu_{ij} = \nu_j - \nu_i + z(\varepsilon_{is} - \varepsilon_{js}) \\
x_{ijm} = \exp(-\beta\omega_{ijm}) - 1, \quad \omega_{ijm} = \varepsilon_{jm} - \varepsilon_{js} - \varepsilon_{im} + \varepsilon_{is}
\end{gathered} \tag{2.38}$$

System (2.38) is the initial one for calculating the probabilities of the configurations of a three-component solution. It is easily generalized to any number of components

$$\begin{gathered}
\theta^i_{\{n\}} = \exp\beta\left(\nu_{ij} - \sum_{m=1}^{s-1}\omega_{ijm}n_m\right)\sum_{k_1=0}^{\tilde{z}}\cdots\sum_{k_{s-1}=0}^{\tilde{z}-l} C_{\tilde{z}}^{\{k\}}\prod_{m=1}^{s-1}(x_{ijm})^{k_m}\theta^j_{\{n+k\}} \\
\tilde{z} = z - \sum_{m=1}^{s-1}n_m, \quad l = \sum_{m=1}^{s-2}k_m, \quad \tilde{k} = \sum_{m=1}^{s-1}k_m, \\
C_{\tilde{z}}^{\{k\}} = \tilde{z}!\Big/\left(\prod_{m=1}^{s-1}k_m!(\tilde{z}-\tilde{k})!\right)
\end{gathered} \tag{2.39}$$

where the three dots stand for the omitted summation signs from the second to the $(s-2)$ one over the species of particles, while $\{n+k\} = n_1 + k_1, \ldots, n_{s-1} + k_{s-1}$. The generalizations of expressions (2.36) and (2.37) for any number of components will be written as follows:

$$\theta^i_{\{n\}} = \sum_{k_1=0}^{\tilde{z}} \cdots \sum_{k_{s-1}=0}^{\tilde{z}-l} C_{\tilde{z}}^{\{k\}} \theta_{\{[n+k]\}}$$

$$\theta^i_{\{[n]\}} = \sum_{k_1=0}^{\tilde{z}} \cdots \sum_{k_{s-1}=0}^{\tilde{z}-l} (-1)^k C_{\tilde{z}}^{\{k\}} \theta^i_{\{n+k\}} \tag{2.40}$$

System of equations (2.39) is not closed—the number of unknown correlators exceeds the number of relations between them. To close it, we shall introduce the lacking bonds between the second-type correlators of different dimension. Let us consider the quasichemical and molecular-field approximations (the polynomial approximation is given in Appendix 1).

Quasichemical Approximation. Multiparticle correlators are expressed in terms of unary and two-body correlators:

$$\theta^s_{\{n\}} = \theta_i \prod_{m=1}^{s-1} (t_{im})^{n_m}, \quad t_{ij} = \theta_{ij}/\theta_i, \quad \theta_{ij} = \langle \gamma_f^i \gamma_g^j \rangle \tag{2.41}$$

where t_{ij} is the conditional probability of finding a particle j adjacent to a particle i. Substitution of approximation (2.41) into system (2.39) yields the expressions

$$\theta^i_{\{n\}} = \exp(\beta \nu_{ij}) \theta_j \prod_{m=1}^{s-1} (t_{jm} e^{-\beta \omega_{ijm}})^{n_m} \left(1 + \sum_{m=1}^{s-1} x_{ijm} t_{jm}\right)^{\tilde{z}} \tag{2.42}$$

from among which we must choose equations for the unambiguous determination of all the functions θ_i and θ_{ij} with consideration of the normalizing relations (1.11) and (1.13).

The equations for the unary functions are obtained with $n_1 = \ldots = n_{s-1} = 0$:

$$\theta_s = \theta_i \exp(\beta \nu_{ij}) \left(1 + \sum_{m=1}^{s-1} x_{sim} t_{im}\right)^z, \quad 1 \leqslant i \leqslant (s-1) \tag{2.43}$$

The remaining equations of the system are obtained for all the zero subscripts n_m except the single one $n_j = 1$, where $j = i, i+1, \ldots, s-1$.

$$\theta_{sj} = \theta_{ij} \exp \beta(\nu_{si} - \omega_{sj}) \left(1 + \sum_{m=1}^{s-1} x_{sim} t_{im}\right)^{z-1} \tag{2.44}$$

The ratio of expressions (2.44) for $s = j$ and when $j = i$ and $s = j$ yields the generally used relation of the couplings in a quasichemical approximation [163]:

$$\theta_{ii}\theta_{jj} = \theta_{ij}^2 \exp \beta\omega_{ij}, \quad \omega_{ij} = \varepsilon_{ii} + \varepsilon_{jj} - 2\varepsilon_{ij} \tag{2.45}$$

Molecular-Field Approximation. The multiparticle correlators are expressed in terms of the product of the particle concentrations:

$$\theta^i_{\{n\}} = \theta_i \prod_{m=1}^{s-1} (\theta_m)^{n_m} \tag{2.46}$$

Inserting (2.46) into the right-hand sides of system (2.39) for $\{n\} = \{0\}$ and performing a limiting process $\beta\varepsilon_{ij} \to 0$: $1 + \sum_m x_{sim}\theta_m \approx \exp\left(-\beta \sum_m \omega_{sim}\theta_m\right)$ and then extrapolating the found expressions to the finite values of the quantities $\beta\varepsilon_{ij}$, we obtain a closed system of equations for the unknown concentrations of the particles:

$$\theta_i = \sigma_i \Big/ \sum_{i=1}^{s} \sigma_i, \quad \sigma_i = \exp \beta \left(-\nu_i + z \sum_{m=1}^{s} \varepsilon_{im}\theta_m\right) \tag{2.47}$$

Having solved the closed systems of equations (2.43), (2.44), or (2.45), (2.47), by formulas (2.40) we determine the correlators of the first type:

$$\theta^i_{\{[n]\}} = \theta_i \prod_{m=1}^{s} (\varphi_{im})^{n_m}, \quad \varphi_{im} = \begin{cases} t_{im} & \text{QCA} \\ \theta_m & \text{MFA} \end{cases} \tag{2.48}$$

where n_s is given by formula (2.4).

The obtained equations correspond to the models of adsorption, absorption, and solid solutions. The first two are treated in the following section, while for the model of solid solutions we must take into account that in equations (2.43) and (2.47) we must assume that $\varepsilon_i = 0$ in the parameters ν_i (1.23) because the interaction of a particle i with the lattice atoms is set by the parameters ε_{ij}.

2.5. Adsorption and Absorption of a Gas Mixture

The available experimental data point to the strong mutual influence of the components of sorption (ad- or absorption) systems and to the complicated nature of their distribution [164-166]. For example, experiments involving the diffraction of low-energy electrons on single crystals of the platinum metals yielded data that the systems contains CO and oxygen as stratifying ordered phases, and also their mixed ordered phases coexisting with an ordered stratifying phase [167]. Sorption isotherms are studied experimentally the most frequently, therefore we shall limit ourselves to their treatment.

The closed equations given in Sec. 2.4 with consideration of the properties of vacant sites: $\nu_v = \varepsilon_{i,v} = 0$, $1 \leqslant i \leqslant s$, are sorption isotherms of $(s - 1)$ components of the gas phase by a homogeneous sorbent with account taken of the interaction of the nearest neighbors. By (1.23), the partial occupancies of a lattice and the equilibrium pressure of the molecules in the gas phase are related by the expressions

$$\theta_i = \frac{y_i(S_i)^z}{1 + \sum_{j=1}^{s-1} y_j(S_j)^z}, \quad S_i = \begin{cases} \left(1 + \sum_{m=1}^{s-1} x_{im} t_{im}\right)^{-1} & \text{QCA} \\ \exp \beta \sum_{m=1}^{s-1} \varepsilon_{im}\theta_m & \text{MFA} \end{cases} \tag{2.49}$$

where $x_{im} = \exp(-\beta\varepsilon_{im}) - 1$, while the functions t_{im} are determined by solving the system of equations (2.45). In the absence of interaction, formula (2.49) transforms into (1.28).

With identical interaction parameters, the concentration dependences of the thermodynamic characteristics in various approximations depend on how accurately the correlation effects are taken into account. In the quasichemical approximation, system (2.45) in the general form cannot be solved relative to θ_{ij}, but in the region of low concentrations of the adsorbed particles, the solution of the system has the form

$$\theta_{ij} = \theta_i\theta_j \exp \beta\varepsilon_{ij}, \quad 1 \leqslant i, j \leqslant (s - 1) \tag{2.50}$$

Examination of formula (2.50) reveals that the repulsion of particles reduces the probability of their being at neighboring sites, while their attraction increases this probability. In the molecular-field approximation, by (2.46), we have $\theta_{ij} = \theta_i\theta_j$.

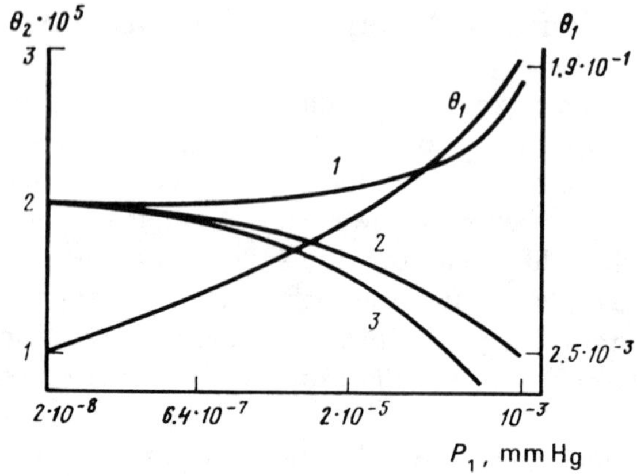

Fig. 2.10. Influence of the interaction between particles of the first and second species on the number of adsorbed particles of the second species:

1—$\beta\varepsilon_{12} = 0.5$; *2*—$\beta\varepsilon_{12} = -0.5$; *3*—$\beta\varepsilon_{12} = -1.5$; $a_1/a_2 = 10^{1.3}$; $\beta\varepsilon_{11} = -1.0$; $\beta\varepsilon_{22} = -2.0$; $P_2 = 10^{-6}$ mm Hg

One must regard with care, however, the concepts of "small" occupancies as relating to a region in which the particles are considered to behave independently of one another, and in which the data on mixtures can be used to determine the parameters of a model of individual components (e.g. the adsorption coefficients). Figure 2.10 shows adsorption isotherms of two species of molecules (the partial pressure of the second component is fixed) for various values of $\beta\varepsilon_{12}$. With an increase in the fraction of adsorbed particles of the first component, the fraction of the second component grows when $\varepsilon_{12} > 0$ and diminishes when $\varepsilon_{12} < 0$ even for the given small occupancies by the particles. On the other hand, this indicates the possibility of obtaining information on the particle interactions of the adsorbed molecules. With a growth in the pressure of the first component at $\varepsilon_{12} > 0$, the values of θ_2 (P_2 = const) always pass through a peak because of displacement of the second component molecules by the first component ones.

The number of different phases appearing in multicomponent sorption systems is large. The classification of the phase transitions even for a two-component mixture (by analogy with a three-component mixture in the theory of alloys [15]) with consideration of the interaction of the nearest neighbors is an involved task and requires independent treatment. We shall confine ourselves to an analysis of the main variants of adsorption isotherms for two species of molecules with a general growth in the pressure of the adsorption system $P = P_1 + P_2$, and P_1/P_2 = const (see also [168]). Depending on the relation between the parameters of particle interaction, the partial occupancies, and the temperature of a system, its single-phase disordered state may become energetically unprofitable, and the system will stratify. The concentrations of the coexisting phases are determined from the equations of the material balance and the equality of the chemical potentials of the system's components [169]:

$$V^- \theta_i^- + V^+ \theta_i^+ = \theta_i, \quad \sum_{i=1}^{s} \theta_i^{\pm} = 1, \quad \mu_i^+ = \mu_i^- \tag{2.51}$$

$$\beta\mu_i = \frac{z}{2}\beta\varepsilon_{ii} + (1 - z)\ln\theta_i + \frac{z}{2}\ln\theta_{ii}$$

where μ_i is the chemical potential of a particle i of the lattice system written in the quasichemical approximation; the signs (+) and (−) signify different phases; $V^{\pm}$ and $\theta_i^{\pm}$ are the volume fraction and mole fraction of the particles i for the phase (+) or (−).

The types of the isothermal cross sections of the phase diagrams for a two-component adsorption mixture are shown in Fig. 2.11. Eight types of diagrams can be singled out by their topological characteristics. They are divided into four groups depending on whether or not the binary solutions (12), (1v), (2v) in the three-component lattice system stratify. In the quasichemical approximation, a binary solution stratifies if the temperature is below the critical one

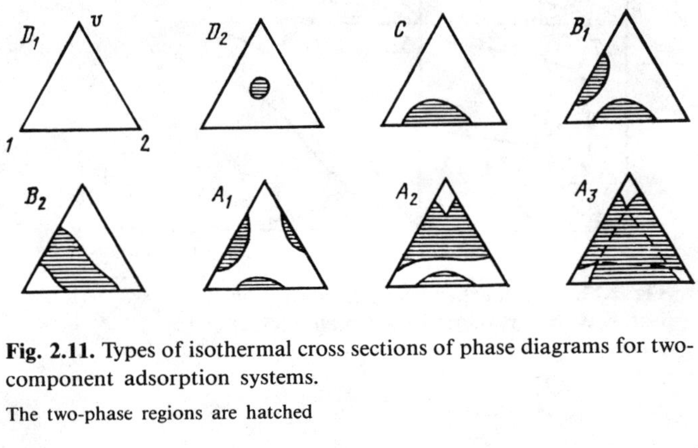

Fig. 2.11. Types of isothermal cross sections of phase diagrams for two-component adsorption systems.

The two-phase regions are hatched

$\beta_c\omega = 2 \ln [z/(z-2)]$ [77, 142]. If none of the binary solutions stratifies, the three-component system is a single-phase one (D_1) or stratifies with attraction of components 1 and 2 (D_2). For chemisorption, ε_{11}, $\varepsilon_{12} < 0$, but depending on the value of the parameter ε_{12} the diagrams D_1 or C are possible—stratification may occur if the repulsion between the particles of different species is greater than the repulsion of the particles of one species. Diagrams B_1 and B_2 are realized when the particles of the first species, for instance, attract one another, while those of the second species repel or only slightly attract one another, but at the same time the solution (12) stratifies. Finally, with strong attraction of the particles of each of the components to one another (physical adsorption) and stratification of the solution (12), three types of diagrams are realized, namely, A_1, A_2, and A_3.

Figure 2.12 shows typical relations of coadsorption isotherms. Motion along a curve from the apex in Fig. 2.11 corresponding to a free surface to the opposite line corresponding to solution (12) shows the surface coverage at P_1/P_2 = const. The ratio θ_1/θ_2 at $\theta_v \to 0$ ($P \to \infty$) is determined from Eq. (2.14). The shape of the curve depends on the ratio a_1/a_2 and on the values of all the interaction parameters. In the two-phase region, the curves of θ_i versus the pressure P have three roots. For one of the components (say 1), the relation θ_1 vs. P is an analogue of the van der Waals loop (Fig. 2.6), while

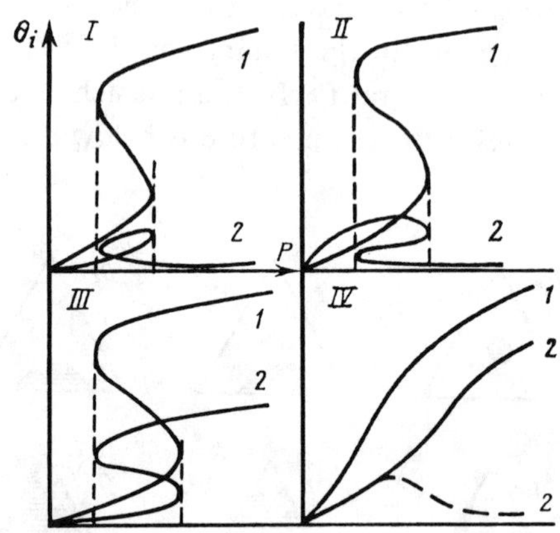

Fig. 2.12. Typical dependences of adsorption isotherms for two species of molecules at P_1/P_2 = const (it is assumed that $a_1P_1 > a_2P_2$):

I—$A_1A_2A_3$, D_2; *II*—B_1B_2; *III*—C; *IV*—D_1

the behavior of θ_2 vs. P depends on all the parameters of the model: *I*—a closed loop, *II*—an inverted loop, *III*—a similar loop. Cases *I* and *II* correspond to stratification of solution (12), while cases *III* and *IV* correspond to the absence of its stratification. The position of the jump in the densities on the partial isotherms is determined from the condition of the equality of the chemical potentials (2.51) or with the aid of Maxwell's rule generalized for a mixture of particles. For component 1, as for $s = 2$, the relation $\theta_1^+ > \theta_1^-$ holds, while for component 2 depending on the position of the secant for the type I isotherms, we have $\theta_2^- < \theta_2^+$ (*I*, *III*) or $\theta_2^- > \theta_2^+$ (*I*, *II*). Similar relations are observed for absorption isotherms.

3 The Equilibrium State of Inhomogeneous Systems

3.1. Inhomogeneous Systems

Let us consider an inhomogeneous lattice system containing t types of sites. The arrangement of the sites of different types is set by the quantities η_f^q, $1 \leqslant q \leqslant t$ (1.15). We shall assume that any site of a lattice can be occupied by a particle i, $1 \leqslant i \leqslant s$, where s is the number of particle species. Inhomogeneous lattice systems correspond to a broad range of models.

Let us single out four types of inhomogeneous systems that are of the main interest for surface processes: (1) if one of the states of occupancy of the lattice sites corresponds to vacant sites, and the lattice sites correspond to regions of adsorbed particle localization, this is a mode of adsorption of $(s - 1)$ components of the gas phase on an inhomogeneous surface; (2) if one of the states of occupancy of the lattice sites corresponds to vacant sites, and the lattice sites are interstices of the lattice of atoms of a solid, this is a model of absorption of $(s - 1)$ components of the gas phase by an inhomogeneous sorbent; (3) if the particles of a solution occupying the lattice sites form periodically repeating "patterns" (with a period greater than the lattice constant), this corresponds to an ordered solution. Each pattern is characterized by its own way of particle arrangement that depends on the concentration of the solution components and the particle interaction; and (4) if sites of one type are arranged in layers forming a set of bonded layers, this is a model of the subsurface region of a phase interface.

Other examples of inhomogeneous systems are a combination of the inhomogeneous systems of the indicated types. For instance, one example is a model of absorption into octa- and tetrahedral interstices with account taken of the ordered arrangement of the particles in these interstices or ordering of the solution components at the phase interface.

The diversity of factors leading to inhomogeneous systems is readily illustrated by the example of inhomogeneous surfaces [34, 170]. Inhomogeneities of a surface may be associated with a disturbance in the regularity of arrangement of the surface atoms of a solid (structural inhomogeneities) and with the difference in the nature of the surface atoms (chemical inhomogeneities).

Point inhomogeneities are the simplest example of inhomogeneity. They include vacancies of surface atoms, autoadsorbed atoms of a solid and atoms of an impurity on the surface, points of emergence to the surface of edge dislocations, and impurities at the sites of a surface lattice or incorporation of impurities into the subsurface layer. Point defects may modify the properties of the neighboring sites of the surface structure, which increases the number of types of surface sites. The linear imperfections of a surface include faces of edges, steps, and dislocations in the plane of the surface.

Fractures on steps and roughness of the singular and vicinal faces of crystals are more complicated examples of structural inhomogeneities. An example of chemical inhomogeneity is the surface of the binary alloy A_xB_{1-x} (for instance, Pd-Ag) having an unlimited solubility. A homogeneous surface of a face, e.g. (100), of this alloy is inhomogeneous at the expense of the different adsorptivity of the A and B atoms. The fraction of each component on the surface may vary from zero to unity depending on the temperature and concentration of the particles in the bulk. In the general case, the chemical and structural inhomogeneities occur simultaneously.

On inhomogeneous lattices, the internal degrees of freedom of the particles and the potentials of particle interaction depend on the types of the sites. We shall consider that the one-particle contributions are described by expressions (1.29). Let ε_{qp}^{ij} stand for the interaction between the particles i and j on neighboring sites of type q and p. Now the function of the distribution of the particles among the sites of an inhomogeneous lattice is determined by expression (1.18) in which the effective Hamiltonian is written as follows [171]:

$$H = \sum_{f,i} \nu_q^i \gamma_f^i \eta_f^q - \frac{1}{2} \sum_{f,g} \sum_{i,j} \varepsilon_{qp}^{ij} \gamma_f^i \gamma_g^j \eta_f^q \eta_g^p \tag{3.1}$$

As for homogeneous lattices, the subscript f runs through all the lattice sites, while the subscript g runs through the nearest neighbors of site f. The sums over i and j are taken over all the states of occupancy of the lattice sites. The zero values of interaction of the particles with vacant sites are taken into consideration in the final expressions of the theory of ad- and absorption.

Let us depict an inhomogeneous lattice on a set of clusters M_1 consisting of a central site and its z_q nearest neighbors [171, 172]. The type of cluster is determined by the types of all its sites. Let us denote by σ_{1q} the number of different types of clusters with a central site of type q differing in the types of sites of its neighbors, $\sigma_{1q} < t^{z_q}$, because part of the types of clusters may be absent. The total number of clusters on which the initial inhomogeneous lattice will be depicted is $L_1 = \prod_{q=1}^{t} \sigma_{1q}$. The probabilities of realization of clusters of various types can be characterized by the cluster correlators similar to the cluster correlators of a homogeneous system (2.3). Their determination is connected with the solution of problems on the distribution of atoms of a solid (Secs. 3.8, 4.4), and here they are considered to be known.

Let us denote by $f(q\{p\})$ the normalized probability of finding on a lattice a cluster at whose center there is a site of type q, the first site of the coordination sphere is site p_1, the second is p_2, and so on up to p_z inclusively; the braces signify the completeness of the set of sites $p_1 \ldots p_z$:

$$f(q\{p\}) = \frac{1}{N} \sum_{f} \eta_f^q \eta_{g_1}^{p_1} \ldots \eta_{g_z}^{p_z}, \quad \sum_{\{p\}} f(q\{p\}) = f_q \tag{3.2}$$

The sum over $\{p\}$ signifies summation over all the types of coordination spheres of a cluster, while the second relation (3.2) determines the normalization of the functions $f(q\{p\})$, where f_q has been determined in (1.34).

If we take into consideration only the direct correlations of the interacting particles, this will enable us to employ a rougher description of the types of the coordination sphere sites, i.e. limit ourselves to a knowledge of the number of sites m_{qp} of various types without going into their arrangement, $0 \leqslant m_{qp} \leqslant z_q$, $\sum_{p=1}^{t} m_{qp} = z_q$. In the given case, the lattice is depicted on a set of clusters M_2, and their number is $L_2 = \prod_{q=1}^{t} \sigma_{2q}$, where $\sigma_{2q} \leqslant \overset{*}{\sigma}_{2q}$, and $\overset{*}{\sigma}_{2q}$ is the number of solutions of the equation $\sum_p m_{qp} = z_q$ for integral values of m_{qp}. We shall denote the normalized probabilities of the clusters of set M_2 by $f(q\{[m]\})$:

$$f(q\{p\})\xi(\{m\}) = f(q\{[m]\}), \quad \sum_{\sigma_{2q}} f(q\{[m]\}) = f_q \tag{3.3}$$

where $\xi(\{m\})$ is the degree of degeneration of the coordination sphere states for the set M_1 differing in the ordinal numbers of the sites with a fixed number of sites of various types $\{[m]\} = [m_{q1}, \ldots, m_{qt}]$, $\xi(\{m\}) < C_{z_q}^{\{m\}}$. Here $C_{z_q}^{\{m\}}$ is the number of combinations of z_q sites taken $m_{q1} \ldots m_{qt}$ types of sites at a time. The set of clusters M_2 is considerably smaller than the set M_1, for example, for $t = 2$ we have $\sigma_{1q} \leqslant 2^{z_q}$, while $\sigma_{2q} \leqslant z_q + 1$.

The energy of the central particles on a cluster of type $q\{[m]\}$ is described by the following expression:

$$h_f(q\{[m]\}) = \sum_{i=1}^{s} h_f^i(q\{[m]\}),$$

$$h_f^i(q\{[m]\}) = \left(\nu_q^i - \sum_{g,j} \varepsilon_{qp}^{ij} \gamma_g^j \eta_g^p\right) \gamma_f^i \tag{3.4}$$

where the set of quantities η_g^p corresponds to the set of numbers of site types $\{[m]\}$.

For each cluster of the set M_2, we shall determine two types of cluster correlators characterizing the configurations of the particles on them. The correlators of the first type, by definitions (1.8) and (1.18), have a form similar to that of expressions (2.3) and (2.5):

$$\theta_{q\{[m]\}}^{i\{[n]\}} = \frac{1}{Q} \sum_{i_1=1}^{s} \cdots \sum_{i_N=1}^{s} \gamma_f^i \prod_{j=1}^{s} (\gamma_g^j)^{n_j} \exp\left(-\beta[h_f(q\{[m]\}) + G_f(q\{[m]\})]\right)$$

$$= \frac{\text{const}\ (\{[n]\} | \{[m]\})}{Q} \exp \beta \left(-\nu_q^i + \sum_{p=1}^{t} \sum_{j=1}^{s} \varepsilon_{qp}^{ij} n_{qp}^j\right) \tag{3.5}$$

The kind of configurations of the particles is given by the set of the numbers of bonds n_{qp}^j of the central particle i at site q with the neighboring particles j on sites of type p, $\{[n]\} = [\bar{n}_{q1} \ldots \bar{n}_{qt}]$, $\bar{n}_{qp} = n_{qp}^1 \ldots n_{qp}^{s-1}$. The braces signify that $1 \leqslant p \leqslant t$, while the square brackets signify that at m_{qp} sites of type p there are exactly n_{qp}^j particles of species j, $1 \leqslant j \leqslant s$. For the unambiguous setting of the configurations of the particles, it is sufficient to set $(s - 1)$ numbers n_{qp}^j; the particles of species s are a complement up to the complete occupancy of all the sites m_{qp} of type p:

$$n_{qp}^s = m_{qp} - \sum_{j=1}^{s-1} n_{qp}^j, \quad 0 \leqslant n_{qp}^j \leqslant m_{qp} \tag{3.6}$$

In formula (3.5), the expression for $G_f(q\{[m]\}) = H.- h_f(q\{[m]\})$ does not depend on the state of occupancy of the central site f. This enables us to rewrite definition (3.5) in accordance with the second equality, where the unknown constant takes into account the contributions to the energy of the system from all the lattice sites except site f.

The correlators of the second type are written as follows:

$$\theta^{i\{n\}}_{q\{[m]\}} = \left\langle \gamma^i_f \prod_{p=1}^{t} \prod_{j=1}^{s-1} (\gamma^j_g)^{n^j_{qp}} \middle| \eta^q_f \prod_{g=1}^{z} \eta^p_g \right\rangle \tag{3.7}$$

In the absence of square brackets, the superscript n^j_{qp} indicates that on m_{qp} sites of types q there are at least n^j_{qp} particles of species j, where $1 \leqslant j \leqslant s - 1$. Equation (3.6) is not observed for them.

The thermodynamic characteristics of a system are expressed in terms of correlators of the second type. To calculate the correlators, in Sec. 3.2 closed systems of equations in a two-component system are obtained, while in Sec. 3.6 they are obtained in a multicomponent system. In concluding this section, we shall discuss the appraisals of the distribution functions for sites of various types with whose aid the local configurations of the particles are "weighted" when describing the characteristics of inhomogeneous surfaces.

Information on the distribution functions for sites of various types can be obtained with the aid of direct and indirect measurements. The direct methods include experiments involving field ion microscopy, low-energy electron diffraction, molecular beam scattering, especially of helium atoms, and Auger electron spectroscopy [95-98, 170, 173]. These measurements provide information on the number of defects and autoadsorbed atoms, structure and chemical composition of a surface, the presence of steps, and the degree of roughness. From the integral characteristics, using model notions, we can obtain stricter appraisals for the surface layer. For example, the signal of Auger electron spectroscopy contains information on several surface layers. Using techniques for calculating the concentration profile of an alloy [81-83, 174] (see Secs. 3.8, 5.4, and 5.5), we can "distribute" its total number of components by layers and obtain information on the composition and structure of the surface layer. The same information can be obtained indirectly from data on the distribution of the components of an alloy in the bulk and on the surface tension (we have in view the achievement of an equilibrium state of an alloy).

Information on the nature of inhomogeneity can also be based on experimental data and model notions on the mechanisms of creating surfaces in the process of crystal growth [22, 175], the deposition of films, in the cleavage of crystals [21], and in the mechanical and thermal processing of a specimen. If an alloy in the course of its preparation beginning from a certain temperature was cooled sharply, it will be characterized for a certain time by a distribution of the particles corresponding to this temperature. Consequently, knowing the prehistory of an adsorbent, we can appraise the functions of distribution of sites of various types on a surface. The dependence on the conditions and duration of thermal treatment of an alloy has been shown on the example of the thermodesorbed spectra of CO from the alloy Pd-Au [176]).

We shall present very simple examples of the distribution functions for sites and linear imperfections of a surface that are convenient to represent in terms of $d(q\{[m]\})$—the conditional probabilities of being on a lattice of a cluster with a coordination sphere of type $\{[m]\}$ if a site of type q is at its center:

$$f(q\{[m]\}) = f_q d(q\{[m]\}), \quad \sum_{\sigma_{2q}} d(q\{[m]\}) = 1 \tag{3.8}$$

Assume that a point defect on the face (111) of a single crystal distorts the adsorption properties of its nearest neighbors. Without going into its nature, we shall consider that the position of the defect coincides with a lattice site. Now there are three types of sites on the surface, namely, the defect, its neighbors, and the remaining sites on the surface. All the types of sites have the same number of neighbors. Assume that the defects are quite far from one another. Denoting their concentration by c, we have $f_1 = c$, $f_2 = zc$, and $f_3 = 1 - (z + 1)c$.

Five types of clusters exist on a face (111). They differ in the type of the central site and its nearest neighbors; here $\{[m]\}$ signifies $[m_{q1}, m_{q2}, m_{q3}]$, $1 \leqslant q \leqslant 3$. The sites of the first type are surrounded only by sites of the second one, i.e. the single configuration of neighboring sites [0, 6, 0] is realized, therefore $d(1[0, 6, 0]) = 1$. A site of the second type always has one neighboring site of the first type, two neighboring sites of the second type, and three neighboring sites of the third type, therefore $d(2[1, 2, 3]) = 1$.

The sites of the third type depending on their position have three different types of surrounding (the plurality of clusters is three). In any case, a site of the third type has no neighbors of the first one—$m_{31} = 0$. A site of the third type can have two, one, or zero neighbors of the second type. Clusters of the types [0, 2, 4], [0, 1, 5], and [0, 0, 6] correspond to them. In the first two cases, the sites of the third type are adjacent to sites of the second one; the number of sites in both cases is equal and, consequently, $d(3[0, 2, 4]) = d(3[0, 1, 5]) = zc/f_3$. The remaining sites on the surface relate in the clusters 3 [0, 0, 6]. To determine $d(3[0, 0, 6])$, we must subtract all the central sites of the clusters listed above on a unit area: $1 - (1 + 3z)c$ and relate them to the number of sites of the third type. As a result, we obtain $d(3[0, 0, 6]) = (1 - 19c)/(1 - 7c)$.

Let us consider as an example of a linear imperfection the surface of a face (100) with a monatomic step (Fig. 3.1); let l be the length of the terrace. We

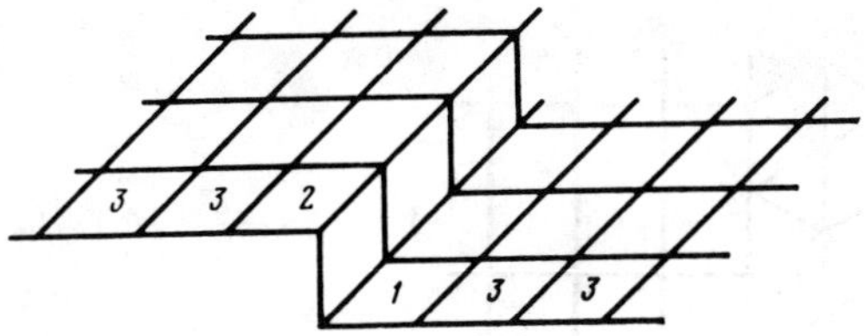

Fig. 3.1. A monatomic step

shall distinguish the sites adjoining the step (first type) and in the first row of the terrace from the step (second type). The remaining sites are of the third type (without consideration of the edges along the length of the step), $t = 3$. The sites of each type occupy a whole row, hence their fractions are characterized by the following values: $f_1 = f_2 = 1/l$, $f_3 = 1 - 2/l$.

For a face (100), the sites of the first and second types are not nearest neighbors (the particles at them are at a distance of second neighbors $a\sqrt{2}$). The structure of the surface is characterized by five types of clusters. Each site of the first type is surrounded by two sites of this type and one site of the third type. Matters are similar for sites of the second type, hence $z_1 = z_2 = 3$, and $d(1\,[2, 0, 1]) = d(2\,[0, 2, 1]) = 1$. Sites of the third type have four neighbors each. They differ depending on whether or not sites of the first and second types are next to each other. For a row of sites of the third type adjoining ones of the first type, the clusters have the surrounding [1, 0, 3]. Similarly, for a row of sites adjoining ones of the second type, the clusters have the surrounding [0, 1, 3]. Their probabilities are equal to the probabilities of getting by chance into one of the rows of the third type, therefore $d(3\,[1, 0, 3]) = d(3\,[0, 1, 3]) = 1/(l - 2)$. The remaining rows of sites of the third type do not have sites of the first and second types as neighbors, hence $d(3\,[0, 0, 4]) = (l - 4)/(l - 2)$.

The boundary between faces (100) and (111) (Fig. 3.2) is another example of linear imperfection. It affects the properties of the sites in neighboring layers at each side. This leads to five types of sites: (1) sites on the face (111), $z_1 = 6$; (2) sites of the face (111) adjacent to the boundary, $z_2 = 6$; (3) the boundary sites, $z_3 = 5$; (4) sites of the face (100) adjacent to the boundary, $z_4 = 4$, and (5) sites of the face (100), $z_5 = 4$. Let us designate the ratios of the number of sites of the boundary to the numbers of them on the faces (111) and (100) by $1/b$ and $1/l$. Hence, $f_1 = (b - 1)/\sigma$, $f_2 = f_3 = f_4 = 1/\sigma$, and $f_5 = (l - 1)/\sigma$, where $\sigma = b + l + 1$. Within the confines of each layer, all the sites have the same surrounding.

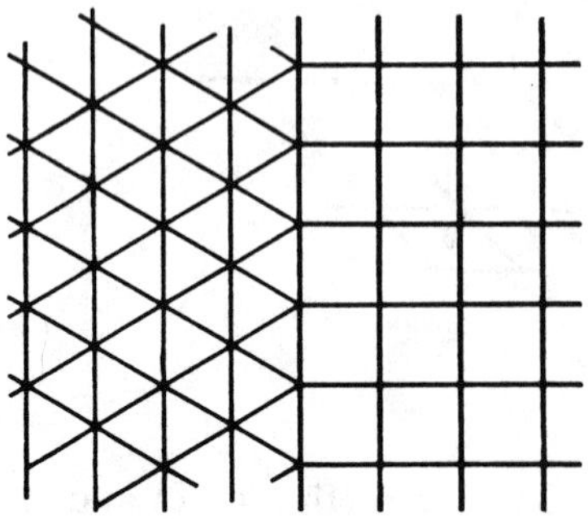

Fig. 3.2. Boundary of faces (100) and (111)

The structure of a surface is characterized by seven types of clusters relating to different layers, namely, the boundary itself and three layers at each side of it; the remaining layers of the faces are equivalent to the layers adjoining the ones with sites of types 1 and 5 in Fig. 3.2. Repeating the reasoning relating to a stepped surface, it is not difficult to obtain that

$$d(1\,[6,\ 0,\ 0,\ 0,\ 0]) = (b-2)/(b-1),$$

$$d(1\,[4,\ 2,\ 0,\ 0,\ 0]) = (b-1)^{-1}$$

$$d(2\,[2,\ 2,\ 2,\ 0,\ 0]) = d(3\,[0,\ 2,\ 2,\ 1,\ 0]) = d(4\,[0,\ 0,\ 1,\ 2,\ 1]) = 1$$

$$d(5\,[0,\ 0,\ 0,\ 1,\ 3]) = (l-1)^{-1}, \quad d(5\,[0,\ 0,\ 0,\ 0,\ 4]) = (l-2)/(l-1)$$

Other examples of point and linear imperfections have been considered by the author elsewhere [172, 177]. The functions of distribution of clusters in more complicated cases are constructed similarly, e.g. consideration of fractures in the steps, surface roughness, and faces with high indices, as well as in cases when such defects are present on a surface simultaneously.

Finding of the distribution functions is simplified if it is known that the distribution of the atoms of a solid obey an equilibrium distribution:

$$d(q\{[m]\}) = C_{z_q}^{\{m\}} \prod_{p=1}^{t} (d_{qp})^{m_{qp}}, \quad \sum_{p=1}^{t} m_{qp} = z_q \tag{3.9}$$

where the functions d_{qp} are the conditional probabilities of finding sites of types p adjacent to a site of type q. These functions are similar to functions (2.41) introduced for multicomponent solutions. The functions d_{qp} can also be used to describe the structure of a surface for nonequilibrium conditions.

In this case, the distribution functions d_{qp} and $d(q\{[m]\})$ are related as follows:

$$z_q d_{qp} = \sum_{\sigma_{2q}} m_{qp} d(q\{[m]\}) \tag{3.10}$$

Both the right-hand and left-hand sides of formula (3.10) contain expressions for the average number of neighbors of type p adjoining sites q. Formula (1.63) is an example of using the functions d_{qp}.

The use of expression (3.9) for nonequilibrium conditions is an additional approximation that simplifies the mathematical model. In a number of situations, the use of the distribution functions $d(q\{[m]\})$ and their approximations (3.9) and (3.10) yields equivalent results, namely, patchwise and chaotic surfaces, as well as surfaces with a regular alternation of sites of different types. In the general case, this is not true.

Let us give for comparison the distribution functions d_{qp} for the point and linear surface imperfections indicated above (the functions f_q retain the same values).

Point Defect on a Face (111). Since the bonds "11" and "13" are absent, while a site of the first type is surrounded by ones of the second type, we have $d_{11} = d_{13} = d_{31} = 0$, and $d_{12} = 1$. Sites of the second type have one neighbor of the first type, two neighbors of the second type, and three of the third type, hence $d_{21} = 1/6$, $d_{22} = 2/6$, and $d_{23} = 3/6$. The quantity d_{32} can be obtained on the basis of the relation $f_q d_{qp} = f_p d_{pq}$ reflecting the obvious equality $f_{qp} = f_{pq}$, i.e. $d_{32} = 3c/f_3$. The same result is obtained by direct calculations. The sites of the second type are surrounded by 12 sites of the third type. Six of them have one bond to sites of the second type, and the other six have two bonds each. The average number of bonds between the sites of the second and third type for each impurity is $\left(6 \times \frac{1}{6} + 6 \times \frac{2}{6}\right) = 3$. Consequently, $d_{32} = 3c/f_3$. We can calculate in a similar way the average number of bonds between sites of the third type. These calculations are verified by employing the normalizing condition (1.64), $d_{33} = 1 - d_{31} - d_{32}$.

Stepped Surface (100). For the sites of the first and second types, the functions d_{qp} are the fraction of the number of neighboring sites of each type in the total number of sites: $d_{11} = d_{22} = 2/3$, $d_{13} = d_{23} = 1/3$, $d_{12} = d_{21} = 0$.

For sites of the third type, the differences in their arrangements must be taken into account—whether or not they are adjacent to sites of the first and second type. If a site of the third type belongs to a layer adjoining sites of the first or second type, then for this site the fraction of neighboring sites of another type is 1/4, and since the probability of getting into such a layer equals the quantity $(l - 2)^{-1}$, then $d_{31} = d_{32} = [4(l - 2)]^{-1}$. The function d_{33} is determined from the following relations. The step has $(l - 4)$ layers where each site of the third type has four neighbors of the third type and two layers where each site has three neighbors of the third type. The total number of bonds in $(l - 2)$ layers is $4(l - 2)$; they include $4(l - 4) + 2 \times 3$ bonds between sites of the third type. Their ratio yields $d_{33} = 1 - [2(l - 2)]^{-1}$; the same result is obtained from formula (1.64).

Boundary of Faces (111) and (100). In each layer, all the sites have an identical surrounding. In layers 2-4, as above, the functions d_{qp} are the ratio of the numbers of adjacent sites of different types to the total number of adjacent sites, $d_{21} = d_{22} = d_{23} = 1/3$, $d_{32} = d_{33} = 2/5$, $d_{34} = 1/5$, $d_{44} = 1/2$, $d_{43} = d_{45} = 1/4$. For sites of the first and fifth types, it is necessary to consider the possibility of finding sites of types 2 and 4 next to one another. Each site of the first type in layer 1 has two neighbors of type 2, i.e. the probability is 1/3. Since the probability of getting into this row while remaining among sites of the third type is $(b - 1)^{-1}$, then $d_{12} = [3(b - 1)]^{-1}$; the function $d_{54} = [4(l - 1)]^{-1}$ is calculated similarly. As previously, it is not difficult to determine the number of adjacent bonds "11" and "55"—the result is $d_{11} = (3b - 4)/(3b - 3)$, and $d_{55} = (4l - 5)/(4l - 4)$.

Other examples of constructing the functions d_{qp} are given in [116, 135].

3.2. Distribution of Particles of a Two-Component Solution

Let us first consider an inhomogeneous lattice consisting of sites of two types. The lattice is depicted on a set of clusters given by the quantities f_1, f_2, $d(1[m_{11}, m_{12}])$, and $d(2[m_{21}, m_{22}])$, where $m_{11} + m_{12} = z_1$, and $m_{21} + m_{22} = z_2$. Let A and B be the components of a binary solution. Assume that the B particles supplement the sites m_{qp} up to complete occupancy, hence in the numbers n_{qp}^{A} of the A particles in the coordination sphere of a cluster the superscript A may be omitted. The procedure for obtaining a system of equations for cluster correlators of the second type consists of the same three stages indicated in Sec. 2.2. When dealing with an inhomogeneous lattice, this procedure is carried out for all the types of clusters. Let us introduce an abridged designation for a cluster type, i.e. $Y = q[m_{q1}m_{q2}]$ and compile a relation between the correlators (3.5) with the central particles of A and B and with n_{q1} particles of A at the sites of the first type and n_{q2} at the sites of the second one:

$$\theta_Y^{\mathrm{B}[n_{q1}n_{q2}]} = \theta_Y^{\mathrm{A}[n_{q1}n_{q2}]} M_Y^{n_{q1}n_{q2}}$$
$$M_Y^{n_{q1}n_{q2}} = \exp \beta[-\nu(Y) + n_{q1}\omega_{q1} + n_{q2}\omega_{q2}] \tag{3.11}$$

$$\nu(Y) = \nu_q^{\mathrm{B}} - \nu_q^{\mathrm{A}} + \sum_{p=1}^{2} m_{qp}(\varepsilon_{qp}^{\mathrm{AB}} - \varepsilon_{qp}^{\mathrm{BB}})$$
$$\omega_{qp} = \varepsilon_{qp}^{\mathrm{AA}} + \varepsilon_{qp}^{\mathrm{BB}} - \varepsilon_{qp}^{\mathrm{AB}} - \varepsilon_{qp}^{\mathrm{BA}} \tag{3.12}$$

The unknown constant and the partition function of the system cancel out. The correlators of the first and second types are related by the expression

$$\theta_Y^{in_{q1}n_{q2}} = \sum_{k_{q1}=0}^{\tilde{m}_{q1}} \sum_{k_{q2}=0}^{\tilde{m}_{q2}} C_{\tilde{m}_{q1}}^{k_{q1}} C_{\tilde{m}_{q2}}^{k_{q2}} \theta_Y^{i[n_{q1}+k_{q1},\, n_{q2}+k_{q2}]}$$
$$\tilde{m}_{qp} = m_{qp} - n_{qp}, \quad C_{m_{qp}}^{k_{qp}} = m_{qp}!/(k_{qp}!(m_{qp} - k_{qp})!) \tag{3.13}$$

Formula (3.13) is a generalization of formula (2.9). Its derivation is based on the same reasoning on the number of transitions between different configurations of particles with a continuous growth in the number of particles of A in the coordination sphere of a cluster provided that each type of sites is occupied independently of another one.

Inversion of formula (3.13) makes it possible to express the correlators of the first type in terms of those of the second type:

$$\theta_Y^{i[n_{q1}n_{q2}]} = \sum_{k_{q1}=0}^{\tilde{m}_{q1}} \sum_{k_{q2}=0}^{\tilde{m}_{q2}} (-1)^{k_{q1}+k_{q2}} C_{m_{q1}}^{k_{q1}} C_{m_{q2}}^{k_{q2}} \theta_Y^{in_{q1}+k_{q1}, n_{q2}+k_{q2}} \tag{3.14}$$

Inserting expressions (3.14) into (3.11), we obtain a system of linear equations for the correlators of the second type. Solving it consecutively from $n_{q1} = m_{q1}$ and $n_{q2} = m_{q2}$ and lowering the values of n_{q1} and n_{q2} independently, we obtain

$$\theta_Y^{\mathrm{B}n_{q1}n_{q2}} = \sum_{k_{q1}=0}^{\tilde{m}_{q1}} \sum_{k_{q2}=0}^{\tilde{m}_{q2}} C_{\tilde{m}_{q1}}^{k_{q1}} C_{\tilde{m}_{q2}}^{k_{q2}} M_Y^{n_{q1}+k_{q1},n_{q2}+k_{q2}}$$

$$\times \sum_{r_{q1}=0}^{l_{q1}} \sum_{r_{q2}=0}^{l_{q2}} (-1)^{r_{q1}+r_{q2}} C_{l_{q1}}^{r_{q1}} C_{l_{q2}}^{r_{q2}} \theta_Y^{\mathrm{A}n_{q1}+k_{q1}+r_{q1},n_{q2}+k_{q2}+r_{q2}} \tag{3.15}$$

where $l_{qp} = \tilde{m}_{qp} - k_{qp}$. System (3.15) generalizes the system of equations (2.12) for the inhomogeneous cluster Y. The number of its unknowns exceeds the number of couplings, therefore to obtain closed expressions the correlators of a larger dimension must be approximated in terms of the correlators of a smaller one.

Let us consider the generalization for an inhomogeneous lattice using the quasichemical approximation. On an inhomogeneous lattice, the approximation of the higher correlators in terms of the lower ones depends on how account is taken of the effects of correlation between the interacting particles and of the influence of lattice inhomogeneity on particle distribution. The state of occupancy of a site f depends on its type and on the state of occupancy of its neighbors g, which, in turn, depend on the types of these sites g and on the states of occupancy of their neighbors, including site f, and so on. As a result, the state of occupancy of site f should depend on the types of the neighboring sites g.

We shall confine ourselves to an approximation without a detailed consideration of how the inhomogeneity of a lattice affects particle distribution. The function θ_q^i for an inhomogeneous lattice has been determined in expression (1.33), while the macroscopic mean quantity θ_{qp}^{ij} characterizing the probability of finding particles i and j at sites of type q and p on a lattice is determined as

$$\theta_{qp}^{ij} = \frac{1}{N_{qp}} \sum_{(f_1 f_2)=1}^{N_{qp}} \theta_{f_1 f_2}^{ij}\,(qp), \quad N_{qp} = \frac{1}{2} \sum_{f=1}^{N} \sum_{g=1}^{z_q} \eta_f^q \eta_g^p \tag{3.16}$$

where N_{qp} is the number of pairs of type qp on an inhomogeneous lattice; the subscript f_1 runs through all the sites of type q, and the subscript f_2 through the neighboring sites p.

The quasichemical approximation is obtained as a result of the following assumptions:

$$\theta_Y^{in_{q1}n_{q2}} = \theta_q^i (t_{q1}^{i\mathrm{A}})^{n_{q1}} (t_{q2}^{i\mathrm{A}})^{n_{q2}}, \quad t_{qp}^{ij} = \theta_{qp}^{ij} / \theta_q^i \tag{3.17}$$

where t_{qp}^{ij} is the conditional probability of finding a particle j at site p adjacent to a particle i at site q. Inserting (3.17) into system (3.15), we obtain

$$\theta_q^{B}(t_{q1}^{BA})^{n_{q1}}(t_{q2}^{BA})^{n_{q2}} = \exp(-\beta\nu(Y))\theta_q^{A}\prod_{p=1}^{2}[t_{qp}^{AA}e^{-\beta\omega_{qp}}]^{n_{qp}}$$

$$\times(1 + x_{qp}t_{qp}^{AA})^{m_{qp}-n_{qp}}, \quad x_{qp} = \exp(-\beta\omega_{qp}) - 1 \tag{3.18}$$

The unknown functions θ_q^{i} and θ_{qp}^{ij} are determined from expressions (3.18) for $n_{q1} = n_{q2} = 0$; $n_{q1} = 1$, $n_{q2} = 0$; and $n_{q1} = 0$, $n_{q2} = 1$. Using normalizing relations (1.11) and (1.13) with consideration of (1.33) and (3.16) for them, we arrive at the following system of equations ($q = 1$ and 2):

$$1 - \theta_q^{A} = \theta_q^{A}\exp(-\beta\nu(Y))\prod_{p=1}^{2}(1 + x_{qp}t_{qp}^{AA})^{m_{qp}} \tag{3.19}$$

$$\begin{aligned} t_{qp}^{AA} &= 2\theta_p^{A}/(b_{qp} + \delta_{qp}), \quad \delta_{qp} = 1 + x_{qp}(1 - \theta_q^{A} - \theta_p^{A}) \\ b_{qp} &= \{\delta_{qp}^{2} + 4x_{qp}\theta_q^{A}\theta_p^{A}\}^{1/2} \end{aligned} \tag{3.20}$$

The expression for t_{qp}^{AA} does not depend on the type of cluster, it holds for a lattice as a whole. The dependence on the type of cluster is included in the first equations (3.19) in terms of the sets of values of m_{qp}, therefore to obtain self-consistent expressions for θ_q^{A} relating to the entire lattice, we shall multiply both sides of this equation by $f(q[m_{q1}, m_{q2}])$ and summate over various types of coordination spheres of clusters containing a site q at their center. The result is the closed system of equations (3.20) and (3.21):

$$1 - \theta_q^{A} = \theta_q^{A}\sum_{\sigma_{2q}} d(q[m_{q1}m_{q2}])\exp(-\beta\nu(q[m_{q1}m_{q2}]))$$

$$\times(1 + t_{q1}^{AA}x_{q1})^{m_{q1}}(1 + t_{q2}^{AA}x_{q2})^{m_{q2}} \tag{3.21}$$

that describes the distribution of the interacting particles among two types of sites of an inhomogeneous lattice. The given equations take into account the effects of correlation of interacting particles and correlation in the arrangement of sites of various types on a lattice.

Let us now go over to the general case of inhomogeneous lattices containing $t > 2$ types of sites. The procedure for obtaining equations for the cluster correlators is similar to that set out for $t = 2$, the distinctions being connected only with the larger number of site types. An inhomogeneous lattice is characterized by the distribution functions f_q and $d(Y)$, where $Y = q\{[m]\}$ and $1 \leqslant q \leqslant t$. For each type of clusters, we compile relation (3.5) for the correlators of the first type:

$$\theta_Y^{B\{[n]\}} = \theta_Y^{A\{[n]\}} M_Y^{\{n\}},$$
$$M_Y^{\{n\}} = \exp\left(-\beta\left[\nu(Y) + \sum_{p=1}^{t} n_{qp}\omega_{qp}\right]\right) \tag{3.22}$$
$$\nu(Y) = \nu_q^{B} - \nu_q^{A} + \sum_{p=1}^{t} m_{qp}(\varepsilon_{qp}^{AB} - \varepsilon_{qp}^{BB})$$

where ω_{qp} has been determined in Eq. (3.12).

The couplings between the correlators of the first and second types have the form

$$\theta_Y^{i\{n\}} = \sum_{k_{q1}=0}^{\tilde{m}_{q1}} \cdots \sum_{k_{qt}=0}^{\tilde{m}_{qt}} \prod_{p=1}^{t} C_{\tilde{m}_{qp}}^{k_{qp}} \theta_Y^{i\{[n+k]\}}$$
$$= \prod_{p=1}^{t}\left(\sum_{k_{qp}=0}^{\tilde{m}_{qp}} C_{\tilde{m}_{qp}}^{k_{qp}}\right)\theta_Y^{i\{[n+k]\}}, \quad \tilde{m}_{qp} = m_{qp} - n_{qp} \tag{3.23}$$

$$\theta_Y^{i\{[n]\}} = \prod_{p=1}^{t}\left(\sum_{k_{qp}=0}^{\tilde{m}_{qp}} (-1)^{k_{qp}} C_{\tilde{m}_{qp}}^{k_{qp}}\right)\theta_Y^{i\{n+k\}} \tag{3.24}$$

Formulas (3.23) and (3.24) naturally generalize expressions (3.13) and (3.14) because with an increase in the number of particles of A in the coordination sphere of a cluster, each type of sites is occupied independently of the others. The symbol $\{[n+k]\}$ signifies that at m_{qp} sites of type p there are $n_{qp} + k_{qp}$ particles of A, where $0 \leqslant k_{qp} \leqslant \tilde{m}_{qp}$. The second equality in (3.23) is the determination of the abridged form of writing all the sums that are taken over the types of adjacent sites p, where $1 \leqslant p \leqslant t$.

By excluding the correlators of the first type from relations (3.22) with the aid of formula (3.24), we obtain a system of equations for correlators of the second type ($l_{qp} = \tilde{m}_{qp} - k_{qp}$):

$$\theta_Y^{B\{n\}} = \prod_{p=1}^{t}\left(\sum_{k_{qp}=0}^{\tilde{m}_{qp}} C_{\tilde{m}_{qp}}^{k_{qp}}\right) M_Y^{\{n+k\}} \prod_{p=1}^{t}\left(\sum_{r_{qp}=0}^{l_{qp}} (-1)^{r_{qp}} C_{l_{qp}}^{r_{qp}}\right)\theta_Y^{A\{n+k+r\}} \tag{3.25}$$

The system of equations (3.25) is closed with the aid of the same approximations as for a lattice with two types of sites.

The quasichemical approximation is

$$\theta_Y^{i\{n\}} = \theta_q^i \prod_{p=1}^{t} (t_{qp}^{iA})^{n_{qp}} \tag{3.26}$$

where the functions t_{qp}^{ij} are defined in (3.17). Their values are determined in terms of the probability of particles of A occupying sites of type q and p [in terms of the normalizing relations and formula (3.20)]. The quantities θ_q^A are calculated from the following system of equations (its dimension is t):

$$1 - \theta_q^A = \theta_q^A \sum_{\sigma_{2q}} d(q\{[m]\})e^{-\beta\nu(q\{[m]\})} \prod_{p=1}^{t} (1 + t_{qp}^{AA} x_{qp})^{m_{qp}} \tag{3.27}$$

where the quantities x_{qp} have the form given in (3.18). Solution of the system (3.27) enables us to obtain all the correlators of the second type by formula (3.26), while the introduction of formula (3.26) into (3.24) enables us to obtain all the correlators of the first type:

$$\theta_Y^{i\{[n]\}} = \theta_q^i \prod_{p=1}^{t} (t_{qp}^{iA})^{n_{qp}} (t_{qp}^{iB})^{m_{qp} - n_{qp}},$$
$$t_{qp}^{iB} = 1 - t_{qp}^{iA} \tag{3.28}$$

The distribution of the particles in all the approximations formulated above for a homogeneous lattice can be considered in a similar way. The polynomial approximation is given in Appendix 1. If the correlation effects are not considered, the correlators of the first and second types are expressed as follows:

$$\theta_Y^{i\{[n]\}} = \theta_q^i \prod_{p=1}^{t} (\theta_p^A)^{n_{qp}} (\theta_p^B)^{m_{qp} - n_{qp}}, \quad \theta_Y^{i\{n\}} = \theta_q^i \prod_{p=1}^{t} (\theta_p^A)^{n_{qp}} \tag{3.29}$$

while the occupancy of the sites by particles of A is determined from the relevant system of equations. In the chaotic approximation, this system has the form of (3.27), where t_{qp}^{AA} is replaced by θ_p^A. In the molecular-field approximation, a system of equations is obtained after an additional limiting process $\beta\varepsilon_{qp}^{ij} \to 0$. It can be written in the form

$$\theta_q^i = \sigma_q^i / (\sigma_q^A + \sigma_q^B),$$

$$\sigma_q^i = \exp \beta \left[-\nu_q^i + z_q \sum_{p=1}^{t} d_{qp} \sum_{j=A}^{B} \varepsilon_{qp}^{ij} \theta_p^j \right] \tag{3.30}$$

where relation (3.10) has been used. The system (3.30) considers the interaction of neighboring particles without correlation effects, but the correlation effects in the distribution of sites of various types on a lattice are retained in it.

The obtained systems of equations enable us to describe the joint influence of lattice inhomogeneities and the interaction of neighboring particles. In the following sections (3.3-3.5), these equations will be used to describe adsorption systems.

3.3. Adsorption Isotherms

For an inhomogeneous surface, the adsorption isotherm determining the fraction of the surface covered at a given temperature depending on the pressure of the gas phase molecules is expressed as follows:

$$\theta(P) = \sum_{q=1}^{t} f_q \theta_q^{\mathrm{A}}(P) \tag{3.31}$$

where $\theta_q^{\mathrm{A}}(P)$ is the fractional coverage by adsorbed particles of sites of type q whose values are found from the solution of the closed systems of equations in Sec. 3.2. Let us rewrite these systems with considerations of the properties of vacant points $\nu_q^{\mathrm{v}} = \varepsilon_{qp}^{i\,\mathrm{v}} = 0$, $i = \mathrm{A}$, v. When treating the adsorption of one component of the gas phase below, the superscript A in θ_q^{A}, y_q^{A}, and $\varepsilon_{qp}^{\mathrm{AA}}$ will be omitted.

An adsorption isotherm in a quasichemical approximation is obtained by formula (3.31) in which θ_q is calculated from the following system [172]:

$$y_q = \frac{\theta_q}{1 - \theta_q} \sum_{\sigma_{2q}} d(q\{[m]\}) \prod_{p=1}^{t} (1 + t_{qp} x_{qp})^{m_{qp}} \tag{3.32}$$

where y_q is determined in Eq. (1.29), while x_{qp} and t_{qp} are determined in (3.18) and (3.20) with $\omega_{qp} = \varepsilon_{qp}$. If the distribution function for clusters of various types can be expressed in the form of Eq. (3.9), system (3.32) is simplified:

$$y_q = \frac{\theta_q}{1 - \theta_q} (\Sigma_q)^{z_q}, \quad \Sigma_q = 1 + \sum_{p=1}^{t} d_{qp} t_{qp} x_{qp} \tag{3.33}$$

A similar system of equations in the molecular-field approximation has the form [178, 179]:

$$y_q = \frac{\theta_q}{1 - \theta_q} \exp\left(-\beta z_q \sum_{p=1}^{t} d_{qp} \varepsilon_{qp} \theta_p\right) \tag{3.34}$$

In the absence of interactions, formulas (3.32)-(3.34) transform into formula (1.32). With any interactions in the region of very small coverages, isotherm (3.31) transforms into the Henry isotherm $\theta = \tilde{a}p^{1/m}$, where $\tilde{a} = \sum_{q=1}^{t} f_q(a_q^{\mathrm{A}})^{1/m}$ is the effective constant of adsorption on an inhomogeneous surface. By (1.29), the adsorption coefficient a_q^{A} can be represented as $a_q^{\mathrm{A}} = \hat{a}_q \exp \beta\varepsilon_q$, where $\hat{a}_q$ is the preexponential factor of the adsorption coefficient depending only slightly on the temperature, ε_q is the heat of adsorption at a site of type q. If we assume that for different sites the preexponential factors depend only slightly on the site type, then presuming that the subscript $q = 1$ relates to the site with the largest adsorption bond, we have

$$\tilde{a} = (a_1)^{1/m} \sum_{q=1}^{t} f_q \exp(\beta\Delta\varepsilon_q/m), \quad \Delta\varepsilon_q = \varepsilon_q - \varepsilon_1 \tag{3.35}$$

The greater the difference between the binding energies of sites of species q and 1, the smaller is the contribution of sites q to the effective adsorption constant. Formula (3.35) reveals that at low coverages the adsorption isotherm depends only on the composition of the adsorbent surface. With an increase in the surface coverage, the particle interactions begin to play a greater and greater role, and not only the surface composition, but also the surface structure becomes important.

We shall limit ourselves below to an analysis of surfaces for which the final results of expressions (3.32) and (3.33) are close to each other and shall investigate the joint influence of surface inhomogeneity and lateral interaction of the adsorbed particles on the behavior of the adsorption isotherms. For a homogeneous surface, consideration of interaction changes the region of pressure corresponding to a change in the surface coverage from zero values to a coverage of about unity, which the pressure P^* corresponds to. An increase in the repulsion of the particles makes this region larger, while an increase in the attraction makes it smaller in comparison with the Langmuir isotherm [147, 156, 157].

For an inhomogeneous surface in the absence of interaction, the considered pressure range depends on the difference in $\Delta\varepsilon_t$: when this difference is greater, the pressure range increases [33, 85]. In this sense, there is a direct analogy with the increase in the repulsion of the adsorbed particles. The joint consideration of the repulsion of particles and an inhomogeneous surface should increase the pressure P^* still more to achieve a given fractional coverage θ^* than is contributed by each factor separately, while the joint consideration of the attraction of particles and surface inhomogeneity should lower the pressure P^*.

The joint effect of both factors manifests itself the most obviously for a surface with two types of sites. It is easy to illustrate the complete spectrum of site arrangement by taking a surface containing both types of sites in the same amount. If the sites of one type form macroscopic homogeneous sections $d_{12} = d_{21} = 0$, this is a case of a patchwise surface (we disregard the contribution of the boundary between the sections). A chaotic distribution of the sites is obtained when they are mixed completely, $d_{qp} = f_p$. Finally, the regular alternation of sites of the two types (a chessboard) is a completely ordered surface $d_{12} = d_{21} = 1$. There may be intermediate ways of the mutual arrangement of the sites between the three extremes indicated above. If $f_1 \neq 0.5$, it is also possible to introduce the same limiting distributions and characterize them by the function d_{12} or $d_{21} = f_1 d_{12}/f_2$. When $d_{12} = 0$, we have a patchwise surface, when $d_{12} = f_2$—a chaotic one, and when $d_{12} = 1$, if $f_1 < 0.5$, each site of the first type is surrounded by ones of the second type, and by analogy with the case when $f_1 = 0.5$ this can be considered as an ordered surface. When $f_1 > 0.5$, $d_{21} = 1$ corresponds to an ordered surface, i.e. $d_{12} = f_2/f_1$.

Isotherms of physical adsorption are shown in Fig. 3.3 [180]. Attraction of the particles causes a more rapid growth of the coverage with increasing pressure than in the absence of interaction. At $\theta < f_1$, attraction of the particles manifests itself differently depending on the structure of the surface: the value of θ grows more rapidly with an increase in the pressure on a patchwise surface than on an ordered one. In the second case, the contribution of interaction

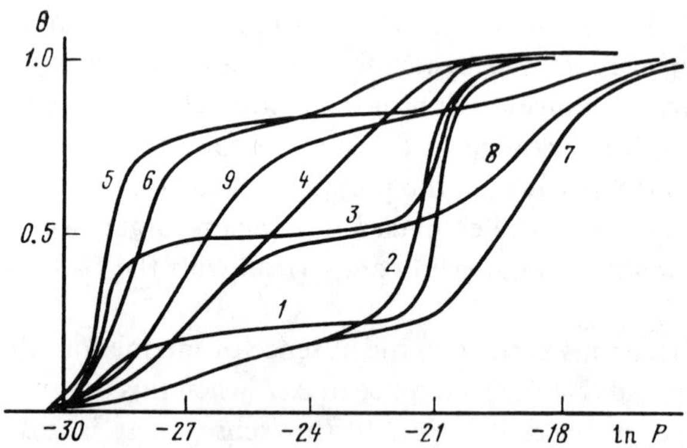

Fig. 3.3. Adsorption isotherms for molecules attracting one another at $\beta\varepsilon = 1$ (curves *1*—*6*) and in the absence of interaction (curves *7*—*9*):

$Q_1 = 113$, $Q_2 = 80$ kJ/mol, $z = 4$, $T = 500$ K, $a_1/a_2 = \exp \beta (Q_1 - Q_2)$; $f_1 = 0.2$ (*1, 2, 7*), $f_1 = 0.5$ (*3, 4, 8*); $f_1 = 0.8$ (*5, 6, 9*). The surface is patchwise (odd numbers) or ordered (even numbers)

plays a smaller role because sites with a higher adsorptivity are surrounded by ones with a lower adsorptivity that have a lower coverage than neighboring sites on a homogeneous section in the first case. Therefore isotherms at $\varepsilon > 0$ on ordered surfaces at $f_1 = 0.2$ and 0.5 virtually coincide with the isotherms at $\varepsilon = 0$.

For $f_1 = 0.8$, a similar region of coincidence of the isotherms is limited by $\theta < 0.07$. At $\theta \approx f_1$, the adsorption isotherms for surfaces with a different structure approximately coincide. This is associated with the condition $\varepsilon_{qp} = \varepsilon$. With a further growth in the pressure, the isotherms for ordered surfaces grow more rapidly than on patchwise ones because the adsorbed molecules on them have a larger number of occupied neighboring sites. The extreme values of P^* in addition to their dependence on the surface composition and the adsorption coefficients depend on the values of the interaction parameters and on the structure of the surface. It follows from system (3.33) that when $\theta_q \to 1$, we have

$$y_q \to \frac{1}{1-\theta_q} (\Sigma_q)^{z_q}, \quad \Sigma_q = \sum_{p=1}^{t} d_{qp} \exp(-\beta \varepsilon_{qp}) \tag{3.36}$$

Typical chemisorption isotherms are shown in Fig. 3.4. For particles that repel one another, the behavior of the curve of θ vs. P for patchwise and or-

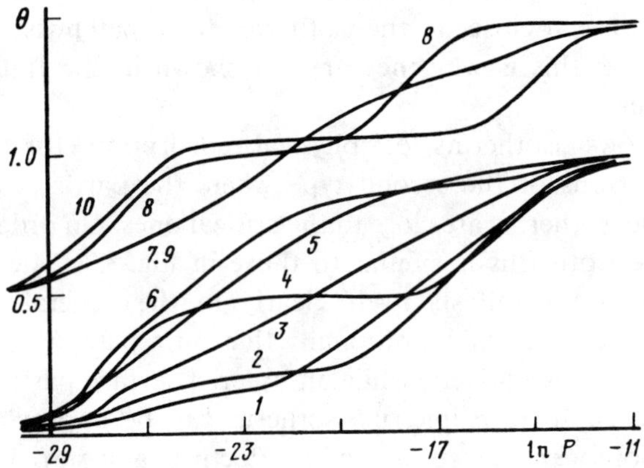

Fig. 3.4. Chemisorption isotherms:

1-6—$\beta\varepsilon = -1$; *7-10*—$\beta\varepsilon_{11} = -1.0$; $\beta\varepsilon_{22} = -1.5$; $\beta\varepsilon_{12} = -0.75$ (*7, 8*), -1.75 (*9, 10*); $f_1 = 0.2$ (*1, 2*), 0.5 (*3, 4, 7-10*); 0.8 (*5, 6*). Curves *7-10* are displaced along the axis of ordinates by $\Delta\theta = 0.5$. The remainder is as in Fig. 3.3

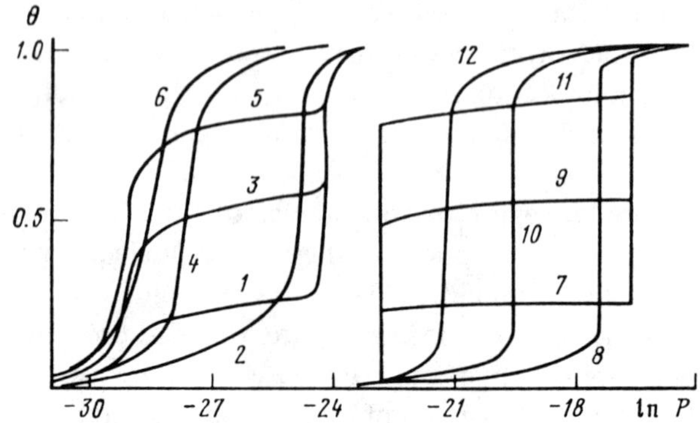

Fig. 3.5. Adsorption isotherms for attracted molecules:
Q_2 = 88 kJ/mol; *1-6*—$\beta\varepsilon_{11}$ = 1.0, $\beta\varepsilon_{12}$ = 1.75; $\beta\varepsilon_{22}$ = 1.5; *7-12*—$\beta\varepsilon_{12}$ = 2.3 (curves *7-12* are displaced along the axis of abscissas by 7 units); f_1 = 0.2 (*1, 2, 7, 8*), 0.5 (*3, 4, 9, 10*), 0.8 (*5, 6, 11, 12*). The remainder is as in Fig. 3.3.

dered surfaces is opposite to that indicated above for particles attracting one another. If the parameters of particle interaction depend on the type of the sites accommodating them, this may greatly change the coordinates of the isotherm intersection points for surfaces with a different structure (curves *9, 10*) or cause them to intersect twice (curves *7, 8*), or both. (The adsorption isotherms for surfaces with a chaotic arrangement of the sites are in the majority of cases much closer to the isotherms for a patchwise surface than for an ordered one; this is why they are not shown in the figures to avoid their overcrowding.)

Figure 3.5 shows isotherms of physical adsorption for $\varepsilon_{qp} = \varepsilon$. On homogeneous sections of the second type where the particles are attracted more strongly, the isotherms are close to the critical ones. On ordered surfaces, the nature of the isotherms is similar to those in Fig. 3.3. The influence of the surface structure manifests itself clearly in the general form of the isotherms. It can be assumed conditionally that on a patchwise surface each section is occupied consecutively, while an ordered surface participates in adsorption completely. In principle, its isotherm can be approximated by an isotherm for a homogeneous surface with effective values of the interaction parameter and the adsorption coefficient. This can be seen the best for curves *7-12* in Fig. 3.5. Under conditions when two phase transitions occur on a patchwise surface, one phase transition occurs on an ordered surface (at the given parameters for curve *12*, no phase transition occurs).

The critical conditions of two-dimensional condensation are determined by the simultaneous solution of system (2.30) and a system describing the occupation of sites of various types [one of the systems (3.1)-(3.3)]. These conditions were analyzed in the molecular-field approximation [181]. The presence within the confines of a monolayer of several phase transitions is one of the chief features of the joint influence of inhomogeneity of a surface and the attraction of adsorbed particles. The number of phase transitions and the values of the critical densities and temperatures are sensitive to the composition and structure of a surface. This is why the experimental determination of the critical densities and temperatures can evidently allow one to appraise the properties of a surface.

In the general case, notwithstanding the indicated strong effect of the properties of an adsorption system on the shape of the adsorption isotherms, it is often difficult to arrive at a conclusion on the composition and structure of a surface on the basis of the typical features of this shape. The effects of inhomogeneity and repulsion of the particles "screen" each other, and their separation makes it necessary to greatly change the temperature and pressure, which is not always possible.

Matters are somewhat simpler in physical adsorption, where the two factors "work" in opposite directions. Experimental isotherms with density jumps within the confines of a monolayer are well known [182, 183]. They are generally interpreted as relating to patchwise surfaces [182]. This includes the adsorption of the inert gases, ethane, nitrogen oxide and other gases on carbon, boron nitride, pure cadmium and its halides, nickel and sodium chlorides, and on other adsorbents.

Analysis of isotherms (3.32)-(3.34) reveals that although the maximum number of phase transitions in a monolayer is determined by the number of macroscopically homogeneous sections, similar density jumps remain with appreciable deviations from a strictly patchwise surface when the contribution of the boundaries between different sections may not be ignored. It should be noted that the molecular-field approximation gives a greater region of values of the parameters of the model, this region containing a plurality of phase transitions, than the approximations considering the correlation effects.

In the supercritical region of temperatures, the shape of the isotherms enables one to judge on the number of types of centers if only the difference $\beta(\varepsilon_q - \varepsilon_{q+1}) \geqslant 5$. For chemisorption, depending on the structure of the surface and the degree of repulsion, this difference does not always allow one to correctly appraise the number of types of sites on a surface. It is therefore good practice to involve independent information on the properties of the surface and, by using the theory being set out, verify the correctness of the appraisals of the distribution functions for sites of various types by the shape of the adsorption isotherms being calculated (and not seek these distribution functions from the adsorption isotherms) or vary them within preliminarily restricted limits. The expediency of such an approach has been confirmed by considering the system argon-zeolite NaX [180].

3.4. Heat of Adsorption

The isosteric heat of adsorption is determined by a general relation of the type of the Clausius-Clapeyron equation (2.31) in which the derivative of the temperature with respect to the pressure is taken for a constant surface coverage (3.31). The surface coverage and the pressure at constant temperature are implicitly related by one of the systems of equations in Sec. 3.3 relative to θ_q. If this system is differentiated with respect to β, we obtain a linear system of equations. In the matrix form, it is written as follows:

$$M\left|\overline{\frac{d\theta_q}{d\beta}}\right| = |\overline{Q_q^*}|, \quad Q_q^* = Q_q - Q + E_q, \quad Q_q = \frac{d \ln a_q}{d\beta} = \varepsilon_q^{\mathrm{A}} \tag{3.37}$$

where Q_q is the heat of adsorption at a site of type q (the preexponential factor of the adsorption coefficient is considered to depend only slightly on the temperature): $|\overline{d\theta_q/d\beta}|$ and $|\overline{Q_q^*}|$ are vector columns composed from the quantities $d\theta_q/d\beta$ and Q_q^*, where $1 \leqslant q \leqslant t$, and M is the matrix of the system of equations, the form of its elements M_{qp} and of the quantities E_q depending on the approximation used to consider interaction. As previously, we shall limit ourselves to the approximation (3.33) [184]:

$$E_q = \frac{z_q}{\Sigma_q} \sum_{p=1}^{t} d_{qp}\varepsilon_{qp}t_{qp}e^{\beta\varepsilon_{qp}}[1 - x_{qp}(1 - \theta_q - \theta_p + \theta_{qp})/b_{qp}]$$

$$\begin{aligned} M_{qp} = {} & \Delta_{qp}([\theta_q(1 - \theta_q)]^{-1} + \frac{z_q}{\Sigma_q} \sum_{p=1}^{t} d_{qp}x_{qp}^2 t_{qp}(1 - t_{qp})/b_{qp}) \\ & + t_{qp}\frac{z_q d_{qp}}{\Sigma_q\theta_p} x_{qp}(1 + x_{qp}\theta_p(1 - t_{pq})/b_{qp}] \end{aligned} \tag{3.38}$$

The solution of system (3.37) has the form [185]

$$\frac{d\theta_q}{d\beta} = \Delta_q(Q_q^*)/\Delta \tag{3.39}$$

where Δ is the determinant of matrix M, $\Delta_q(Q_p^*)$ is the determinant of matrix M in which the q-th column has been replaced by the column of the right-hand sides. In view of the constancy of the surface coverage with a change in the temperature, we have $\sum_q f_q(d\theta_q/d\beta) = 0$, and we obtain an expression for the heat of adsorption [185]:

$$Q(\theta) = \sum_{q=1}^{t} f_q \Delta_q(Q_p + mE_p) \Big/ \sum_{q=1}^{t} f_q \Delta_q(1) \tag{3.40}$$

where $\Delta_q(1)$ is the determinant of matrix M in which the q-th column has been replaced by a column of ones. Formula (3.40) determines the implicit dependence of the heat of adsorption on the coverage θ in terms of the value of θ_q.

In the absence of interaction, formula (3.40) transforms into the expression [185, 186]

$$Q(\theta) = \sum_{q=1}^{t} f_q Q_q \theta_q (1 - \theta_q) \Big/ \sum_{q=1}^{t} f_q \theta_q (1 - \theta_q) \tag{3.41}$$

For zero coverages, formulas (3.40) and (3.41) are rewritten as follows:

$$Q(0) = \sum_{q=1}^{t} f_q Q_q (a_q)^{1/m} \Big/ \sum_{q=1}^{t} f_q (a_q)^{1/m} \tag{3.42}$$

The heat of adsorption at zero coverages is the weighted value of all the heats of adsorption for various types of sites. It depends on the fraction of the sites and their adsorption coefficients whose magnitude differs from the heat of adsorption at sites with the maximal adsorptivity. This can be seen quite well taking the example of a surface with two types of sites. For the conditions when equation (3.35) is observed, formula (3.42) becomes:

$$Q = (Q_1 + Q_2 \varkappa)/(1 + \varkappa), \quad \varkappa = f_2 e^{-\beta\delta}/f_1, \ \delta = Q_1 - Q_2 \tag{3.43}$$

When the difference δ is larger, the quantity $\varkappa$ is smaller, and Q is closer to Q_1. On the other hand, if $Q_2 \to Q_1$, then we also have $Q \to Q_1$. Consequently, there is a certain value of δ^* that corresponds to the largest difference between Q and Q_1. The value of δ^* depends on the surface composition of the adsorbent and on the temperature and can reach values of the order of 10 kJ/mol [184].

For the complete coverage of a surface, the heat of adsorption is:

$$Q(1) = \sum_{q=1}^{t} f_q(Q_q + mz_q\langle\varepsilon_q\rangle)L_q \Big/ \sum_{q=1}^{t} f_q L_q$$
$$\langle\varepsilon_q\rangle = \sum_{p=1}^{t} d_{qp}\varepsilon_{qp} \exp(-\beta\varepsilon_{qp})/\Sigma_q, \quad L_q = (\Sigma_q)^{z_q}/(a_q)^{1/m} \tag{3.44}$$

where Σ_q has been determined in (3.36). The heat of adsorption depends on all the parameters of the model, namely, the composition and structure of the surface, the adsorption coefficient, and the interaction between particles. If $z_q = z$ and $\varepsilon_{qp} = \varepsilon$, the contributions to the heat of adsorption from the interaction of the particles and from an inhomogeneous surface are separated:

$$Q(1) = mz\varepsilon + \sum_{q=1}^{t} f_q Q_q (a_q)^{-1/m} \Big/ \sum_{q=1}^{t} f_q (a_q)^{-1/m} \tag{3.45}$$

The first addend coincides with the contribution of the interacting particles on a homogeneous surface [see (2.32), (2.34)]. The second addend coincides with the limit expression of formula (3.41) when $\theta_q \to 1$. This expression is the weighted value of the contributions of all the types of sites, and its value can differ appreciably from the heat of adsorption at sites with a minimum adsorptivity (by analogy with what was shown above for zero coverages). It follows from the difference of expressions (3.42) and (3.44) that the repulsion of the particles increases, and the attraction decreases the total increment of the heat of adsorption in comparison with the absence of interaction.

Prior to going over to the complete concentration dependences of the heat of adsorption, let us consider an example illustrating the significant role of surface structure. For a surface with a strictly ordered arrangement of sites of two types with $f_1 = 0.5$ (a chessboard) in the region of low coverages, it can be shown [184] that an increase in the repulsion of the particles increases the heat of adsorption with a growth in the coverage so that it is profitable for the particles to be at sites of the first type ($Q_1 > Q_2$) with $\theta_1(\varepsilon_{12} < 0) > \theta_2(\varepsilon_{12} = 0)$. Such a growth in the heat of adsorption cannot lead to values exceeding the value of Q_1, i.e. we are speaking of an increase by a value of the order of magnitude of δ (Fig. 3.6). (For attraction of the particles, the effect is opposite—with a growth in the coverage the heat of adsorp-

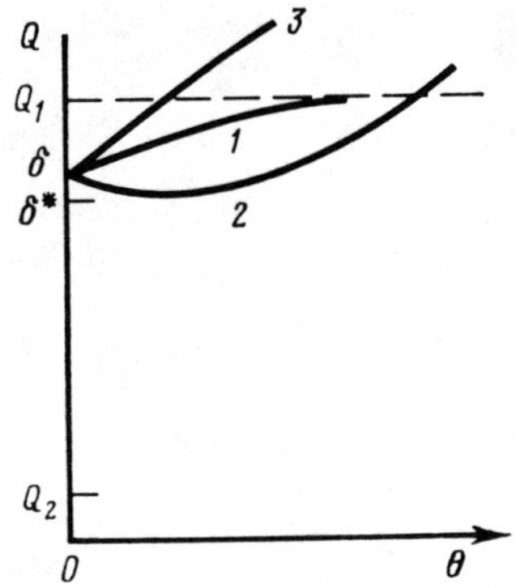

Fig. 3.6. Heats of adsorption at low surface coverages of the "chess-board" type:

1—repulsion of particles; *2*—attraction of particles; *3*—attraction of particles on a patchwise surface

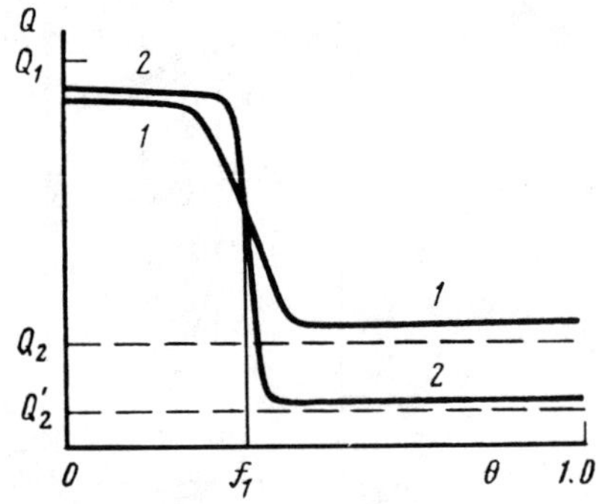

Fig. 3.7. Heat of adsorption in the absence of interaction for a surface with two types of sites, $Q_2 > Q_2'$

tion diminishes.) Here there comes to light the difference in the behavior of particles repelling one another on an ordered surface from that of particles attracting one another on patchwise surfaces for which the heat of adsorption may exceed Q_1.

The concentration dependences of the heats of adsorption involve all the properties of an adsorption system [187]. It is the simplest to treat the joint influence of surface inhomogeneity and particle interaction for surfaces with two types of sites. In the absence of interaction, the dependence of Q on θ has a stepped shape (Fig. 3.7): at $\theta < f_1$ the heat is constant and equals Q for $\theta = 0$, at $\theta > f_1$ the heat is also constant and equals Q for $\theta = 1$, while in the transition region the heat monotonically and virtually linearly decreases with an increasing coverage. The width of the transition region is narrower when the quantity $\beta(Q_1 - Q_2)$ is larger.

Consideration of the interaction depending on the surface structure changes Q vs. θ in various ways (Fig. 3.8). On a patchwise surface, if the fraction of sites of the first type is high, the heat of adsorption at low coverages is similar to that on a homogeneous surface. When the fraction of sites of the first type diminishes, repulsion of the particles leads to an increase in the fraction of the particles at sites of the second type, and the region of the practical constancy of Q vs. θ diminishes. The width of the transition region grows with repul-

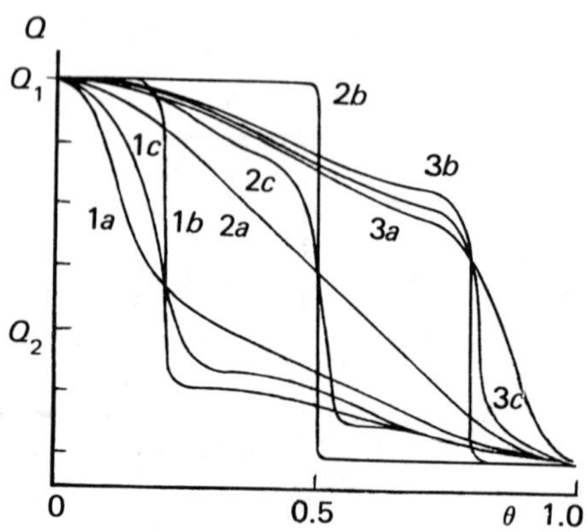

Fig. 3.8. Variation of composition and structure of surface, $\beta(Q_1 - Q_2) = 8$:

$\beta\varepsilon = -1.0$; $z = 4$: *1*—$f_1 = 0.2$; *2*—$f_1 = 0.5$; *3*—$f_1 = 0.8$. The surfaces are: *a*—patchwise; *b*—ordered; *c*—chaotic; $Q_1 = 113$ kJ/mol, $T = 500$ K. Along the axis of ordinates, $\Delta Q = 8.2$ kJ/mol

sion of the particles. For $f_1 = 0.5$, the curve of Q vs. θ is antisymmetrical relative to the point [$\theta = 0.5$; $Q(\theta = 0.5)$], but it differs from similar curves for a homogeneous surface [in the slope of the linear portion at $\theta \approx 0.5$ (about two times)] and for a uniformly inhomogeneous surface without interaction (in the slope of the linear portion and the shape of the curve for high and low coverages).

For an ordered surface, the curves of Q vs. θ depend differently on the composition for $f_1 < 0.5$, $d_{12} = 1$, $d_{11} = 0$ and $f_1 > 0.5$, $d_{21} = 1$, $d_{22} = 0$. In the second case with low coverages, there is only a slight difference from a patchwise surface because the sites of the first type form the major part of the surface. Distinctions appear at $\theta \approx f_1$, when practically all the sites of the first type are occupied, and when a particle settles on a site of the second type, all its occupied neighbors yield a repelling contribution. The heat of adsorption drops sharply to a value of the order of magnitude of $Q_1 - Q_2 + z\varepsilon$ instead of a smooth change on a patchwise surface. If $f_1 < 0.5$, at low coverages, as indicated above, the heat grows to a value of Q_1 and remains constant almost up to complete coverage of sites of the first type. The width of the transition region in comparison with the absence of interaction diminishes. The jump in the heat of adsorption at $\theta \approx f_1$ depends on the difference $Q_1 - Q_2$ and on the contribution of the particles repelling one another that

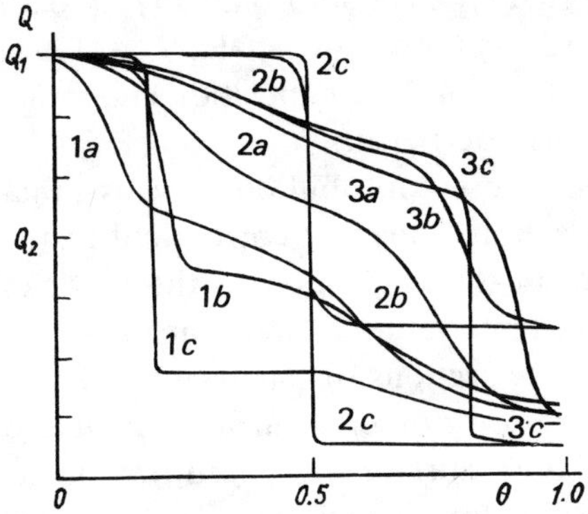

Fig. 3.9. Influence of the parameter $\beta\varepsilon_{12}$: $\beta(Q_1 - Q_2) = 6$, $\beta\varepsilon_{11} = -1.0$, $\beta\varepsilon_{22} = -1.5$, $\beta\varepsilon_{12} = -0.75$; the letter *d* signifies an ordered surface at $\beta\varepsilon_{12} = -1.75$. The remainder is as in Fig. 3.8.

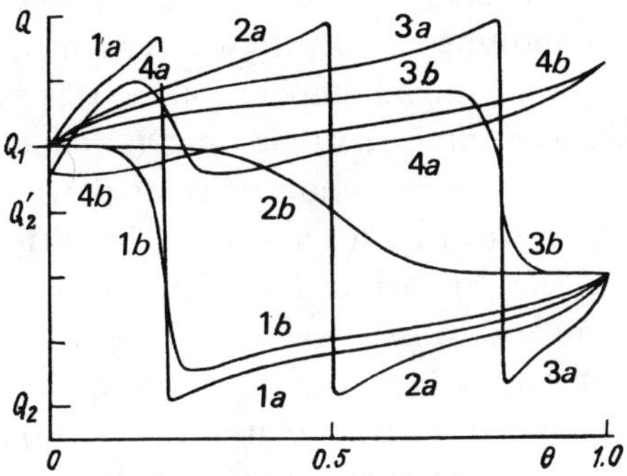

Fig. 3.10. Variation of composition and structure of a surface, $\beta(Q_1 - Q_2) = 8$, $\beta\varepsilon = 1.0$.
The curves *4* are for $f_1 = 0.2$ at $\beta(Q_1 - Q_2) = 2$. The remainder is as in Fig. 3.8

tend to take up positions so as to have no occupied sites of the first type as neighbors. The latter is possible for $(0.5 - f_1)$ sites of the second type (the plateau on curve 1*b*); at $\theta > 0.5$ the heat diminishes monotonically as the sites of the second type become occupied.

The influence of the differences in the parameters of particle interaction on the relation between Q and θ is shown in Fig. 3.9. For an ordered surface at $f_1 = 0.5$, the behavior of the curve of Q vs. θ is determined only by the values of ε_{12}. With an increase in repulsion, ε_{12} lowers the width of the transition region. For $f_1 < 0.5$, the curve of Q vs. θ is also affected by the interactions between particles at sites of the second type, and for $f_1 > 0.5$—at sites of the first type. Their contributions also change the limiting values of $Q(1)$.

The heats of physical adsorption are shown in Fig. 3.10. The influences of attraction and inhomogeneity of the surface on the change in the heat of adsorption are opposite: attraction increases Q with increasing θ, while inhomogeneity lowers it. For a patchwise surface, the contribution of each homogeneous section occurs additively depending on its coverage. The width of the transition region diminishes with an increasing attraction of the particles. On an ordered surface, the particles at isolated sites of the first type (curves 1*b* and 2*b*) do not practically have neighboring occupied sites, and the heat of adsorption at low coverages does not change or diminishes (curve

4*b*), as indicated above. If the sites of the first type are not isolated ($f_1 > 0.5$), the heat of adsorption increases with the coverage by analogy as for a homogeneous surface whose properties depend on the composition and structure of an inhomogeneous surface and the coefficients a_q.

The extreme values of $Q(1)$ depend on whose contribution prevails—that of surface inhomogeneity or particle attraction. Curves *4* correspond to the case when $z\varepsilon > Q_1 - Q_2$. The differences in the interaction of the particles at sites of various types leads to various extreme values of the heats at $\theta \to 1$, to various differences in the heat on separate sections of patchwise surfaces, and to various shapes of the curves of Q vs. θ for ordered surfaces. It should be noted that the relations $Q = f(\theta)$ for chaotic surfaces that occupy an intermediate position between the similar relations for patchwise and ordered surfaces are displaced in the direction of patchwise surfaces [187].

The transition to a large number of site types sharply increases the number of different surface states and it is natural to attempt to answer the question as to the extent to which information on the dependence of Q on θ can indicate the number of site types. The author in a different publication [187] has considered the general situation on the example of surfaces with four types of sites.

An analysis of various arrangements of sites has shown that in the majority of cases the probability of determining the exact number of site types is quite low. The observed curves of Q vs. θ can be approximated by relations with a smaller number of site types, including one, if modified parameters of a model are employed. The probability of determining t diminishes with an increase in its value and with a more regular arrangement of sites of various types. On the other hand, this probability grows for particles that attract one another, especially with a drop in the temperature of the system and with a patchwise arrangement of sites of various types. Nevertheless, in a number of specific situations, a knowledge of the dependence of Q on θ enables one to make reasonable appraisals of the state of inhomogeneity of a surface, especially in combination with data obtained independently by other methods.

The considered concentration dependences relate to the region of relatively small interactions. These relations may change their nature when phase transitions of stratification or ordering occur in an adsorption system.

Care must be taken in the indicated cases because the heterogeneity of particle distribution may be produced not only by the difference in the properties of the lattice sites, but also by interaction between the particles. This manifests itself the most illustratively in a system of particles repelling one another, which will be treated in the following section.

3.5. Ordering of Adsorbed Particles

The interaction between particles is due to the cooperative nature of behavior of the system. If the magnitude of this interaction exceeds the kinetic energy of the particles, then a qualitatively new distribution of the particles is possible in a system in comparison with an ideal one. When particles repel one another and the temperature is lowered, a transition occurs from chaotic distribution of the particles to their regular arrangement by lattice sites so that the predominately occupied sites form a new regular lattice with a period larger than the lattice constant of the surface structure. In other words, the initial lattice breaks up into a set of sublattices with a various probability of being occupied by particles. The phenomenon of ordering occurs in many chemisorbed systems [170, 188-192], and also in solid substitutional and interstitial solutions [12-15, 47, 48].

We shall limit ourselves to a consideration of a very simple ordered system that forms on a face (100) with moderate surface coverages—a superstructure $c(2 \times 2)$ (using the terminology of studies of low-energy electron diffraction).

To describe other superstructures, account must be taken of the contributions from the interaction of more remote neighbors (see Sec. 6.4). The given superstructure is characterized by all the lattice sites breaking up into two sublattices. We shall designate the first one whose sites are occupied by α, and the second one whose sites are mainly vacant by γ. Every site of an α-sublattice is surrounded by sites of a γ-sublattice, and vice versa. This allows us to set the ordered state of the particles by using the functions of distribution of sites of various types introduced above, i.e. $f_\alpha = f_\gamma = 0.5$ and $d(\alpha[0, z]) = d(\gamma[z, 0]) = 1$. The same adsorption equations are obtained when simplifying the description with the aid of the distribution functions $d_{\alpha\gamma} = d_{\gamma\alpha} = 1$.

Let $\theta_\sigma(\sigma = \alpha, \gamma)$ be the probability of occupation of the sites of types σ. Adsorption isotherm (3.31) will be rewritten as $\theta = (\theta_\alpha + \theta_\gamma)/2$, where θ_σ are determined from the system of equations

$$y = \frac{\theta_\alpha}{1 - \theta_\alpha}(1 + t_{\alpha\gamma}x)^z, \quad y = \frac{\theta_\gamma}{1 - \theta_\gamma}(1 + t_{\gamma\alpha}x)^z$$

$$x = \exp(-\beta\varepsilon) - 1, \quad \theta_{\alpha\gamma} = 2\theta_\alpha\theta_\gamma/(\delta + b),$$

$$t_{\lambda\sigma} = \theta_{\alpha\gamma}/\theta_\lambda, \quad \lambda \neq \sigma \tag{3.46}$$

$$\delta = 1 + x(1 - \theta_\alpha - \theta_\gamma), \quad b = (\delta^2 + 4x\theta_\alpha\theta_\gamma)^{1/2}$$

Here definitions (3.18), (3.20) and (1.23), (2.23) are used because the particles are on a homogeneous surface, and their internal degrees of freedom and interaction parameters do not depend on the type of the sublattice site. (It should be noted that the initial equations in Sec. 3.2 allow one to consider the difference in the parameters of a model depending on the type of superstructure.)

If we exclude y from Eqs. (3.46), we obtain an equation relating the relative occupancy of the sites θ_α and θ_γ. Having in view that $\theta_\gamma = 2\theta - \theta_\alpha$, this equation allows us to calculate θ_α vs. θ and $\beta\varepsilon$. It has one or three roots. In the first case, ordering of the particles is absent in the system, while in the second the root with the higher value belongs to θ_α, and with the lower one to θ_γ. The temperature below which the formation of an ordered state is possible is known as the order-disorder transition temperature. Its value is determined by expanding into a series the equation relating the occupancy of sites α and γ by the small values of the order parameter ϱ that characterizes the fraction of particles at the sites of the α-sublattice [12]: $\theta_\alpha = \theta + \varrho(1 - \theta)$, $\theta_\gamma = \theta - \varrho(1 - \theta)$. If all the particles are at the sites of the α-sublattice, then $\varrho = 1$ and $\theta_\alpha = 1$. The values $\varrho = 0$ and $\theta_\alpha = \theta_\gamma = \theta$ correspond to the disordered state.

Limiting ourselves to the first terms of the expansion at $\varrho = 0$, we obtain a relation between the critical values of the particle concentration and the order-disorder transition temperature: $1 + x + b = 2zx\theta(1 - \theta)$. For the particular case of half coverage, we have $\beta\varepsilon = -2 \ln (1 - 2/z)$. In the region of low temperatures ($\beta\varepsilon \to -\infty$) the extreme values of the coverages limiting the region of existence of the ordered state of $c(2 \times 2)$ are $\theta = 1/z$ (Fig. 3.11).

Consideration of ordering changes the shape of the adsorption isotherms. Figure 3.12 shows the occupancies of the sites of both sublattices and the overall coverage of a surface depending on the external pressure. When the critical coverage of the surface is reached, rapid occupancy of the α-sublattice begins chiefly as a result of the particles leaving sites of the γ-sublattice. The quantity θ_γ passes through a minimum and begins to grow with increasing pressure. When the second critical coverage is approached, the change in the occupancy of sites γ mainly occurs at the expense of the particles leaving sites α. When equal occupancy of both sublattices is achieved, the system again passes into a disordered state.

The nature of distribution of the particles repelling one another with lowering of the temperature is also illustrated by the curve of θ_{AA} vs. $\beta\varepsilon_{AA}$ for a half surface coverage (see Fig. 2.9, curve *5*). Consideration of ordering sharply alters curve *4* and brings it closer to the exact Onsager solution within virtually

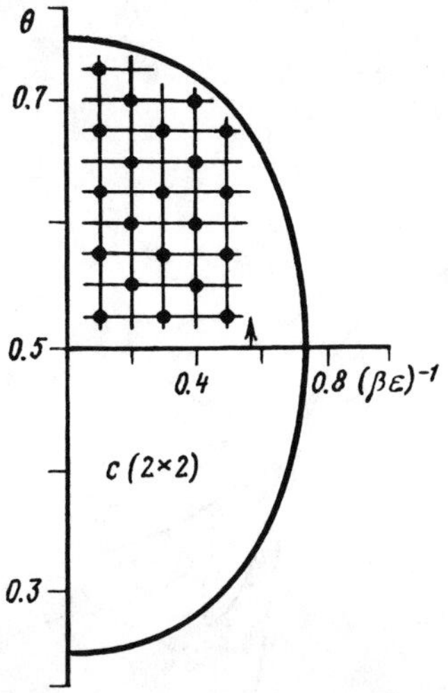

Fig. 3.11. Region of existence of the ordered phase c (2 × 2), $z = 4$.

The arrow indicates the exact value of the order-disorder transition temperature

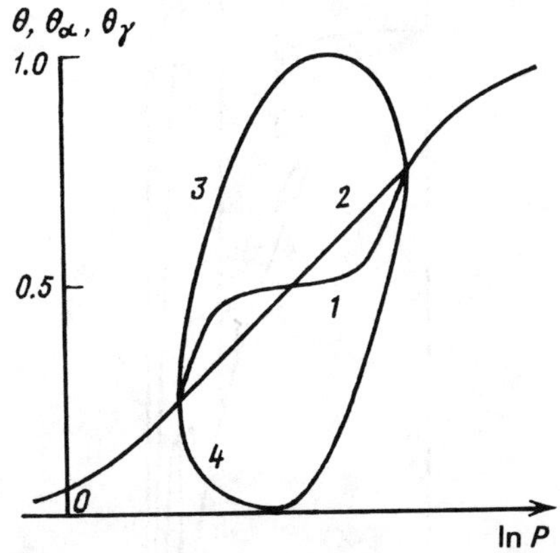

Fig. 3.12. Adsorption isotherm for ordered particles (*1*), isotherm of disordered system (*2*), and occupancy of sites of α- and γ-sublattices (*3*, *4*)

the entire region of temperatures (the maximal difference is in the vicinity of the critical point of the order-disorder phase transition).

The heats of adsorption of particles repelling one another on a homogeneous surface with account taken of ordering are calculated by formula (3.40). The curves of Q vs. θ for the region of ordering differ sharply from those for the disordered phase. Figure 3.13 shows the heats of adsorption for various values of $\beta\varepsilon$. After the point of ordering, the curve breaks and the heat of adsorption sharply grows because the particles mainly occupying sites of the α-sublattice have no neighboring occupied sites. In the region with $\theta = 0.5$, the heat of adsorption drops sharply: the particles being adsorbed now have all their neighboring sites occupied. The curve of Q vs. θ remains antisymmetrical relative to the point $[\theta = 0.5,\ Q(\theta = 0.5)]$. The maximum (at $\beta\varepsilon \approx -1.7$) value of the heat of adsorption above Q for $\theta < 0.5$ and the drop in the heat of adsorption relative to $Q(0) + z\varepsilon$ for $\theta > 0.5$ do not exceed one-fourth of the difference between the curves of Q vs. θ for the ordered and disordered states. The fact that the values of Q protrude from the region limit-

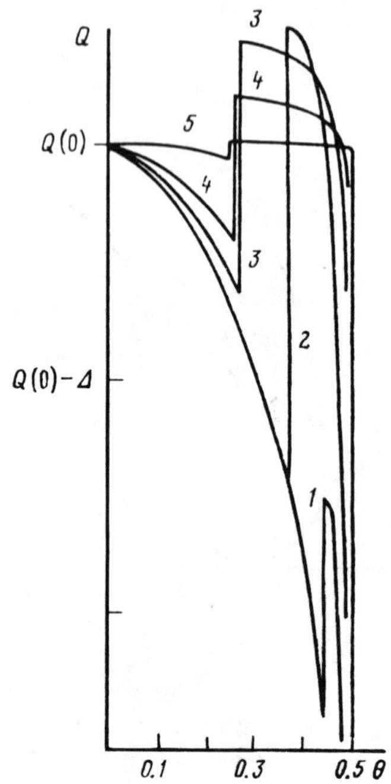

Fig. 3.13. Heat of adsorption of ordered particles on a homogeneous surface:

$z = 4$; *1*—$\beta\varepsilon = -1.4$; *2*—-1.7; *3*—-2.2; *4*—-3.0; *5*—$\beta\varepsilon = -6.0$; along the ordinate axis $\Delta = 4.2$ kJ/mol

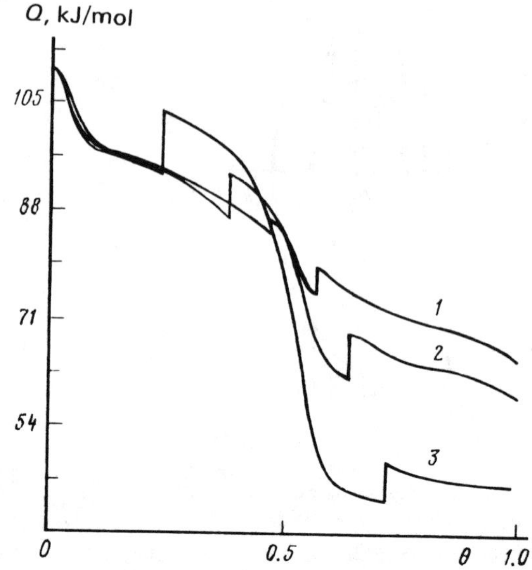

Fig. 3.14. Heat of adsorption:

stepped surface: $l = 10$, $t = 6$, $Q_1 = Q_2 = 113$, $Q_3 = Q_4 = 97$, $Q_5 = Q_6 = 80$ kJ/mol. The nonzero distribution functions are: $d_{mk} = 2/3$, $mk = 12, 21, 56, 65$; $d_{mk} = 1/3$, $mk = 14, 23, 54, 63$; $d_{mk} = [4(l-2)]^{-1}$, $mk = 41, 32, 45, 36$; $d_{34} = 1 - [2(l-2)]^{-1} = d_{43}$, $f_q = 1/2l$, $q = 1, 2, 5, 6$; $f_3 = f_4 = (l-2)/2l$; $\beta\varepsilon = -1.6$ (*1*), -2 (*2*), -3 (*3*)

ed by $Q(0)$ and $Q(0) + z\varepsilon m$ is associated with the approximation used, which overstates the correlation effects in comparison with the exact solution.

Ordering of the particles occurs not only on a homogeneous surface. Figures 3.14 and 3.15 show the heat of adsorption for a stepped surface and the occupancy of its sites. For a stepped surface, the number of sites of various types is doubled (compare with Sec. 3.1) because of the differentiation of sites α and γ on a terrace and next to a step. Although the curves of Q vs. θ become more complicated than for a homogeneous surface, sections can be singled out for which inhomogeneity of the sites (low and high coverages) and ordering of the particles (moderate coverages) are "responsible". This cannot always be done for more involved situations.

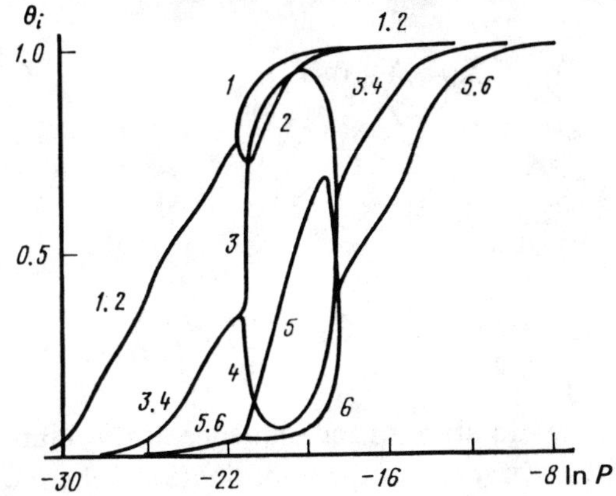

Fig. 3.15. Occupancy of sites on a stepped surface by ordered particles at $\beta\varepsilon = -2$. The remainder is as in Fig. 3.14.

The sublattices in a simple cubic lattice ($z = 6$) and volume-centered cubic lattice ($z = 8$) of the type of β-brass are a volume analogue of the superstructure $c(2 \times 2)$. All the expressions indicated above (only the number z changes) remain in force if the absorption process is being considered in single crystals of pure components, or the analogous expressions for binary solutions (see Sec. 3.2) if solid substitutional solutions are being considered.

3.6. Distribution of Multicomponent Solution Particles and Adsorption Isotherms of a Gas Mixture

Let us summarize the results obtained in Sec. 3.2 for a multicomponent system. The procedure of obtaining a system of equations for cluster correlators consists of a consideration of (1) the relations between functions (3.5) with different particles at the central sites of the clusters, (2) the relation between the correlators of the first [Eq. (3.5)] and second [Eq. (3.7)] types, and (3) exclusion of the correlators of the first type.

For each of the clusters $Y = q\{[m]\}$ on which an inhomogeneous lattice is depicted, the cluster correlators are expressed in terms of one another as follows:

$$\theta_Y^{i\{n\}} = \prod_{p=1}^{t} \left(\sum_{k_{qp}^1=0}^{\tilde{m}_{qp}} \cdots \sum_{k_{qp}^{s-1}=0}^{=m_{qp}-\varkappa_{qp}} C_{\tilde{m}_{qp}}^{\{k_{qp}\}} \right) \theta_Y^{i\{[n+k]\}}$$

$$\theta_Y^{i\{[n]\}} = \prod_{p=1}^{t} \left(\sum_{k_{qp}^1=0}^{\tilde{m}_{qp}} \cdots \sum_{k_{qp}^{s-1}=0}^{=m_{qp}-\varkappa_{qp}} (-1)^{\tilde{k}_{qp}} C_{\tilde{m}_{qp}}^{\{k_{qp}\}} \right) \theta_Y^{i\{n+k\}} \tag{3.47}$$

$$\tilde{m}_{qp} = m_{qp} - \sum_{j=1}^{s-1} n_{qp}^j, \quad \varkappa_{qp} = \sum_{j=1}^{s-2} k_{qp}^j, \quad \tilde{k}_{qp} = \sum_{j=1}^{s-1} k_{qp}^j$$

Here we have used a compact notation of the sequence of the sums describing the occupancy of m_{qp} sites of type p in the coordination sphere of a cluster with a center q similar to the notation (3.23). Each type of sites p is occupied independently. For a multicomponent system, the state of occupancy of these sites is given by sets of the numbers $n_{qp}^j + k_{qp}^j$, where $1 \leqslant j \leqslant s-1$, as in (2.40) for a homogeneous lattice.

The relations of pairs of the correlators of the first type (3.5) yield:

$$\theta_Y^{i\{[n]\}} = \theta_Y^{j\{[n]\}} \exp \beta \left[\nu_q^j - \nu_q^i + \sum_{p=1}^{t} \sum_{l=1}^{s} (\varepsilon_{qp}^{il} - \varepsilon_{qp}^{jl}) n_{qp}^l \right] \tag{3.48}$$

Excluding correlators of the first type from relations (3.48), we obtain a generalization of system of equations (3.25) for $s > 2$. By additionally regrouping the addends on the right-hand sides of the equations, we obtain a generalization of system of equations (2.39) for an inhomogeneous lattice:

$$\theta_Y^{i\{n\}} = \exp \beta \left(\nu_q^j - \nu_q^i - \sum_{p=1}^{t} \sum_{l=1}^{s} (\varepsilon_{qp}^{jl} - \varepsilon_{qp}^{il}) n_{qp}^l \right)$$

$$\times \prod_{p=1}^{t} \left(\sum_{k_{qp}^1=0}^{\tilde{m}_{qp}} \cdots \sum_{k_{qp}^{s-1}=0}^{m_{qp}-\varkappa_{qp}} \prod_{l=1}^{s-1} (x_{qp}^{ijl})^{k_{qp}^l} C_{\tilde{m}_{qp}}^{\{k_{qp}\}} \right) \theta_Y^{j\{n+k\}} \tag{3.49}$$

$$x_{qp}^{ijl} = \exp(\beta \omega_{qp}^{ijl}) - 1, \quad \omega_{qp}^{ijl} = \varepsilon_{qp}^{il} - \varepsilon_{qp}^{jl} - \varepsilon_{qp}^{is} + \varepsilon_{qp}^{js}$$

The equations are obtained in a quasichemical approximation when the following approach is employed:

$$\theta_Y^{i\{n\}} = \theta_q^i \prod_{p=1}^{t} \prod_{j=1}^{s-1} (t_{qp}^{ij})^{n_{qp}^j}; \quad \theta_Y^{i\{[n]\}} = \theta_q^i \prod_{p=1}^{t} \prod_{j=1}^{s} (t_{qp}^{ij})^{n_{qp}^j} \tag{3.50}$$

where the functions t_{qp}^{ij} have been determined in (3.17). In this case, system (3.49) at $\{n\} = \{0\}$ yields $(s-1)$ equations for the functions θ_q^i, where $1 \leqslant i \leqslant s-1$:

$$\theta_q^s = \theta_q^i \exp \beta(\nu_q^i - \nu_q^s) \prod_{p=1}^{t} [S_{qp}^i(s)]^{m_{qp}}$$

$$S_{qp}^i(s) = \sum_{j=1}^{s} t_{qp}^{ij} \exp \beta(\varepsilon_{qp}^{sj} - \varepsilon_{qp}^{ij}) \tag{3.51}$$

The equations for the functions θ_{qp}^{ij} are obtained from system (3.49) for all the zero indices of n_{qp}^l except the one $n_{qp}^j = 1$. These equations can be rewritten in an equivalent form [177] that generalizes the relations between two-body correlators (2.45) for a homogeneous lattice for their application to an inhomogeneous lattice:

$$\theta_{qp}^{ij} \theta_{qp}^{kl} = \theta_{qp}^{il} \theta_{qp}^{kj} \exp \beta(\varepsilon_{qp}^{ij} + \varepsilon_{qp}^{kl} - \varepsilon_{qp}^{il} - \varepsilon_{qp}^{kj}) \tag{3.52}$$

Equations (3.51) and (3.52) have been obtained for a specific cluster Y. For the functions θ_q^i and θ_{qp}^{ij} to relate to an entire inhomogeneous lattice, it is necessary to perform averaging with the aid of the distribution functions f_q and $f(q\{[m]\})$. The equations for the pairs (3.52) do not depend on the type of cluster, therefore they retain their form for a lattice as a whole, while the averaging of equations (3.51) yields the following system of equations:

$$\theta_q^s = \theta_q^i \exp \beta(\nu_q^i - \nu_q^s) \sum_{\sigma_{2q}} d(q\{[m]\}) \prod_{p=1}^{t} [S_{qp}(s)]^{m_{qp}} \tag{3.53}$$

In closing system of equations (3.52) and (3.53), we shall take into account the normalizing relations

$$\sum_{i=1}^{s} \theta_q^i = 1, \quad \sum_{i=1}^{s} \theta_{qp}^{ij} = \theta_p^j, \quad \sum_{j=1}^{s} \theta_{qp}^{ij} = \theta_q^i \tag{3.54}$$

The polynomial approximation is considered in Appendix 1. In the absence of correlation between the interacting particles, formulas (3.50) will be rewritten as follows:

$$\theta_Y^{i\{n\}} = \theta_q^i \prod_{p=1}^{t} \prod_{j=1}^{s-1} (\theta_p^j)^{n_{qp}^j}, \quad \theta_Y^{i\{[n]\}} = \theta_q^i \prod_{p=1}^{t} \prod_{j=1}^{s} (\theta_p^j)^{n_{qp}^j} \tag{3.55}$$

In this case, the equations in a chaotic approximation have the form of (3.53) in which the functions t_{qp}^{ij} have been replaced by θ_p^j, while the equations in the molecular-field approximation are obtained with the additional limiting process $\beta\varepsilon_{qp}^{ij} \to 0$, as has been indicated previously:

$$\theta_q^i = \sigma_q^i \Big/ \sum_{i=1}^{s} \sigma_q^i, \quad \sigma_q^i = \exp \beta \left(-\nu_q^i + z_q \sum_{p=1}^{t} d_{qp} \sum_{j=1}^{s} \varepsilon_{qp}^{ij} \theta_p^j \right) \tag{3.56}$$

where relation (3.10) has been used. System of equations (3.56) was obtained by the author in a different publication [193] with the aid of Bogolyubov's variation principle.

If we take into account the properties of vacant sites $\nu_q^{\mathrm{v}} = \varepsilon_{qp}^{i\mathrm{v}} = 0$, where $1 \leqslant i \leqslant s$, Eqs. (3.53) and (3.56) are adsorption isotherms of an $(s-1)$-component gas mixture for an inhomogeneous surface with consideration of the interaction of the nearest neighbors.

In the quasichemical approximation, the adsorption isotherms have the form

$$y_q^i = \frac{\theta_q^i}{\theta_q^{\mathrm{v}}} \sum_{\sigma_{2q}} d(q\{[m]\}) \prod_{p=1}^{t} \left(1 + \sum_{j=1}^{s-1} t_{qp}^{ij} x_{qp}^{ij}\right)^{m_{qp}}$$

$$x_{qp}^{ij} = \exp(-\beta\varepsilon_{qp}^{ij}) - 1, \quad \theta_q^{\mathrm{v}} = 1 - \sum_{i=1}^{s-1} \theta_q^i \tag{3.57}$$

where the functions t_{qp}^{ij} are determined by solving system (3.52) and (3.54), while y_q^i has been determined in (1.29). If the distribution function of the clusters can be represented in form (3.9), equations (3.57) are simplified:

$$\theta_q^{\mathrm{v}} y_q^i = \theta_q^i \left[1 + \sum_{p=1}^{t} d_{qp} \sum_{j=1}^{s-1} t_{qp}^{ij} x_{qp}^{ij}\right]^{z_q} \tag{3.58}$$

In the molecular-field approximation, the adsorption isotherms are expressed by a formula similar to formula (1.32):

$$\theta_q^i = \tilde{y}_q^i \Big/ \left(\sum_{i=1}^{s-1} \tilde{y}_q^i + 1\right),$$

$$\tilde{y}_q^i = y_q^i \exp \beta z_q \sum_{p=1}^{t} d_{qp} \sum_{j=1}^{s-1} \varepsilon_{qp}^{ij} \theta_p^j \tag{3.59}$$

In the absence of interaction, formulas (3.57)-(3.59) transform into (1.32).

In the region of low coverages, we can obtain the following appraisal from system (3.52), (3.54):

$$\theta_{qp}^{ij} = \theta_q^i \theta_p^j \exp \beta \varepsilon_{qp}^{ij} \tag{3.60}$$

which shows that attraction of the particles increases the probability of finding them at neighboring sites in comparison with a chaotic arrangement, while repulsion diminishes this probability.

In the general case, the systems of equations in a quasichemical approximation have to be solved relative to the partial occupancies of sites of various types [how to reduce the dimension of systems (3.52), (3.57) or (3.58) is indicated in Appendix 1]. The system of equations in the molecular-field approximation can be simplified substantially if we assume that the interactions between neighboring particles do not depend on the types of the sites at which they are ($\varepsilon_{qp}^{ij} = \varepsilon_{ij}$), and sites of various types are distributed chaotically ($f_{qp} = f_q f_p$). Notwithstanding the roughness of these assumptions, it is good to use them as the first approximation for describing an involved problem such as the adsorption of interacting particles on an inhomogeneous surface. The chief idea of an approximate description consists in the transition from a system for θ_q^i to a system for the average values of the partial surface coverages θ_i (1.34):

$$\theta_i = \sum_{q=1}^{t} \frac{f_q \tilde{y}_q^i}{\left(1 + \sum_{j=1}^{s-1} \tilde{y}_q^i\right)}, \quad \tilde{y}_q^i = y_q^i \exp \beta z_q \sum_{j=1}^{s-1} \varepsilon_{ij}\theta_j \tag{3.61}$$

At the given values of the surface composition of the adsorbent, system (3.61) determines the implicit dependence of the partial coverages on the partial pressure of the molecules in the gas phase. Further simplifications are possible on the basis of system (3.61). They are associated with additional conditions for the distribution function f_q and for the dependences of the binding energies of various particles on the type of site q. The author [193] has generalized the concepts of the simbatic and antybatic behavior of the adsorption coefficients of various particles [194]. With a uniform function of distribution of various types of sites, they make it possible to conduct summation on the right-hand part of (3.61) and express the system of equations for the partial adsorption isotherms in terms of elementary functions.

3.7. Absorption of Gas-Phase Molecules

It has been stressed on a previous page that lattice models of adsorption and absorption from the viewpoint of the molecular-statistical theory are identical and differ in the number of nearest neighbors. The adsorption equations obtained above in Secs. 3.3, 3.5, and 3.6, if by z_q we understand the number of nearest neighbors of an absorption center of type q, correspond to the absorption isotherms of interacting particles. This is why we shall define more specifically the absorption coefficients and the distribution functions of sites of various types contained in these equations. Let us consider pure metals and solid solutions.

There are two types of absorption centers in pure metals, namely, octahedral (e.g. the first type) and tetrahedral (the second type) interstices. In some cases, centers of the first type are occupied (H_2/Pd, Ni, C/Ti, O/Ta), in others, of the second type (H_2/Nb, Ta), and in still others, of both types (H_2/V, Ce) [47]. If absorption proceeds into one type of centers, the lattice of the absorption sites is homogeneous; the number of its sites N is related to the number of metal atoms N_M by the expression $N = \sigma N_M$, where $\sigma = 3$ $(q = 1)$ and 6 $(q = 2)$ for a body-centered cubic (bcc) lattice and $\sigma = 1$ $(q = 1)$ and 2 $(q = 2)$ for a face-centered cubic (fcc) lattice. If absorption proceeds into both types of centers, the lattice is inhomogeneous: $f_1 = 1/3$ and $f_2 = 2/3$, while $d(q[m_{q1}, m_{q2}]) = 1$, where the values of m_{q1} and m_{q2} depend on the type of lattice of the metal and the type of lattice of the interstices shown in Figs. 3.16 and 3.17 (for example, for an fcc lattice $m_{12} = 8$, $m_{21} = 4$, and $m_{11} = m_{22} = 0$). In the latter case, a major role can be played by repulsion of the nearest neighbors. This results in blocking of neighboring types of interstices because of exchange interaction, which facilitates ordering of the particles at the expense of differences in the energies of the bond to the metal lattice.

Matters become more involved in absorption by multicomponent solid solutions. Let us illustrate them for a binary alloy A-B. We shall consider that the interaction of particle i in interstice σ with the neighboring atoms j is of a two-body nature—$\varepsilon_{ij}(\sigma)$, j = A, B. Here the number of site types and the energy binding particle i to the lattice depend on the type of interstice σ and on the number of atoms of A and B forming as absorption center. The number of absorption site types is $\sum_{\sigma} \tilde{\varkappa}_\sigma$, where $\tilde{\varkappa}_\sigma = 1 + \varkappa_\sigma$. Here $\varkappa_\sigma$ is the number of alloy atoms forming an absorption center, $\varkappa_1 = 6$ and $\varkappa_2 = 4$. The binding energy for an fcc lattice is written as $\varepsilon^i_{\sigma n} = n\varepsilon_{iA}(\sigma) + (\varkappa_\sigma - n) \times \varepsilon_{iB}(\sigma)$; n is

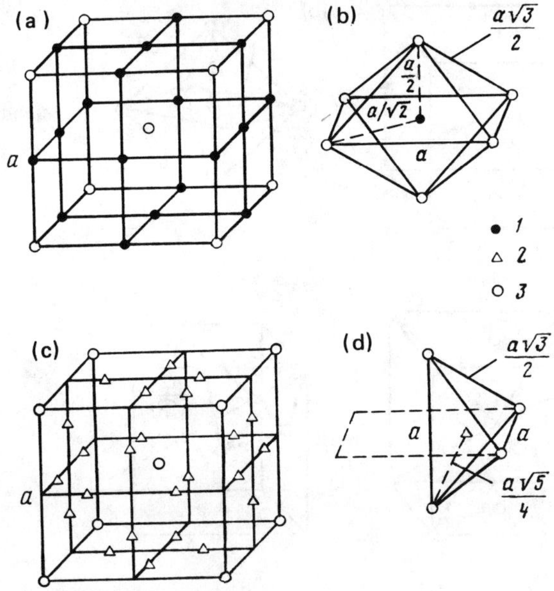

Fig. 3.16. A bcc lattice:

The octahedral (*1*) (*a*) and tetrahedral (*2*) (*c*) interstices in a cubic cell of metal atoms (*3*), and metal atoms neighboring with the relevant interstice (*b*, *d*) [48]

the number of A atoms among $\varkappa_\sigma$ sites. The distances between i and j are $a/2$ for $\sigma = 1$ and $a/\sqrt{2}$ for $\sigma = 2$. For a bcc lattice, these distances are $a\sqrt{5}/4$ for $\sigma = 2$, and ε^i_{2n} has the same form, while for $\sigma = 1$ these distances differ, namely, four atoms are at a distance of $a/\sqrt{2}$ and two at a distance of $a/2$. If we denote by n_1 and n_2 the number of atoms of A at these distances, we have $\varepsilon^i_{1n_1n_2} = n_1\varepsilon_{iA}(1) + (4 - n_1)\varepsilon_{iB}(1) + n_2\tilde{\varepsilon}_{iA}(1) + (2 - n_2)\tilde{\varepsilon}_{iB}(1)$, where the tilde indicates the atoms at a distance of $a/2$. The number of various types of sites in a bcc lattice at $\sigma = 1$ grows to 15.

The distribution functions of sites of various types f_q and $d(q\{[m]\})$ for the absorption of gas-phase molecules by absorbents that do not reorganize themselves are determined from the solution of problems on the distribution of atoms of A and B. Assume that an alloy was annealed at a temperature T_1 and sharply cooled to a temperature T_2. The local distributions of the alloy atoms are described by equations of Sec. 2.2 if the alloy is disordered and by equations of Secs. 3.2 and 3.5 if the alloy was ordered at the temperature

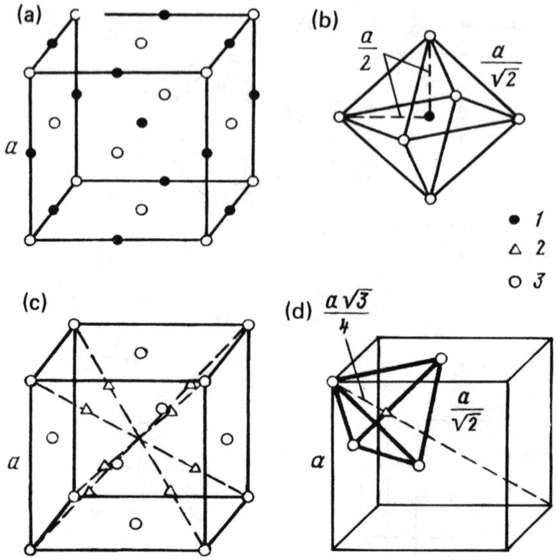

Fig. 3.17. An fcc lattice:

The octahedral (*1*) (*a*) and tetrahedral (*2*) (*c*) interstices in a cubic cell of metal atoms (*3*), and metal atoms neighboring with the relevant interstice (*b, d*) [48]

T_1. The functions f_q are the probability of finding n particles of A among $\varkappa_\sigma$ sites because definite values of σ and n unambiguously correspond to each q.

For simplicity, we shall consider octahedral interstices of an fcc lattice. The probability of a configuration of six atoms of the alloy can be expressed in terms of two-body probabilities. If we denote the probability of finding particles of types i and j at neighboring sites f and g by θ_{fg}^{ij}, the probability of a closed configuration consisting of six atoms (Fig. 3.17) will be written as $\theta_{12\ldots6}^{ij\ldots k} = \prod_{f=1}^{6} \theta_f^i \prod_{\langle fg\rangle} \xi_{fg}^{ij}$, where $\xi_{fg}^{ij} = \theta_{fg}^{ij}/(\theta_f^i\theta_g^j)$, and $\langle fg\rangle$ signifies a set of all the pairs between neighboring numbers of sites: 1g, 6g, g = 2-5, 23, 25, 34, 45; among the indices of types $ij\ldots k$ there are n particles of A. The probabilities θ_{fg}^{ij} are determined from a solution of equations of Secs. 2.2, 3.2, and 3.5. Therefore, the distribution functions f_q equal $C_6^n\theta_{1\ldots6}^{i\ldots k}$.

To calculate the functions $d(q\{[m]\})$, it is necessary to consider the probabilities of configurations of a considerably larger number of atoms of A and B. If we are interested only in octahedral interstices, the number of neighboring

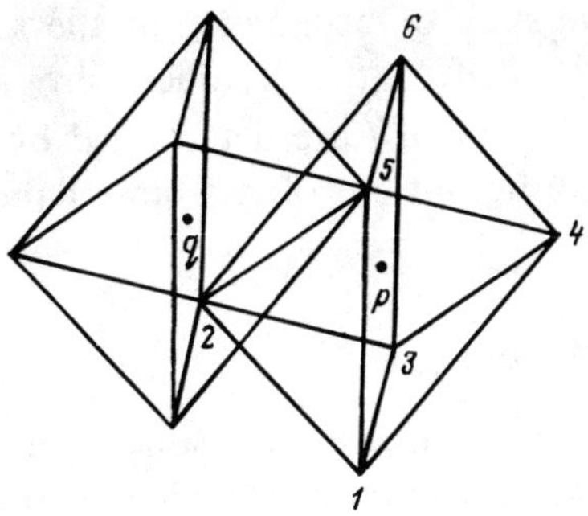

Fig. 3.18. Two neighboring octahedral interstices in an fcc lattice and the metal atoms surrounding them

adsorption centers in an fcc lattice is 12. The type of neighboring absorption center p is characterized by the number of particles of A and B from the six sites surrounding it, two of them also being included in the absorption center of site q (Fig. 3.18).

The dimension of the function $d(q\{[m]\})$ equals the number of lattice sites of the alloy enveloping all the absorption centers, namely, 6 (center q) + 12×4 (centers p) = 54. The probability of such a configuration is also expressed in the form $\prod_{f=1}^{54} \theta_f^i \prod_{\langle fg \rangle} \xi_{fg}^{ij}$, where $\langle fg \rangle$ signifies the set of all the pairs between neighboring sites. When calculating the weighting factor of configurations with different values of the numbers n_A at various absorption centers, we must have in view that a change in the state of occupancy of a lattice site of the alloy around the absorption center q simultaneously changes the type of center for the four neighboring centers p. Moreover, for each center p of the four "remaining" sites of the alloy enveloping it, the states of occupancy change independently only for the two most remote sites. The change in the state of the two other sites changes the state of occupancy at two other neighboring centers p. As a result, the weighting factor of the configuration of the atoms of a solid depends on the arrangement of the particles of A at 12×2 lattice sites of the alloy with a fixed arrangement of particles A and B around center q.

Matters are substantially simplified if we use the distribution function d_{qp}. Now for each absorption center p, the state of occupancy of its four sites is considered independently of the state of occupancy of the sites at other absorption centers: $d_{qp} = \theta^{i...k}_{1...6} / (\theta^{j}_{2}\theta^{l}_{5}\xi^{jl}_{25})$ (the numbers are indicated in Fig. 3.18), the species of the particles j and l are determined by the type of center q, while the weights of the configurations obey a binomial distribution.

3.8. Solid-Vacuum Interface

The distribution of the components of a solid in the subsurface region differs from that in the bulk. Lattice models enable us to consider the subsurface regions with a flat and rough phase interfaces (Fig. 3.19). In both cases, the subsurface region is divided into a number of parallel monatomic layers. The set of the quantities θ^{i}_{k}, where $1 \leqslant i \leqslant s$, $1 \leqslant k \leqslant \varkappa$, $\varkappa$ is the width of the subsurface region, is the concentration (depth) profile of a multicomponent solution showing how the local densities of the components change with the depth. The inhomogeneity of distribution of the atoms of a solid in the subsurface region, as in an ordered system, is due to interaction between the particles. The equations describing the distributions of the particles obtained above in Sec. 3.6 hold for them, and now it is only necessary to specify the distribution functions for sites of various types.

If the number of each layer k is compared with the type of the site in it, we have $f_k = 1/\varkappa$. Let z_{kq} be the number of nearest neighbors of a site in layer k, these neighbors being in layers $q = k, k + 1$, and let the total number of these neighbors be z_k, i.e. $z_k = z_{kk+1} + z_{kk} + z_{kk-1}$. Generally lattices with

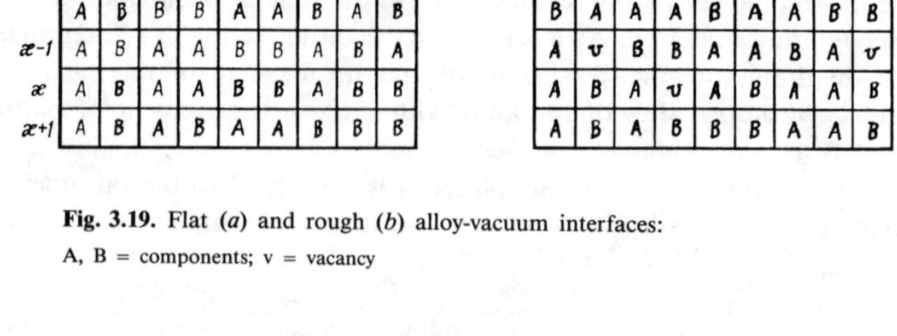

Fig. 3.19. Flat (*a*) and rough (*b*) alloy-vacuum interfaces:
A, B = components; v = vacancy

the following numbers of neighbors are used: $6 = 1 + 4 + 1$, $8 = 4 + 0 + 4$, and $12 = 3 + 6 + 3$. Since each particle of layer k interacts with its nearest neighbors, the structure of an inhomogeneous lattice is set with the aid of the functions $d(k[z_{kk+1}, z_{kk}, z_{kk-1}]) = 1$, $k \neq 1$. For a surface layer of a flat interface $k = 1$, neighbors at the side of the layer $k = 0$ are absent in vacuum, therefore the total number of sites is $z_1 = z_{11} + z_{12}$. The function $d(1[z_{12}, z_{11}, 0]) = 1$ has been determined for it. The system of equations for the concentration profile in the quasichemical approximation, by (3.51), will be written as follows:

$$\theta_k^s = \theta_k^i \exp \beta(\nu_k^i - \nu_k^s) \prod_{q=k+1}^{k-1} [S_{kq}^i(s)]^{z_{kq}}, \quad 2 \leqslant k \leqslant \varkappa \tag{3.62}$$

where the functions $S_{kq}^i(s)$ are calculated with a view to the system of equations for two-body correlators (3.52) (see also Appendix 1).

System (3.62) is closed by equations for the layer $k = 1$:

$$\theta_1^s = \theta_1^i \exp \beta(\nu_1^i - \nu_1^s)[S_{12}^i(s)]^{z_{12}}[S_{11}^i(s)]^{z_{11}} \tag{3.63}$$

Equations (3.62) and (3.63) by analogy with the model of adsorption on an inhomogeneous surface retain the dependence of the model parameters ν_k^i and ε_{kq}^{ij} on the number of the lattice layer. Traditionally in calculations for solid alloys [16-20] and liquid solutions of nonelectrolytes [81-83], it is assumed that these quantities do not depend on the layer number. If we adopt this assumption, the numbers of the layers should be omitted, although it is general knowledge that, for instance, for the surface of NaCl [73] the relaxation displacements of the anions and cations differ, and this results in an additional contribution to the polarization of the surface layer. This is why the step-by-step consideration of the presence of an interface makes it necessary to take into account in a lattice model the dependence of the internal degrees of freedom of the particles and their interaction parameters on the layer number [174].

Recall (see the end of Sec. 2.1) that here the quantities ν_k^i do not contain the binding energies of a particle i with its neighbors, but show the ratio of the internal degrees of freedom of a particle i in a solid and in a gas. If $\nu_k^i = \nu_i$, the exponential term in (3.62) may be excluded by using the equations for the bulk layer $\varkappa + 1$:

$$\frac{\theta_k^s}{\theta_{\varkappa+1}^s} = \frac{\theta_k^i}{\theta_{\varkappa+1}^i} \prod_{q=k+1}^{k-1} [S_{kq}^i(s)/S_{\varkappa+1q}^i(s)]^{z_{kq}} \tag{3.64}$$

When solving system (3.62), (3.63), the values of $\theta^i_{\varkappa+1}$ corresponding to the bulk of a solid solution are considered to be known. Such a statement of the problem presumes that the size of the subsurface region is much smaller than the volume of the solid, $\varkappa \ll L$, where L is the linear dimension of the solid phase. This allows us to consider the change in the state of the subsurface region for the condition of the state of the bulk being constant. For $\varkappa/L \leqslant 1$ (thin films, fine crystals), it is necessary to additionally consider the material balance for each component of the solution in the bulk and subsurface region. Solution of system (3.62), (3.63) enables one to determine the surface composition of a solid and the total accumulation of each component in the subsurface region:

$$\Delta\theta_i = \sum_{k=1}^{\varkappa} (\theta^i_k - \theta^i_{\varkappa+1})/\varkappa \qquad (3.65)$$

Concentration profiles of binary alloys are presented in Table 3.1 [192]. Their form depends on the quantity ω (3.12) (the properties of the particles are considered to be identical for all the layers). When $\omega > 0$ (a tendency of stratification), a readily volatile component accumulates on the surface, and its concentration monotonically diminishes with an increasing layer number. A growth in the value of $\beta\omega$ causes the width of the subsurface region to increase. When $\omega < 0$ (a tendency of ordering), the concentration profile fluctuates relative to $\theta^A_{\varkappa+1}$. With a growth in $|\beta\omega|$, the width of the subsurface region increases considerably more slowly than for stratifying alloys. With a growth in the layer number, the amplitude of the fluctuations rapidly diminishes. For alloys becoming ordered, situations are possible when the less volatile component A accumulates on the surface (a change in the nature of the surface segregation): the tendency of an atom of B being surrounded by atoms of A predominates over the tendency of the readily volatile component accumulating in the layer $k = 1$. These situations depend on the bulk concentration of the components, the lattice structure, and the value of $\beta\omega$.

If a solid solution is ordered in the bulk, the distribution functions of sites of various type acquire a different form. We shall restrict our treatment to the case of the two sublattices α and γ considered in Sec. 3.5. Within the limits of each layer, the sites divide into two types. For clarity, we shall designate the site type by two indices k and σ, $\sigma = \alpha, \gamma$. The functions $f_{k\sigma} = 1/2\varkappa$. Each particle in layer k at a site of sublattice σ has neighbors in sublattice λ, where $\lambda \neq \sigma$, and in layers k and $k \pm 1$. Consequently, $d(k\sigma[\lambda z_{kk+1}, \lambda z_{kk}, \lambda z_{kk-1}]) = 1$, $2 \leqslant k \leqslant \varkappa$, and for the layer $k = 1$ we have $d(1\sigma[\lambda z_{12}, \lambda z_{11},$

Table 3.1. Concentration of Component A in the Subsurface Region

k	v	g	v	g	v	g	v	g
	$\beta\omega = -5.5$		$\beta\omega = -3.0$		$\beta\omega = -1.5$		$\beta\omega = -0.5$	
1	0.7423	0.7498	0.7140	0.7312	0.6883	0.7179	0.6418	0.6945
2	0.6944	0.6934	0.6982	0.6960	0.7014	0.6979	0.7041	0.7004
3	0.7008	0.7009	0.7002	0.7005	0.6998	0.7002	0.6997	0.7000
4	0.6999	0.6999	0.7000	0.6999	0.7000	0.7000	0.7000	
5	0.7000	0.7000		0.7000				
	$\beta\omega = 0$		$\beta\omega = 0.25$		$\beta\omega = 0.50$		$\beta\omega = 0.65$	
1	0.5860	0.6682	0.5357	0.6461	0.4483	0.6046	0.3554	0.5663
2	0.7000	0.7000	0.6889	0.6964	0.6534	0.6827	0.5893	0.6598
3			0.6993	0.6998	0.6916	0.6969	0.6670	0.6883
4			0.7000	0.7000	0.6985	0.6995	0.6904	0.6966
5					0.6997	0.6999	0.6972	0.6990
6					0.7000	0.7000	0.6992	0.6997
7							0.6998	0.6999
8							0.6999	0.7000

Note. For all the layers ($k = 1, 2, \ldots, 8$) following the ones for which $\theta^{A}_{x+1} = 0.7$, the relevant quantity also equals 0.7; v = vacuum, g = gas phase, $\beta\varepsilon_{AA} = 5.0$, $\beta\varepsilon_{BB} = 4.0$, $\beta\varepsilon_{XA} = 2.5$, $\beta\varepsilon_{YA} = 3.0$, $\beta\varepsilon_{XB} = 2.0$, $\beta\varepsilon_{YB} = 2.5$, $\beta\nu_X = 0$, $\beta\nu_Y = 1.0$, $\beta\nu_v = -2.0$, $\beta\varepsilon_{XX} = 1.0$, $\beta\varepsilon_{XY} = 1.5$, $\beta\varepsilon_{YY} = 1.75$.

0)) = 1. As a result, we shall have the following system of equations in a quasichemical approximation:

$$\theta^{s}_{k\sigma} = \theta^{i}_{k\sigma} \exp \beta(\nu^{i}_{k} - \nu^{s}_{k}) \prod_{q=k+1}^{k-1} [S^{i}_{k\sigma,q\lambda}(s)]^{z_{kq}} \tag{3.66}$$

For the surface layer $k = 1$, Eqs. (3.66) hold if we consider that $z_{10} = 0$. System (3.66) is supplemented with the relevant equations (3.52) for the two-body functions $\theta^{ij}_{k\sigma,q\lambda}$ characterizing the probability of finding particles i and j in layers k and q on sublattices σ and λ.

The model of a rough surface is a more general model of a solid-vacuum interface. At the interface of two immiscible phases, there is always a transition region containing components of both phases. If by these phases we understand a solid solution and vacuum (a phase of vacancies), the transition region will be a rough surface [195]. The concentrations of the components in both immiscible phases are related by the conditions of equality of the chemical potentials of all the components and by the equations of the material balance [equations similar to Eqs. (2.51)]. Having determined these concentrations, we shall find the values of θ_k^i from the solution of system (3.62), where $1 \leqslant k \leqslant \varkappa$, by including the vacancies into the number of components of the system. Such an approach makes it possible to determine the role of the vacancies in the processes of surface segregation in alloys and the nature of the roughness of the surface of a solid.

The roughness of a surface characterizes the distribution function of columns [196] consisting of atoms of the solid over the height h, where $0 \leqslant h \leqslant \varkappa - 1$. Let P_h be the probability of a column having the height h. The height of a column h signifies that in a vertical sequence of sites in the layers $\varkappa, \varkappa - 1, \ldots, \varkappa - h$ there are atoms of the solid, whereas in the layer $\varkappa - h - 1$ there is a vacancy (this is the traditional definition for crystals without vacancies [197]). Knowing the solution of system (3.62), we can calculate the value of P_h quite readily. For a single crystal of A, we have

$$P_h = \prod_{k=1}^{h} t^{\mathrm{AA}}_{\varkappa-k+1,\varkappa-k} (1 - t^{\mathrm{AA}}_{\varkappa-h,\varkappa-h-1}) \tag{3.67}$$

An expression for a multicomponent system similar in its structure is given in [195]. The traditional methods [196-199] of determining the surface roughness are a difficult task to cope with even for single crystals, while the approach being considered also enables one to solve it for multicomponent alloys.

The influence of vacancies on the concentration profile of a binary alloy has been studied by the author earlier [192]. It was found that when $\beta\varepsilon_{ii} \geqslant 3.0$, where i = A, B, the fraction of self-adsorbed atoms of an alloy does not exceed values of 10^{-2} to 10^{-4}, while when $\beta\varepsilon_{ii} > 6.5$, the solid practically has a flat surface. The concentration of the vacancies in the layer $k = 1$ is generally higher by 1.5-2 orders of magnitude than in the layer $k = 2$. Notwithstanding the low values of the vacancy concentrations in an alloy (it is usually considered that in the vicinity of the melting point $\theta_v \approx 10^{-4}$-10^{-3} and their fraction diminishes with lowering of the temperature [200]), their presence may affect the concentration profile of the alloy. When $\omega > 0$, the vacancies increase the degree of segregation and the accumulation of the readily volatile component somewhat. Their concentration drops monotonically with an increase in the number of the layer of the subsurface region. When $\omega < 0$, the vacancies lower the amplitude of fluctuations of the concentration profile of the alloy components. Their concentration diminishes with a growth of the layer k and may also fluctuate in the subsurface region. These features of vacancy distributions can have a major significance for describing the processes of particle migration in the subsurface region by a vacancy mechanism.

We shall note in concluding that the found distribution of the atoms of a solid in the subsurface region can be used for constructing functions of distribution of sites of various types in problems of the absorption of gas phase particles in exactly the same way as this was discussed in Sec. 3.7. Nonuniformity in the distribution of the atoms of the solid increases the number of types of sites of absorption in this region, and also plays a substantial role in the processes of migration of the absorption particles [201].

3.9. Solid-Gas Phase Interface

As we noted in Chap. 1, at a solid-gas phase interface there always occurs adsorption of gas phase molecules. At low contact times or a low binding energy between the adsorbed particles and the atoms of the solid, or both, the distribution of the solid's atoms remains virtually unchanged. The theory of adsorption for surfaces that do not rearrange themselves has been considered in Secs. 3.3-3.6. With prolonged contact times or a large binding energy, or both, the adsorption of molecules changes the distribution of the solid's atoms in the subsurface region in comparison with their distribution in vacuum (Sec. 3.8).

Let us consider a flat solid-gas phase interface (Fig. 3.20). The layer $k = 0$ (not shown in Fig. 3.19) is a layer of adsorbed particles. The structure of the layer $k = 0$ may differ from that of the layers of the subsurface region of a solid. This manifests itself in that z_{00}, z_{01} do not equal z_{11}, z_{12} and, in addition, $z_{01} \neq z_{10}$; as in Sec. 3.8, z_{kq} is the number of neighbors of a site of layer k in layer q. When the interface is flat, the layer $k = 0$ contains no self-adsorbed atoms of the solid, while in the layers $k \geqslant 1$ we presume that there are no dissolved adsorbed particles. Consequently, another distinction of the present problem from the models treated in the present chapter is that the components themselves and their number in the layers $k = 0$ and $k \geqslant 1$ differ.

Let s_a and s_g be the number of components of the alloy and adsorption layer including vacant sites. Both indicated distinctions, owing to the independence of occupation of neighboring sites of various types by particles of different species (see Secs. 3.2 and 3.6), are "covered" by the system of equations in Sec. 3.6. They describe the distributions of the interacting particles on an

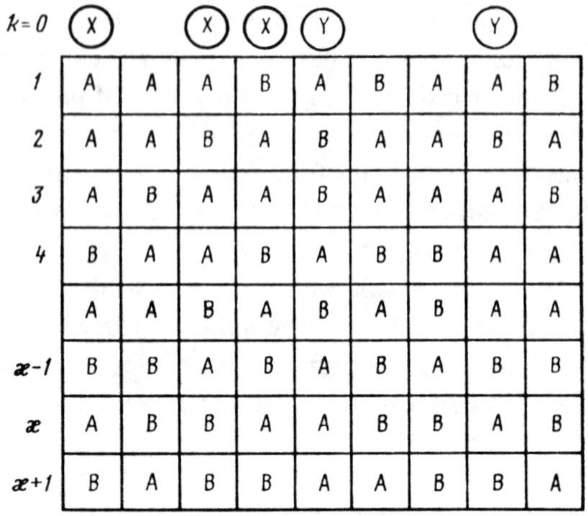

Fig. 3.20. Flat binary alloy-gas interface:
A, B = alloy components; X, Y—adsorbed particles

arbitrary inhomogeneous lattice if we consider that depending on the type of a site it may have different states of occupancy and, therefore, a different number of particles may be contained at it. Formally, the same equations are obtained if we consider a system with $s_a + s_g$ components and take into account the zero values of ν_k^i and ε_{kq}^{ij} for it when the presence of the relevant types of sites is forbidden. (The assumption that different species of particles can be at different sites is employed quite frequently in models of catalysis [86], ionic solutions [202, 203], nonstoichiometric compounds [204, 205], and so on (see [206]).)

Everything said above results in system of equations (3.51) retaining its structure, and only its multipliers are rewritten as follows:

$$S_{kq}^i(s) = \sum_{j=1}^{s_q} t_{kq}^{ij} \exp \beta(\varepsilon_{kq}^{sj} - \varepsilon_{kq}^{ij}) \tag{3.68}$$

where s_q is the number of species of particles in layer q, and s is the number of species of particles in layer k. In formula (3.51), $s_q = s$. (Recall that in Secs. 3.8 and 3.9 the types of sites q and p relate to different layers k and q.) The functions t_{qp}^{ij}, as above, are determined from the solution of system of equations (3.52) in which the indices i and k (j and l) relate to the species of particles at sites of types $q(p)$.

For a solid-gas phase interface, the equations of the concentration profile, by system (3.51), will have the form:
for the layer $k = 0$

$$\theta_0^{s_g} = \theta_0^i \exp \beta(\nu_0^i - \nu_0^{s_g})[S_{00}^i(s_g)]^{z_{00}}[S_{01}^i(s_g)]^{z_{01}} \tag{3.69}$$

for the layer $k = 1$

$$\theta_1^{s_a} = \theta_1^i \exp \beta(\nu_1^i - \nu_1^{s_a}) \prod_{q=2}^{0} [S_{1q}^i(s_a)]^{z_{1q}} \tag{3.70}$$

The equations for the layers $2 \leqslant k \leqslant \varkappa$ coincide with those in Sec. 3.8 and are not given herein. The normalizing relations for the functions θ_{01}^{ij} are written as follows:

$$\sum_{i=1}^{s_g} \theta_{01}^{ij} = \theta_1^j, \quad \sum_{j=1}^{s_a} \theta_{01}^{ij} = \theta_0^i, \quad \sum_{j=1}^{s_a} \theta_1^j = 1, \quad \sum_{i=1}^{s_g} \theta_0^i = 1 \tag{3.71}$$

The system of equations for the concentration profile $1 \leqslant k \leqslant \varkappa$ and the adsorption layer is solved simultaneously. For its solution, the boundary conditions in the bulk phases must be given. In a solid, the concentrations of the components $\theta_{\varkappa+1}^i$, where $1 \leqslant i \leqslant s_a$, are generally considered to be known. In the gas phase, it is more convenient to set the partial pressures of the components $1 \leqslant i \leqslant s_g - 1$ (in terms of the quantities ν_0^i).

It should be noted that the conventional definition of the adsorption coefficient (1.23), (1.29) includes the energy binding an adsorbed particle to the solid ε_q^i. In the equations of the present section, the energy binding an adsorbed particle i to an atom j of the solid corresponds to the parameter ε_{01}^{ij} that "is contained" in multipliers (3.68) of Eqs. (3.69) and (3.70). This is why here the quantities ν_0^i characterize the ratio of only the internal degrees of freedom of the particles in the adsorbed and gaseous states.

The influence of the adsorption of gas phase molecules on the concentration profile of a binary alloy was treated by the author elsewhere [192]. The presence of adsorbed particles of X and Y on a surface (see Table 3.1) changes the state of the subsurface region in comparison with its state in vacuum. This change is minimal for highly ordered alloys and increases with a decreasing distance to the region of alloy stratification. As in vacuum, at $\omega > 0$, the readily volatile component B accumulates on the surface, and its concentration diminishes monotonically with a growth in the layer number. When $\omega < 0$, the concentrations of the alloy components fluctuate, and depending on the structure of the lattice and fraction of adsorbed particles, as well as on all the parameters of particle interaction, a change in the nature of surface segregation of the alloy is possible. Here adsorption of the molecules may change the nature of the surface segregation of the alloy in comparison with vacuum (see $\beta\omega = -1.5$).

The influence of the surface coverage (Table 3.1 was obtained for practical constancy of the adsorption system components $\theta_X = 0.293$, $\theta_Y = 0.427$, except for $\beta\omega = 0.5$, for which other parameters have been used, namely, $\beta\varepsilon_{XX} = 0.5$, $\beta\varepsilon_{XY} = 0.6$, and $\beta\varepsilon_{YY} = 0.4$) and of interaction between the adsorbed particles on the surface composition of an alloy has also been considered in [192]. It was found that the equilibrium distributions of the adsorbed particles on a surface and of the alloy atoms in the subsurface region are quite sensitive to particle interactions and require detailed analysis in each specific case.

The adsorption of gas phase molecules may not only cause redistribution of the atoms of the solid in the subsurface region with a flat interface, but also facilitate its transition to a rough interface. The influence of adsorption may substantially exceed that of a change in the system's temperature on the probability of formation of a rough surface. In this case, the distribution of the particles is described by system of equations (3.62) for $s_a + s_g$ components, including vacancies. As in Sec. 3.8, the layer $k = 0$ corresponds to the bulk of the gas phase containing $s_g - 1$ species of molecules, while the layer $k = \varkappa + 1$ corresponds to the bulk of the solid solution. Hence, in the transition region, each site may be occupied by atoms of the alloy, particles of the gas phase, or be vacant.

The nature of particle distribution in the transition region is determined by particle interaction and the states of the bulk phases. If the gas phase molecules fail to dissolve in the bulk of the solid, the bulk values of the concentrations of the alloy components and the partial pressures of the gas-phase components unambiguously determine the state of the transition region. If s^* components of the gas phase dissolve in the alloy, where $s^* \leqslant s_g - 1$, it is first necessary to solve the problem of determining the bulk concentrations of s^* species of particles in both phases. These concentrations are related by the conditions of coexistence of the phases by equations similar to (2.51).

4 Rates of Surface Reactions. Fast Mobility of Reactants

4.1. The Fundamental Kinetic Equation

A change in the state of a condensed phase, which we are modeling by a lattice structure at a molecular level, is associated with a change in the state of its individual sites as a result of the realization of elementary processes. The state of the sites changes at the expense of two main types of elementary process, namely, migration of the particles and their participation in chemical transformations. In the first case, the particles change their position in space while remaining unchanged in a chemical respect. In the second case, the particles transform into their other chemical compounds without changing their position. (The description of the process when a particle simultaneously changes its coordinates and partly changes its chemical nature by entering into the composition of a complex or an intermediate does not virtually differ from the description of the two basic types of processes.)

Let us consider the evolution in time of the states of a condensed phase described by the overall distribution function $P(\{\gamma_f^i\}, \tau)$ (see Sec. 1.3). The total energy of a lattice system does not contain an explicit time dependence [see, for example, (2.1)], therefore the dynamics of the states of a system's sites is set by the probabilities of realizing elementary processes $W_\alpha(\{I\} \rightarrow \{II\})$ of a transition of a system from its initial state $\{I\}$ to its final state $\{II\}$. If the elementary process α proceeds at one site, the lists of the states $\{I\}$ and $\{II\}$ of occupancy of the system's sites differ only for this site. One-site processes are associated with a change in the internal degrees of freedom of a particle, with the adsorption-desorption of nondissociating molecules, and with a reaction by an impact mechanism. If the elementary process α proceeds at two neighboring sites, the lists of the states $\{I\}$ and $\{II\}$ differ in the states of occupancy of these two sites. Two-site processes include reactions of exchange, adsorption-desorption of dissociating molecules, processes of migration by a vacancy and exchange mechanisms, and reactions by an adsorption mechanism. The relation between states $\{I\}$ and $\{II\}$ depends on the mechanism of the process determining the set of elementary steps α. The probabilities of realizing the elementary processes W_α obey the condition of detailed equilibrium

$$W_\alpha(\{\mathrm{I}\} \to \{\mathrm{II}\}) \exp(-\beta H(\{\mathrm{I}\})) = W_\alpha(\{\mathrm{II}\} \to \{\mathrm{I}\}) \exp(-\beta H(\{\mathrm{II}\})), \tag{4.1}$$

where $H(\{\mathrm{I}\})$ is the total energy of a system in state $\{\mathrm{I}\}$. Condition (4.1) is needed to ensure a correct description of the evolution of a system to the state of equilibrium.

The fundamental kinetic equation for the overall distribution function is an equation of the balance of all possible flows changing the state of occupancy of each site of a system. It is written as follows [207-210]

$$\frac{d}{d\tau} P(\{\gamma_f^i\}, \tau) = \sum_{\{\mathrm{II}\}} [W_\alpha(\{\mathrm{II}\} \to \{\mathrm{I}\}) P(\{\mathrm{II}\}, \tau) - W_\alpha(\{\mathrm{I}\} \to \{\mathrm{II}\}) P(\{\mathrm{I}\}, \tau)] \tag{4.2}$$

where the sum over the states $\{\mathrm{II}\}$ corresponds to a change in the states of occupancy of all the lattice sites because of various elementary processes. Expressions (24.1) and (24.2) have been written in Markov's approximation [207], for which it is considered that the processes of relaxation of the internal degrees of freedom of all the particles proceed more rapidly than the processes of a change in the state of occupancy. Should this condition be violated, one must go over to a shorter time interval and single out a discrete set of states of a particle differing in its internal degrees of freedom, which can be interpreted as particles of different species. (This procedure is employed in the kinetics of gas-phase reactions, isomerization reactions, and in the analysis of conformation transformations of complex molecules [211].

The large dimension of system (4.2) does not allow it to be used for direct integration, therefore from system (4.2) we go over to a system of equations for the correlators (1.8). The latter are obtained by multiplying the left-hand and right-hand sides of expression (4.2) by $\prod_{n=1}^{m} \gamma_{f_n}^{i_n}$ and averaging over all the states of the system. The species of the particles at the sites f_n leading to nonzero contributions to the change of the correlators (1.8) depend on the elementary process α [209, 210]:

$$\frac{d}{d\tau}\theta_{f_1\ldots f_m}^{i_1\ldots i_m}(\tau) = \sum_{\{II\}} \left[\left\langle \prod_{n=1}^{m} \gamma_{f_n}^{\tilde{i}_n} W_\alpha(\{II\} \to \{I\}) \right\rangle - \left\langle \prod_{n=1}^{m} \gamma_{f_n}^{i_n} W_\alpha(\{I\} \to \{II\}) \right\rangle \right] \quad (4.3)$$

where $\tilde{i}_n$ is the species of a particle at site f_n corresponding to state $\{II\}$ and obtained from the particle i_n in the realization of an elementary process α in the forward direction $(i_n \to \tilde{i}_n)$.

Equations (4.3) are the initial ones for obtaining the kinetic equations of processes occurring in condensed phases. Their specific form is determined by the expressions for W_α. As for ideal lattice systems, we shall describe the rate of elementary processes by a model of a transition state [212]. The quantity W_α is determined by the initial state of a particle, its interaction with the neighboring particles and with the thermal vibrations of the system. The latter play the role of a heat reservoir initiating stochastic transitions from one state of a system to another one. In Chaps. 4 and 5, we shall take account of the interactions between the nearest neighbors, and in Chap. 7 — other potentials of particle interaction.

4.2. One-Site Reactions

The transition state model for elementary processes in condensed systems [212] employs the traditional tenets of Eyring's model [134] (see Sec. 1.5), although it follows from the nature of the condensed state that the driving force of the process is ensured by thermofluctuation excitations of the medium, and not by the translational motion of the particles. In the transition state, an activated complex reacts with the neighboring particles. The parameter of their interaction ε_{ij}^* differs from the parameter of interaction ε_{ij} between the initial reactant i which the activated complex is formed from and the neighboring particle j.

In the absence of an activated complex of the one-site reaction A → B, the total energy of a system is described by expression (2.1). The formation of an activated complex changes the magnitude of a system's energy, and it becomes equal to

$$H_1 = \sum_{f=1}^{N}{}' \sum_{i=1}^{s} \gamma_f^i \nu_i - \frac{1}{2} \sum_{f,g}{}' \sum_{i,j} \varepsilon_{ij} \gamma_f^i \gamma_g^j + \nu_A^* \gamma_k^A - \sum_{n,j} \varepsilon_{Aj}^* \gamma_k^A \gamma_n^j \quad (4.4)$$

The first two addends describe the contributions of the particles in their ground states; the primes on the sum signs indicate that the sums are taken over sites not containing an activated complex that occupies site k ($f \neq k$). The third addend determines the contribution of the internal degrees of freedom of the activated complex (1.35), while the fourth one determines its interaction with the neighboring particles. The last two addends contain no summation over the species of particles at site k because an activated complex forms from a particle of A.

Let us rewrite expression (4.4) in the form $H_1 = H + \delta h_k^A$, where

$$\delta h_k^A = (\overset{*}{\nu}_A - \nu_A - \sum_{n,\, j} \delta \varepsilon_{Aj} \gamma_n^j) \gamma_k^A, \quad \delta \varepsilon_{ij} = \overset{*}{\varepsilon}_{ij} - \varepsilon_{ij} \tag{4.5}$$

The probability of formation of an activated complex of the reaction A → B at site k is the probability of a set of two events, namely, the finding of a particle of A at site k and a change in the energy of the system by the value δh_k^A and is expressed as

$$W([\gamma_k^A] \to [\gamma_k^B]) = \gamma_k^A U_A(k), \quad U_A(k) = \exp(-\beta \delta h_k^A) \tag{4.6}$$

The exponential relation in (4.6) and the use of the concept of the chemical potential $\overset{*}{\mu}_A$ (1.35) are associated with the assumption that the activated complex is in equilibrium with a reaction subsystem consisting of reactant A and a fixed arrangement of all the neighboring particles around A, interaction with which determines the magnitude of the activation barrier. In other words, we are dealing with local equilibrium (reagent-activated complex), although the distribution itself of the particles among the lattice sites may differ greatly from the equilibrium distribution.

For the one-site reaction A ⇄ B, it is easy to verify that expressions (4.6) satisfy condition (4.1). For the given reaction, $H(\{I\}) = h_f^A + G_f$, where h_f^i and G_f have been determined in (2.2) and (2.3), corresponds to state $\{I\}$, while $H(\{II\}) = h_f^B + G_f$ corresponds to state $\{II\}$. Inserting formulas (4.6) for particles of A and B into (4.1) and having in view that: (1) the parameters of an activated complex do not depend on the direction of the process, i.e. $\overset{*}{\nu}_A = \overset{*}{\nu}_B$ and $\overset{*}{\varepsilon}_{Aj} = \overset{*}{\varepsilon}_{Bj}$ and (2) for the given process in the initial and final states a particle of $A(\gamma_k^A = 1)$ and one of B $(\gamma_k^B = 1)$ are authentically at site k, we find that formula (4.1) transforms into the identity $1 \equiv 1$.

Let us consider the procedure for obtaining the right-hand sides of the kinetic equations for one-site processes. The equations for the local concentrations

($m = 1$) are the first ones of system of equations (4.3). Limiting ourselves to the step A ⇄ B, we shall insert expressions (4.6) into formula (4.3):

$$\frac{d}{d\tau}\theta_f^{\mathrm{A}}(\tau) = \dot{\theta}_f^{\mathrm{A}} = U_f^{\mathrm{B}} - U_f^{\mathrm{A}}, \; U_f^i = \langle \gamma_f^i U_i(f) \rangle_\tau \tag{4.7}$$

The addends of (4.7) are calculated on the basis of expressions (1.4)-(1.7) and the more general expressions following from them:

$$(\gamma_f^i)^m = \gamma_f^i, \; m \geqslant 1; \; \exp(b\gamma_f^i\gamma_g^j) = 1 + \gamma_f^i\gamma_g^j(\exp b - 1) \tag{4.8}$$

Formula (4.7) is written as follows:

$$U_f^i = \langle \gamma_f^i(1 + \gamma_f^i x_i) \prod_{g=1}^{z} \prod_{j=1}^{s} (1 + \gamma_f^i \gamma_g^j x_{ij}) \rangle_\tau$$
$$x_i = \exp\beta(\nu_i - \nu_i^*) - 1, \; x_{ij} = \exp\beta\delta\varepsilon_{ij} - 1 \tag{4.9}$$

The first two multipliers in (4.9) yield expression (1.39), while the following transformations hold for the remaining multipliers:

$$\prod_{g=1}^{z} \prod_{j=1}^{s} (1 + \gamma_f^i\gamma_g^j x_{ij}) = \prod_{g=1}^{z} (1 - \gamma_f^i + \gamma_f^i \sum_{j=1}^{s} \gamma_g^j \exp\beta\delta\varepsilon_{ij})$$
$$= 1 - \gamma_f^i + \gamma_f^i \prod_{g=1}^{z} \sum_{j=1}^{s} \gamma_g^j \exp\beta\delta\varepsilon_{ij} \tag{4.10}$$

As a result, we obtain

$$U_f^i = \exp\beta(\nu_i - \nu_i^*) \left\langle \gamma_f^i \prod_{g=1}^{z} \sum_{j=1}^{s} \gamma_g^j \exp\beta\delta\varepsilon_{ij} \right\rangle_\tau \tag{4.11}$$

The same result is obtained by simplified calculations if we use the condition that the energy of a system changes by the amount δh_f^i only when a particle i is at site f. Inserting $\gamma_f^i = 1$ into (4.6) and employing consecutively formula (1.7) and the first relation (1.25), we have:

$$U_i(f \mid \gamma_f^i = 1) = \exp\beta\left(\nu_i - \nu_i^* + \sum_{j=1}^{s} \sum_{g=1}^{z} \delta\varepsilon_{ij}\gamma_g^j\right)$$
$$= \exp\beta(\nu_i - \nu_i^*) \prod_{g=1}^{z} \sum_{j=1}^{s} \gamma_g^j \exp\beta\delta\varepsilon_{ij} \tag{4.12}$$

Consequently, $\gamma_f^i U_i(f) = \gamma_f^i U_i(f \mid \gamma_f^i = 1)$, and we again arrive at expression (4.11).

Each addend in angular brackets in (4.11) is the product $\gamma_f^i \gamma_{g_1}^{i_1} \ldots \gamma_{g_z}^{i_z}$, i.e. the right-hand side of Eq. (4.7) contains cluster correlators of the order of magnitude $(z + 1)$ [see (1.8)]. These are correlators of the first type generalizing correlators (2.3) (the first equation) for nonequilibrium distribution of the particles.

By separating the translational degree of freedom from the partition function of the activated complex along the coordinate of the reaction (1.41) and expressing the velocity along it in terms of the characteristics of the activation barrier (1.37), we obtain an expression for the rate of an elementary process:

$$U_f^i = K_i V_f^i,$$

$$V_f^i = \prod_{g=1}^{z} \sum_{j_g=1}^{s} \theta_{fg_1 \ldots g_z}^{ij_1 \ldots j_z} \exp \beta \delta \varepsilon_{ij_g} \tag{4.13}$$

where K_i is the rate constant of a one-site reaction, its form is determined by formula (1.42) [see also (1.43)-(1.45)], and V_f^i is the concentration component of the rate of a one-site reaction that takes into account the change in the height of the activation barrier depending on the local composition (in the absence of interaction $V_f^i = \theta_f^i$).

For the local concentration of particles of B, an equation holds similar to Eq. (4.7). In this equation, however, the addends have opposite signs because the formation of particles of B is associated with the consumption of particles of A. For other components $i \neq$ A, B, the proceeding of the reaction A $\rightleftarrows$ B does not change their local concentrations, $d\theta_f^i/d\tau = 0$.

System of equations (4.7), (4.13) is not closed — its right-hand side contains unknown correlators of a higher dimension. To close it, we must construct new kinetic equations for the unknown correlators or express them in terms of correlators with a lower dimension. (Such a situation is a general one for a system of interacting particles, in particular, we have already encountered it when deriving equations for equilibrium correlators).

A general analysis of the procedure of closing kinetic equations when taking the interaction of nearest neighbors into account will be set out in Chap. 5. Now we shall limit ourselves to the case of rapid surface mobility of the adsorbed particles when they are distributed uniformly among the sites of a homogeneous surface and the limiting steps are adsorption, desorption, and chemical transformations with the participation of the adsorbed particles.

If the adsorbed particles form a disordered phase, the probabilities of various configurations of the particles do not depend on the numbers of the sites — the macroscopic means coincide with the local means, and they may be omitted ($V_f^i \rightarrow V_i$). The cluster correlators of the first type in expression (4.13) can be regrouped according to the number of particles of species n_j around the central particle i. This makes it possible to rewrite V_i as follows:

$$V_i = \sum_{n_1=0}^{z} \sum_{n_2=0}^{z-n_1} \cdots \sum_{n_{s-1}=0}^{z-l} C_z^{\{n\}} \theta^i_{\{[n]\}} \exp \beta \sum_{j=1}^{s} \delta\varepsilon_{ij} n_j \tag{4.14}$$

where definitions (2.3) and (2.39) are employed. Correlators (2.3), in turn, are determined by solving system of equations (2.39) in various approximations. This allows us to express $\theta^i_{\{[n]\}}$ in terms of θ_i and close the system of kinetic equations with respect to θ_i.

In a general case, when a particle of A participates in different steps $A \rightleftarrows B_\alpha$, the kinetic equation for component A will have the form

$$\dot{\theta}_A = \sum_\alpha [U_B(\alpha) - U_A(\alpha)],$$

$$U_i(\alpha) = K_i(\alpha) V_i(\alpha) \tag{4.15}$$

where the constants and concentration components of the reaction rates contain an index α indicating the step number α [each step is characterized by its values of $F_i^*(\alpha)$, $E_i(\alpha)$, and $\varepsilon_{ij}^*(\alpha)$].

The expressions for $V_i(\alpha)$ depend on the approximation employed to account for interaction. Inserting formula (2.48) into (4.14), we obtain

$$V_i(\alpha) = \theta_i [S_i(\alpha)]^z,$$

$$S_i(\alpha) = \begin{cases} \sum_{j=1}^{s} t_{ij} \exp \beta\delta\varepsilon_{ij}(\alpha) & \text{QCA} \\ \exp \beta \sum_{j=1}^{s} \delta\varepsilon_{ij}(\alpha)\theta_j & \text{MFA} \end{cases} \tag{4.16}$$

Using the expressions in a quasichemical approximation as an example, we shall show that the expressions obtained for the rates of reactions are self-consistent with the description of the equilibrium state of reaction systems. That is, by equating the rates of the forward and reverse directions of each step of the general process, we obtain equations for the equilibrium distribution of the particles. Let us consider a step α: A $\rightleftarrows$ B, and let $U_A(\alpha) = U_B(\alpha)$. The ratio of the rate constants determined by formulas (4.15) and (1.42) yields the equilibrium constant K_{eq} and excludes the partition function of the activated complex:

$$K_{eq} = \frac{K_A(\alpha)}{K_B(\alpha)} = \frac{V_B(\alpha)}{V_A(\alpha)} = \frac{\theta_B}{\theta_A}\left[\frac{S_B(\alpha)}{S_A(\alpha)}\right]^z \tag{4.17}$$

To show that expression (4.17) does not depend on the parameters $\varepsilon^*_{ij}(\alpha)$, let us rewrite expressions (2.43) and (2.44) in an equivalent form:

$$\begin{aligned} &\theta_A = \theta_B \exp\beta(\nu_B - \nu_A)[S_B(A)]^z, \quad S_i(n) = \sum_{j=1}^{s} t_{ij} \exp\beta(\varepsilon_{nj} - \varepsilon_{ij}) \\ &\theta_{AC} = \theta_{BC} \exp\beta(\nu_B - \nu_A + \varepsilon_{AC} - \varepsilon_{BC})[S_B(A)]^{z-1} \end{aligned} \tag{4.18}$$

The ratio of these expressions allows us to write

$$t_{AC} = t_{BC} \exp\beta(\varepsilon_{AC} - \varepsilon_{BC})/S_B(A) \tag{4.19}$$

The following identity is easily proved with the aid of Eq. (4.19):

$$S_n(l) \equiv S_m(l)/S_m(n) \tag{4.20}$$

that enables one to go over from one species of central particle to another one. To prove this, it is sufficient to assume in (4.19) that A $= n$, B $= m$, C $= j$, and insert it into the left-hand side of (4.20). Using identity (4.20) and having in view that $\varepsilon^*_{Aj} = \varepsilon^*_{Bj}$, the ratio in (4.17) can be rewritten as $S_A(\alpha) = S_A(B)S_B(\alpha)$. Hence, with $K_{eq} = \exp\beta(\nu_A - \nu_B)$, the ratio of the concentrations of the particles does not depend on how equilibrium is achieved — at the expense of a chemical reaction (4.17) or at the expense of the interphase redistribution of the particles (4.18).

A similar result is obtained in other approximations [212]. The expressions for the rates of reactions in a chaotic approximation are formally "derived" from the expressions in a quasichemical approximation when t_{ij} is replaced by θ_j; the formulas in the polynomial approximation have been given by the author in [212].

For the rapid surface mobility of adsorbed particles, the expressions for the rates of one-site reactions (4.13), (4.14), and (4.16) can be obtained in a more obvious way. Let us illustrate this on the example of the adsorption of nondissociating molecules. Let $E_i(n)$ and $E_i^*(n)$ be the energies of interaction of the central particle i with its neighbors in the ground and transition states, where n is the number of particles of A in the coordination sphere of particle i. The difference $E_i^*(n) - E_i(n)$ determines the change in the magnitude of the activation barrier of desorption (i = A) or adsorption (i = v, we shall take the properties of the vacant sites into consideration in the final expressions) because of interaction of the nearest neighbors in comparison with an ideal system. For two-body interaction, we have $E_i(n) = n\varepsilon_{iA} + (z - n)\varepsilon_{iv}$, and $E_i^*(n) = n\varepsilon_{iA}^* + (z - n)\varepsilon_{iv}^*$.

The rate of a process at the central site of a cluster containing n particles of A in its coordination sphere will be written as follows:

$$U_i(n) = K_i\theta_{[n]}^i \exp \beta[E_i^*(n) - E_i(n)] \tag{4.21}$$

where K_i is the rate constant of an elementary reaction, and its form has been determined in (1.43) and (1.44), while $\theta_{[n]}^i$ is the normalized probability of there being a cluster with n particles of A around the central particle i on the surface. Expression (4.21) can be considered as a modified expression for the rate of a reaction within the framework of the law of mass action (1.42), where the place of the rate constant has been taken by the product $K_i(n) = K_i \exp \beta[E_i^*(n) - E_i(n)]$, while the place of the concentration of the particles has been taken by the concentration of the quasiparticles, which are the following configurations of the real particles: at the center of a cluster there is particle i, and around it there are n particles of A and $(z - n)$ particles v [the quantity $K_i(n)$ can be interpreted as the rate constant for a quasiparticle].

Consequently, by going over to larger quasiparticles, we can formally retain the structure of the expressions of the law of mass action, i.e. the product of the rate constant of a reaction and the reactant concentration. But one must have in view that the number of such quasiparticles sharply grows with increasing s and z (and the radius of interaction).

The rate of a reaction relating to a surface as a whole is obtained by averaging expression (4.21) over all the local configurations differing in the number of particles of A in the coordination sphere around particle i:

$$U_i = K_i \sum_{n=0}^{z} C_z^n \theta_{[n]}^i \exp \beta[n\delta\varepsilon_{iA} + (z - n)\delta\varepsilon_{iv}] \tag{4.22}$$

where $\delta\varepsilon_{ij}$ has been determined in (4.5), and C_z^n is the statistical weight of the configuration with n particles of A. Formula (4.22) is a particular case of formula (4.14). The ways of an approximate calculation of the functions $\theta^i_{[n]}$ were discussed in Sec. 2.2. In a quasichemical approximation, they are expressed in terms of the function θ_i and θ_{ij} in (2.13) and (2.18), where $\theta_i = N_i/N$, N_i is the number of particles of species i, and $\theta_{ij} = (1 + \Delta_{ij})N_{ij}/zN$ (N_{ij} is the number of pairs of particles ij). Insertion of formula (2.18) into (4.22) yields

$$U_i = K_i\theta_i \sum_{n=0}^{z} C_z^n[t_{iA} \exp \beta\delta\varepsilon_{iA}]^n[t_{iv} \exp \beta\delta\varepsilon_{iv}]^{z-n}$$
$$= K_i\theta_i[t_{iA} \exp \beta\delta\varepsilon_{iA} + t_{iv} \exp \beta\delta\varepsilon_{iv}]^z \tag{4.23}$$

which is a particular case of formula (4.16). Multicomponent systems are considered in a similar way. For these systems, the statistical weights of the configurations of the particles in the coordination sphere of a cluster are expressed in terms of the polynomial coefficients of the number of combinations of z taken $n_1 \ldots n_{s-1}$ particles of different species at a time [see Eq. (2.39)].

The process of adsorption of nondissociating particles is the simplest surface process. We shall use it as an example to discuss the role of correlation effects between interacting particles for the kinetics of elementary processes. Considering explicitly the properties of vacant sites v, the kinetic equation for the process of adsorption of particles of A in a quasichemical approximation, for example, will be written as follows:

$$\dot{\theta}_A = K_v(1 - \theta_A)(1 + t_{vA}x_{vA})^z - K_A\theta_A(1 + t_{AA}x_{AA})^z \tag{4.24}$$

where the rate constants have the form given by (1.43) and (1.44); the relation between t_{AA} and t_{vA} with θ is given by formulas (2.14) and (2.18); by (4.9), $x_{AA} = \exp \beta\delta\varepsilon_{AA} - 1$, $x_{vA} = \exp \beta\varepsilon^*_{AA} - 1$. Equations in other approximations are given in references [145, 213].

For the rate of desorption $x_{AA} \gg 1$ in the region of moderate and high coverages, a power dependence on θ is obtained, namely, $U_A = K_A\theta_A^{z+1} \exp \beta z\delta\varepsilon_{AA}$, which is generally known as the Bagham-Bart-Kwan kinetics [85, 214]. Here the true activation energy of desorption is renormalized to the quantity $z\delta\varepsilon_{AA}$. In the region of low coverages, the rate of desorption is expressed as $U_A = K_A\theta_A$. At the same time, the expression for the rate of adsorption differs from the Langmuir relation: $U_v = K_v(1 - \varkappa\theta)$, where $\varkappa = 1 + z - z \exp \beta\varepsilon^*_{AA}$. When $\varepsilon^*_{AA} < 0$, the rate of adsorption diminishes with increasing coverage more rapidly than in the Langmuir model, while at $\varepsilon^*_{AA} > 0$ it diminishes more slowly; if $\beta\varepsilon^*_{AA} > \ln(1 + z^{-1})$, we have $\varkappa < 0$, and the rate of adsorption increases with a growth in the number of adsorbed particles.

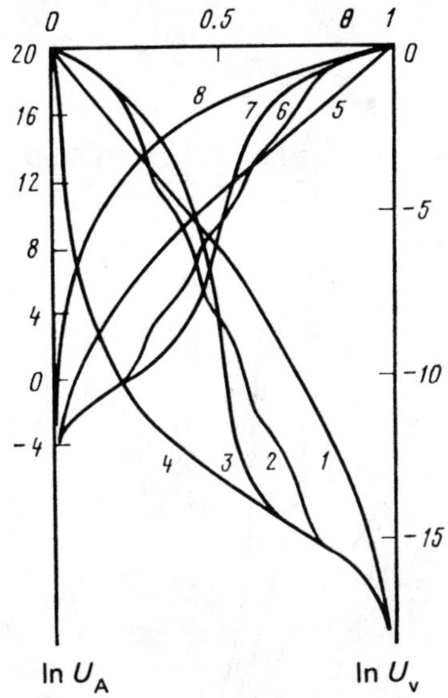

Fig. 4.1. Influence of the approximation used to take account of interaction on the concentration dependence of the adsorption rates U_v ($K_v = 1$, $\beta\delta\varepsilon_{AA} = 1.5$) and desorption rates U_A ($K_A = 1$, $\beta\delta\varepsilon_{AA} = 5$) at $\beta\varepsilon_{AA} = -5$ and $z = 4$:

1, 5—MFA; *2, 6*—PA; *3, 7*—QCA; *4, 8*—CA

The concentration components of the rates of adsorption and desorption in the entire region of coverages in various approximations are shown in Fig. 4.1. The failure to consider the correlation effects sharply changes the rates in the region of low coverages, while in the approximations considering the correlation effects, the values of the rates are close to one another. The "wavy" nature of the curves in the polynomial approximation and the inflection in the quasichemical one are similar to the curves of Q vs. θ (see Fig. 2.7). In the region of moderate coverages, the logarithms of the rates of adsorption and desorption depend linearly on θ. This enables one to explain the Becker-Langmuir empirical relation for adsorption and the Zeldovich-Roginsky one for desorption [85, 150]. The slopes of the straight lines depend on the type of approximation. When the repulsion of the particles grows, the logarithms of the rates of adsorption and desorption can be approximated by several

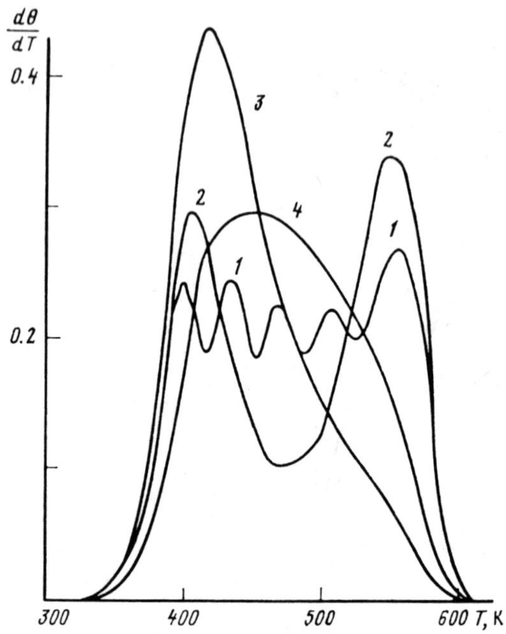

Fig. 4.2. Thermodesorption spectra for an initial coverage of 0.9; $\hat{K}_A = 10^{12}\ s^{-1}$ ($E_A = 126$, $\varepsilon_{AA} = -16.3$, $\varepsilon^*_{AA} = -8.4$ kJ/mol, $b = 50$ K/s, $z = 4$) in the polynomial (*1*), quasichemical (*2*), chaotic (*3*), and molecular-field (*4*) approximations

linear sections [213]. Such relations are often interpreted as a proof of a change in the mechanism of a process with different coverages [215]. Theory shows that such conclusions must be considered with care — there may be a single mechanism of a process. The sharp drop in the rate of adsorption at $\theta \geqslant 0.5$ at the expense of the repulsion of particles enables one to explain the processes of rapid and slow chemisorption [150].

A considerably greater difference between the rates of desorption in various approximations manifests itself when calculating the thermodesorption spectra (TDS), i.e. the dependence of θ on τ at a constant rate b of heating a surface, $T = T_0 + b\tau$, where T_0 is the initial temperature (Fig. 4.2). Consideration of the correlation effects splits the TDS, while the failure to consider them, notwithstanding the high nonlinearity of Q vs. θ in the chaotic approximation, does not split the TDS [145]. In addition to the correlation effects,

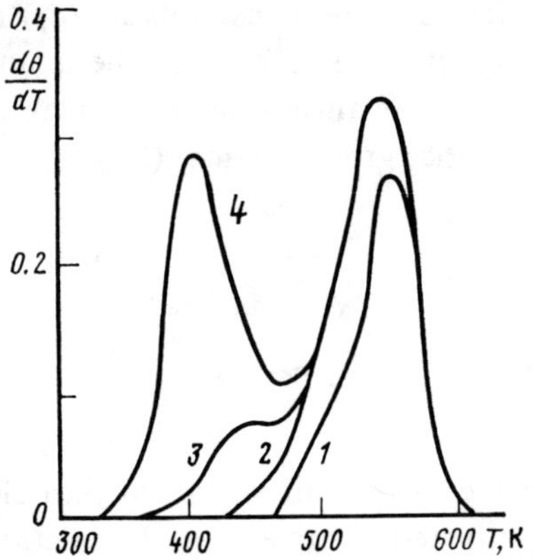

Fig. 4.3. Variation of the initial coverages in the quasichemical approximation (the parameters are the same as for Fig. 4.2): θ_A $(\tau = 0) = 0.3$ (*1*), 0.5 (*2*), 0.65 (*3*), and 0.9 (*4*)

splitting of the TDS requires high initial coverages, $\theta_A(0) = \theta_A(\tau = 0) > 0.6$ (Fig. 4.3) and a definite difference between ε^*_{AA} and ε_{AA} (at low values of $\delta\varepsilon_{AA}$ there is no splitting). The quantity ε^*_{AA} plays the same substantial role as ε_{AA}: at high initial coverages the width of the TDS is proportional to the product $z\delta\varepsilon_{AA}$.

4.3. Two-Site Reactions

Let us consider the reaction A + C → B + D, where A, B, C, and D are the components of a system, and its activated complex occupies two neighboring sites. The total energy of the system containing the activated complex at sites k_1 and k_2 is written as

$$H_2 = \sum_{f,\, i}{}' \nu_i \gamma_f^i - \frac{1}{2}\sum_{f,\, g}{}' \sum_{i,\, j} \varepsilon_{ij}\gamma_f^i \gamma_g^j + \nu^*_{AC}\gamma_{k_1}^A \gamma_{k_2}^C - \sum_{n_1,\, j} \varepsilon^*_{Aj}\gamma_{k_1}^A \gamma_{n_1}^j - \sum_{n_2,\, j} \varepsilon^*_{Cj}\gamma_{k_2}^C \gamma_{n_2}^j \qquad (4.25)$$

The primes in the first addends signify the exclusion of the sites accommodating the activated complex from summation. The contributions from it are described by the last three addends. The subscripts n_1 (n_2) relate to the nearest neighbors of site k_1 (k_2), except for site k_2 (k_1). The quantity ν^*_{AC} has been defined in (1.47). The formation of an activated complex is associated with a change in the energy of the ground state (2.1) by the quantity $\delta h^{AC}_{k_1k_2} = H_2 - H$:

$$\delta h^{AC}_{k_1k_2} = -\nu_A\gamma^A_{k_1} - \nu_C\gamma^C_{k_2} + (\nu^*_{AC} + \varepsilon_{AC})\gamma^A_{k_1}\gamma^C_{k_2} - \sum_{n_1,\ j} \delta\varepsilon_{Aj}\gamma^A_{k_1}\gamma^j_{n_1} - \sum_{n_2,\ j} \delta\varepsilon_{Cj}\gamma^C_{k_2}\gamma^j_{n_2} \tag{4.26}$$

The quantity ε_{AC} appears because of the interaction of particles of A and C in the ground state. For an activated complex, the interaction of A and C relates to its internal bonds. The probability of its formation at the pair of sites k_1k_2 is the probability of a set of events consisting in that a particle of A is at site k_1 and one of C is at site k_2, while the energy of the system changes by the amount $\delta h^{AC}_{k_1k_2}$, hence

$$W([\gamma^A_{k_1}\gamma^C_{k_2}] \to [\gamma^B_{k_1}\gamma^D_{k_2}]) = \gamma^A_{k_1}\gamma^C_{k_2}U_{AC}(k_1,\ k_2)$$
$$U_{AC}(k_1k_2) = \exp(-\beta\delta h^{AC}_{k_1k_2}) \tag{4.27}$$

To prove that the constructed probabilities (4.27) satisfy the principle of detailed equilibrium, we shall write out the cluster Hamiltonian for the two central sites and their nearest neighbors (see Fig. 1.6). By expression (2.1), we shall have

$$h_{fg} = \sum_{ij} h^{ij}_{fg},\quad h^{ij}_{fg} = \nu_i\gamma^i_f + \nu_j\gamma^j_g - \varepsilon_{ij}\gamma^i_f\gamma^j_g - \sum_{n_1,\ l} \varepsilon_{il}\gamma^i_f\gamma^l_{n_1} - \sum_{n_2,\ l} \varepsilon_{jl}\gamma^j_g\gamma^l_{n_2} \tag{4.28}$$

For the reaction A + C $\rightleftarrows$ B + D, the difference of the energies in expression (4.1) is $H(\{I\}) - H(\{II\}) = h^{AC}_{k_1k_2} - h^{BD}_{k_1k_2}$. Inserting the probabilities (4.27) into formula (4.1), we see that it becomes an identity, the same as for one-site reactions.

If we use expression (4.27) and (4.3), the kinetic equations for the local density of the particles of A at site f will be written in the form

$$\dot{\theta}^A_f = \sum_g [U^{BD}_{fg} - U^{AC}_{fg}],\quad U^{ij}_{fg} = \langle\gamma^i_f\gamma^j_g U_{ij}(fg)\rangle_\tau \tag{4.29}$$

where the sites g are the nearest neighbors of site f.

We shall calculate the mean value in (4.29) on the basis of a simplified method because it is not difficult to show that $\gamma_f^i\gamma_g^j U_{ij}(fg) = \gamma_f^i\gamma_g^j U_{ij}(fg \mid \gamma_f^i = \gamma_g^j = 1)$ [212].

$$U_{ij}(fg \mid \gamma_f^i = \gamma_g^j = 1) = \exp \beta(\nu_i + \nu_j - \varepsilon_{ij} - \overset{*}{\nu}_{ij} + \sum_{n_1,\, l} \delta\varepsilon_{il}\gamma_{n_1}^l + \sum_{n_2,\, l} \delta\varepsilon_{jl}\gamma_{n_2}^l)$$

$$= \exp \beta(\nu_i + \nu_j - \overset{*}{\nu}_{ij} - \varepsilon_{ij}) \prod_{n_1} \sum_{l=1}^{s} \gamma_{n_1}^l \exp \beta\delta\varepsilon_{il} \prod_{n_2} \sum_{l=1}^{s} \gamma_{n_2}^l \exp \beta\delta\varepsilon_{jl} \tag{4.30}$$

In our calculations, we have used the first of expressions (4.10). By inserting (4.30) into (4.29) and taking into account the definition of the rate constant for a two-site reaction (1.48)-(1.50), we shall rewrite formula (4.29) as follows:

$$U_{fg}^{ij} = K_{ij} \exp(-\beta\varepsilon_{ij}) V_{fg}^{ij}$$

$$V_{fg}^{ij} = \langle \gamma_f^i\gamma_g^j \prod_{n_1} \sum_{l=1}^{s} \gamma_{n_1}^l \exp \beta\delta\varepsilon_{il} \prod_{n_2} \sum_{l=1}^{s} \gamma_{n_2}^l \exp \beta\delta\varepsilon_{jl}\rangle_T \tag{4.31}$$

where V_{fg}^{ij} is the concentration component of the rate of a two-site reaction. The dimension of correlators (4.31) equals the number of sites of a cluster consisting of two central sites and their nearest neighbors. Consequently, Eq. (4.29) remains unclosed. As in Sec. 4.2, we shall limit ourselves to the equilibrium distribution of the particles over the surface and consider approximate ways of calculating the correlators (4.31) in terms of the macroscopic means θ_i. The correlators constructed on a cluster with two central sites are expressed in terms of the correlators constructed on a cluster with one central site as follows [212]:

$$\left\langle \gamma_f^i\gamma_g^j \prod_{n_1} \gamma_{n_1}^l \prod_{n_2} \gamma_{n_2}^l \right\rangle = \left\langle \gamma_f^i\gamma_g^j \prod_{n_1} \gamma_{n_1}^l \right\rangle \left\langle \gamma_g^j\gamma_f^i \prod_{n_2} \gamma_{n_2}^l \right\rangle \Big/ \langle \gamma_f^i\gamma_g^j \rangle \tag{4.32}$$

while the ways of calculating the latter were treated in Sec. 2.4. Inserting (4.32) into (4.31) and regrouping the addends $\theta^i_{\{[n]\}}$ with respect to the number of particles of species n_l around the central particles i and j, by analogy with (4.14), and then using formulas (2.48), we obtain expressions for the concentration components given in (4.25):

$$V_{ij}(\alpha) = [S_i(\alpha)S_j(\alpha)]^{z-1} \cdot \begin{cases} \theta_{ij} & \text{QCA} \\ \theta_i\theta_j & \text{MFA} \end{cases} \tag{4.33}$$

where the functions $S_i(\alpha)$ in the indicated approximations are determined in (4.16). As in formula (4.15), in formula (4.33) we have additionally introduced the number of the step α and omitted the numbers of the sites at which the reactants and their neighbors are. Here Eq. (4.29) instead of the sum over g should contain the multiplier z.

Let us use the step $A + C \rightleftarrows B + D$ as an example and verify the self-consistence of the description of the kinetics and equilibrium for nonideal reaction systems. It follows from the equation $U_{AC}(\alpha) = U_{BD}(\alpha)$ that the equilibrium constant of the reaction does not depend on the parameters of the activated complex:

$$K_{eq} = \frac{K_{AC}(\alpha)}{K_{BD}(\alpha)} = \frac{V_{BD}(\alpha)}{V_{AC}(\alpha)} = \frac{\theta_{BD}}{\theta_{AC}} \left[\frac{S_B(\alpha)S_D(\alpha)}{S_A(\alpha)S_C(\alpha)}\right]^{z-1}$$

$$\times \exp\beta(\varepsilon_{AC} - \varepsilon_{BD}) = \frac{\theta_{BD}}{\theta_{AC}} [S_B(A)S_D(C)]^{z-1} \exp\beta(\varepsilon_{AC} - \varepsilon_{BD}) \qquad (4.34)$$

Identity (4.20) has been employed in the last equation because $\varepsilon^*_{Aj} = \varepsilon^*_{Bj} \neq \varepsilon^*_{Cj} = \varepsilon^*_{Dj}$, where $1 \leqslant j \leqslant s$. The indicated equations reflect the differences in the interaction of "fragments" of the activated complex A (B) and C (D) with its neighbors and the fact that the interaction of each fragment does not depend on the direction of the reaction.

Let us multiply the right-hand and left-hand sides of relation (4.19) by $\theta_A\theta_B$ and consider a similar relation for the pairs θ_{CF} and θ_{DF}. Having in view that $\theta_{ij} = \theta_{ji}$ and $\varepsilon_{ij} = \varepsilon_{ji}$ and changing the species of the central particle in accordance with identity (4.20), we obtain:

$$\theta_{AC} = \theta_{BD} \frac{\theta_A\theta_C}{\theta_B\theta_D} [S_B(A)S_D(C)]^{-1} \exp\beta(\varepsilon_{AC} - \varepsilon_{BD}) \qquad (4.35)$$

Inserting it into formula (4.34), we have:

$$K_{eq} = \frac{\theta_B\theta_D}{\theta_A\theta_C} [S_B(A)S_D(C)]^z \qquad (4.36)$$

i.e. formally the equilibrium of the reaction $A + C \rightleftarrows B + D$ can be considered as the proceeding of two independent one-site reactions $A \rightleftarrows B$ and $C \rightleftarrows D$, each of which is described in accordance with (4.17) or by the first of Eqs. (4.18). Consequently, the condition of equilibrium of the reaction coincides with the condition of equilibrium of the system as a whole.

In the approximations disregarding the correlation effects, e.g. in the molecular-field approximation, the second equations (4.18) are absent, and, as follows from (2.46), $\theta_{ij} = \theta_i\theta_j$. The insertion of this relation into formula (4.34) and comparison with expression (2.47) or (4.17), where the functions $S_i(\alpha)$ have been determined in (4.16), reveal that the condition of a self-consistent description of the kinetics and equilibrium is not observed.

The expression for the rate of a bimolecular reaction of adsorbed particles when the latter have a rapid surface mobility can be obtained in the traditional way for chemical kinetics if the law of mass action is used for large quasiparticles. Let us consider the reaction AZ + BZ on a homogeneous surface containing three species of particles: A, B, and v (we shall take the properties of the vacant sites v into account in the final expressions). Particles of A and B are at neighboring sites of a lattice and enter the coordination spheres of one another. Assume that at the $(z - 1)$ site a particle of A is surrounded by n_A particles of A, n_B particles of B, and $n_v = z - 1 - n_A - n_B$ particles of v, while at the $(z - 1)$ site a particle of B is surrounded by m_A particles of A, m_B particles of B, and $m_v = z - 1 - m_A - m_B$ particles v.

Let us denote the energy of interaction of particles of A and B with their neighbors in the ground and transition states by $E_{AB}(\{[nm]\})$ and $E^*_{AB}(\{[nm]\})$, where $\{[nm]\}$ is an abridged notation of the numbers of neighboring particles of various species $[n_An_Bn_v;\, m_Am_Bm_v]$, while the square brackets signify fixed values of all the numbers. We shall consider that the central A and B particles and their specific set of nearest neighbors $\{[nm]\}$ form a quasiparticle. Now the rate of an elementary process will be written as

$$\begin{aligned} U_{AB}(\{[nm]\}) &= K_{AB}(\{[nm]\})\theta^{AB}_{\{[nm]\}} \\ K_{AB}(\{[nm]\}) &= K_{AB}\exp\beta[E^*_{AB}(\{[nm]\}) - E_{AB}(\{[nm]\})] \end{aligned} \tag{4.37}$$

where K_{AB} is the rate constant of the reaction of particles of A and B in an ideal model, $K_{AB}(\{[nm]\})$ is the rate constant of the reaction of a quasiparticle, differing from K_{AB} in the change in the activation energy of the reaction when the reactant interacts with neighboring particles; $\theta^{AB}_{[nm]}$ is the probability of finding neighboring particles of A, B, and v given by the set of values $\{[nm]\}$ around the reacting A and B particles. On the other hand, the quantity $\theta^{AB}_{\{[nm]\}}$ is the concentration of the quasiparticles.

The rate of a reaction relating to an entire surface is obtained by averaging expression (4.37) over all the local configurations of the neighboring particles:

$$U_{AB} = \sum_{n_A=0}^{z-1}\ \sum_{n_B=0}^{z-1-n_A}\ \sum_{m_A=0}^{z-1}\ \sum_{m_B=0}^{z-1-m_A} C^{n_An_B}_{z-1}C^{m_Am_B}_{z-1}\theta^{AB}_{\{[nm]\}}K_{AB}(\{[nm]\}) \tag{4.38}$$

where $C_{z-1}^{n_A n_B}(C_{z-1}^{m_A m_B})$ is the statistical weight of configurations with n_A, n_B (m_A, m_B) particles of A and B around the central particle of A (B) differing in their location at $z - 1$ sites.

The energies of interaction of particles forming a quasiparticle in the ground and transition states with consideration of paired interaction of the nearest neighbors are:

$$
\begin{aligned}
E_{AB}(\{[nm]\}) &= \varepsilon_{AB} + \sum_{j=A}^{V} \varepsilon_{Aj} n_j + \sum_{j=A}^{V} \varepsilon_{Bj} m_j \\
E^{*}_{AB}(\{[nm]\}) &= \sum_{j=A}^{V} \varepsilon^{*}_{Aj} n_j + \sum_{j=A}^{V} \varepsilon^{*}_{Bj} m_j
\end{aligned}
\tag{4.39}
$$

In the first formula, the parameter ε_{AB} reflects the interaction of the reactants in the ground state; this parameter is absent in the second formula because an activated complex is a single "molecule" and the interaction between fragments of A and B relates to its internal interactions.

To calculate the rate of a reaction, we must use an approximate expression for the concentrations of the quasiparticles. In the quasichemical approximation, the probabilities of all the multiparticle configurations are approximated in terms of the probabilities of two-body configurations. Here it is possible to use the same division into pairs that was employed in the first formula (4.39) for the energy of two-body interaction, i.e.

$$
\theta^{AB}_{\{[nm]\}} = \theta_{AB} \prod_{j=A}^{V} (t_{Aj})^{n_j} \prod_{j=A}^{V} (t_{Bj})^{m_j} \tag{4.40}
$$

Formula (4.40) is a particular case of relation (4.32). The functions θ_{ij} and t_{ij} have been defined in (2.41), they are found for given values of θ_A and θ_B by solving system of equations (2.45).

Inserting expressions (4.39) and (4.40) into (4.38), we obtain:

$$
\begin{aligned}
U_{AB} &= K_{AB} \exp(-\beta\varepsilon_{AB}) \Phi_A \Phi_B \\
\Phi_A &= \sum_{n_A=0}^{z-1} \sum_{n_B=0}^{z-1-n_A} C_{z-1}^{n_A n_B} \prod_{j=A}^{V} [t_{Aj} \exp \beta\delta\varepsilon_{Aj}]^{n_j}
\end{aligned}
\tag{4.41}
$$

There is a similar expression for Φ_B. Summation in formulas (4.41) yields $\Phi_i = (S_i)^{z-1}$, where S_i has been defined in (4.16), i.e. the obtained expression for the rate of a reaction of adsorbed particles of A and B is a particular case of formulas (4.31) and (4.33).

The use of the cluster approach also provides a standard method for constructing the kinetic equations in more involved cases: for inhomogeneous lattices, when considering indirect correlations, for a large radius of interaction of a two-body potential, in the absence of equilibrium distribution of the adsorbed particles, etc. This is why in the following we shall not stop to consider the possibilities of interpretation of the results obtained on the basis of the notions on quasiparticles.

The rates of elementary adsorption-desorption processes of dissociating molecules A_2 will be written as follows [216]:

quasichemical approximation

$$U_{vv} = K_{vv}\theta_{vv}(1 + t_{vA}x_{vA})^{2(z-1)}$$
$$U_{AA} = K_{AA}\theta_{AA}e^{-\beta\varepsilon_{AA}}(1 + t_{AA}x_{AA})^{2(z-1)} \tag{4.42}$$

molecular-field approximation

$$U_{vv} = K_{vv}\theta_v^2 \exp\,[2\beta(z - 1)\varepsilon_{AA}^{*}\theta_A]$$
$$U_{AA} = K_{AA}\theta_A^2 \exp\beta[2(z - 1)\delta\varepsilon_{AA}\theta_A - \varepsilon_{AA}] \tag{4.43}$$

where the rate constants have been determined in (1.51) and (1.52), while the functions θ_{vv}, t_{vA}, and t_{AA}, as in Eq. (4.24), are related to θ_A by formulas (2.14) and (2.18). In the region of low coverages, the rate of desorption in the quasichemical approximation, by (2.27), has the form $U_{AA} = K_{AA}\theta_A^2$, while in the molecular-field approximation $U_{AA} = K_{AA}\theta_A^2 \exp(-\beta\varepsilon_{AA})$, i.e. the failure to consider the correlation effects leads to artificial renormalization of the activation energy [216]. Matters are similar for any two-site reaction, hence it is necessary to use approximations considering the correlation effects.

The concentration dependences of the rates of adsorption and desorption of dissociating molecules are similar qualitatively to those in Sec. 4.2. The quantitative changes are associated with an increase in the number of nearest neighbors [$2(z - 1)$ instead of z] and with the presence of the functions θ_{vv} and θ_{AA}.

4.4. Kinetic Equations of Nonideal Reaction Systems

Surface processes generally proceed in many steps, therefore the same component of a reaction system is consumed and forms in various steps of the general mechanism of a process. If Q_A stands for the number of steps of the one-site reactions $A \rightleftarrows B_\alpha$, and Q_{AC} for the number of steps of the two-site reactions $A + C_\alpha \rightleftarrows B_\alpha + D_\alpha$, the kinetic equation for the concentration of component A is written as follows:

$$\dot{\theta}_A = \sum_{\alpha=1}^{Q_A} [U_B(\alpha) - U_A(\alpha)] + z \sum_{\alpha=1}^{Q_{AC}} [U_{BD}(\alpha) - U_{AC}(\alpha)] \tag{4.44}$$

where the addends of the first sum have been determined in (4.15) and (4.16), while the components of the second sum — in (4.31) and (4.33). The reaction rate constants in them have the form given by (1.42) and (1.50). Similar equations hold for the other components of the reaction system. With a view to the normalization of the functions θ_i for an s-component system, $(s - 1)$ equations of the type of (4.44) are sufficient.

We shall consider the catalytic process of oxidation of CO molecules on the surface of metals of the platinum group as an example of constructing the kinetic equations for nonideal reaction systems. The process consists of basic elementary steps of adsorption-desorption of nondissociating and dissociating molecules and a reaction between the adsorbed particles. Operating with these steps, we can easily construct the kinetic equations of processes with a larger number of steps. For instance, the contribution of the impact mechanism step of the reaction $AZ + B \rightarrow CZ + D$ is constructed by analogy with the step of nondissociative adsorption with redetermined rate constants [compare (1.44) and (1.45)] and interaction parameters (see also Appendix 3). The contribution of the step $AB + 2Z \rightarrow AZ + BZ$ is similar to that of dissociative adsorption with redetermined molecular parameters. For this reason, the above process will be considered everywhere below. By the modern notions of its mechanism [167], the number of steps of the process at least in the region of low pressures is three: (1) $CO + Z \rightleftarrows ZCO$, (2) $O_2 + 2Z \rightleftarrows 2ZO$, and (3) $ZCO + ZO \rightarrow 2Z + CO_2$, where Z is a vacant site on the surface, and ZCO and ZO are sites on the surface occupied by the molecule CO and the atom O, respectively. We shall designate the particles ZCO by the subscript A, and the particles ZO and Z by the subscripts B and v, respectively. Now the system of kinetic equations will have the following form:

$$\begin{aligned} \dot{\theta}_A &= U_v(1) - U_A(1) - zU_{AB}(3) \\ \dot{\theta}_B &= z[U_{vv}(2) - U_{BB}(2) - U_{AB}(3)] \end{aligned} \tag{4.45}$$

where

$$U_v(1) = K_v(1)\theta_v[S_v(1)]^z,$$
$$U_{vv}(2) = K_{vv}(2)\theta_{vv}[S_v(2)]^{2(z-1)}$$
$$U_A(1) = K_A(1)\theta_A[S_A(1)]^z \qquad (4.46)$$
$$U_{BB}(2) = K_{BB}(2)e^{-\beta\varepsilon_{BB}}\theta_{BB}[S_B(2)]^{2(z-1)}$$
$$U_{AB}(3) = K_{AB}(3)\exp(-\beta\varepsilon_{AB})\theta_{AB}[S_A(3)S_B(3)]^{z-1}, \quad \sum_{i=A}^{v}\theta_i = 1$$

The functions $S_i(\alpha)$ in the quasichemical approximation have been determined by formula (4.16). Recall that $\varepsilon_{iv} = \overset{*}{\varepsilon}_{iv} = 0$, but $\overset{*}{\varepsilon}_{vi}(1) = \overset{*}{\varepsilon}_{Ai}(1)$ and $\overset{*}{\varepsilon}_{vi}(2) = \overset{*}{\varepsilon}_{Bi}(2)$, and i = A or B. The partial pressures of the molecules CO and O_2 are included in the constants $K_v(1)$ and $K_{vv}(2)$. To solve system (4.45), it is necessary to determine the function θ_{ij} from system (2.45) at given values of θ_i (see Appendix 1, Sec. 2).

The parameters of nonideal models can be divided into three groups: 1 — activation energies and preexponential factors of the rate constants, 2 — parameters of interaction between particles in the ground state, and 3 — parameters of interaction of particles in a transition state with their neighbors. The determination of the parameters of the first group by their selection from experimental data is a traditional task of the chemical kinetics of heterogeneous processes [85-90]; the experimental data must relate to extremely low surface coverages. The self-consistency of the description of the kinetics and equilibrium of reaction systems in approximations considering the effects of correlation of particles makes it possible to determine the parameters of the second and third groups independently. For example, the parameters ε_{ij} can be determined from equilibrium characteristics, e.g. from adsorption isotherms or heats.

The parameters $\overset{*}{\varepsilon}_{ij}(\alpha)$ are kinetic ones, they are determined from the kinetic characteristics. They are identical for the forward and reverse directions of each step, but differ for different steps, and it is desirable to determine them from independent kinetic characteristics: in the description of multistep processes, part of the steps can be investigated separately. For example, for surface processes, the parameters of the adsorption-desorption steps of individual components can be determined from their TDS. On the other hand, the data of the TDS can also be employed to determine the parameters ε_{ij} and use them for calculating the adsorption isotherms and heats.

The model (4.45) and (4.46) has been employed [217] for describing the experimental data of the steady-state conditions of oxidizing CO/Pt [218]. The desorption rate constants $K_A(1)$ and $K_{BB}(2)$, as well as the interaction parameters ε_{AA}, $\varepsilon^*_{AA}(1)$, ε_{BB}, $\varepsilon^*_{BB}(2)$ were determined preliminarily [219] from data on the TDS of CO and O_2 molecules [220, 221]. The remaining parameters were considered as varying ones. A satisfactory description of the observed temperature dependence of the rate of formation of CO_2 molecules was obtained for given partial pressures of the CO and O_2 molecules. The model also qualitatively explains the data obtained by Pacia *et al.* [222].

Real matters are undoubtedly more involved. Experimental data witness the influence of migration of the adsorbed particles [223], rearrangement of a surface in the course of a reaction [40, 41, 224], and so on. But the results obtained in ref. [217] and an analysis of investigations of models considering the influence of lateral interactions on the kinetics of surface processes (see Appendix 2) indicate that the nonideal nature of the reaction systems must be taken into account for describing real processes. In this case, the numerical values of the preexponential factors and activation energies of elementary reactions are close to their "conventional" values used in the literature for appraising the rates of mono- and bimolecular reactions.

A numerical analysis of the model (4.45) and (4.46) has revealed that the parameters of the steps of desorption of CO and O_2 obtained from the TDS can be used as a reasonable approximation when describing a multistep process (with a large number of system components), and also that the interaction parameters $\varepsilon^*_{ij}(\alpha)$ greatly affect the height of the activation barrier, and they cannot be treated as small perturbations. Figure 4.4 shows how $U_{AB}(3)$ depends on the temperature when $\varepsilon^*_{ij}(3)$ is varied. It has been assumed for simplicity that $\alpha_{11} = \alpha_{12}$ and $\alpha_{21} = \alpha_{22}$, where $\alpha_{ij} = \varepsilon^*_{ij}(3)/\varepsilon_{ij}$.

The lateral interaction of particles greatly affects the dynamics of surface processes and may cause a whole series of new effects. In particular, it increases the number of steady states of the reaction system. A very simple example of this is the two-step process $A + Z \rightleftarrows AZ$ and $B + AZ \rightarrow Z + C$ [225]. When there is no interaction, the process has a single steady state. The nonideal nature of an adsorption system may result in two or three steady states because of a highly nonlinear change in the rate of the elementary reactions with an increasing coverage. Numerical analysis has shown the conditions of realization of a plurality of steady states to relate to the region of actually existing interactions $\varepsilon^*_{AA}(\alpha)$, $\varepsilon_{AA} \leqslant (3\text{-}5)RT$. The steric difficulties associated with the formation of the product C or the attraction of particles that increases the rate of adsorption with a growth in θ at low coverages are the physical

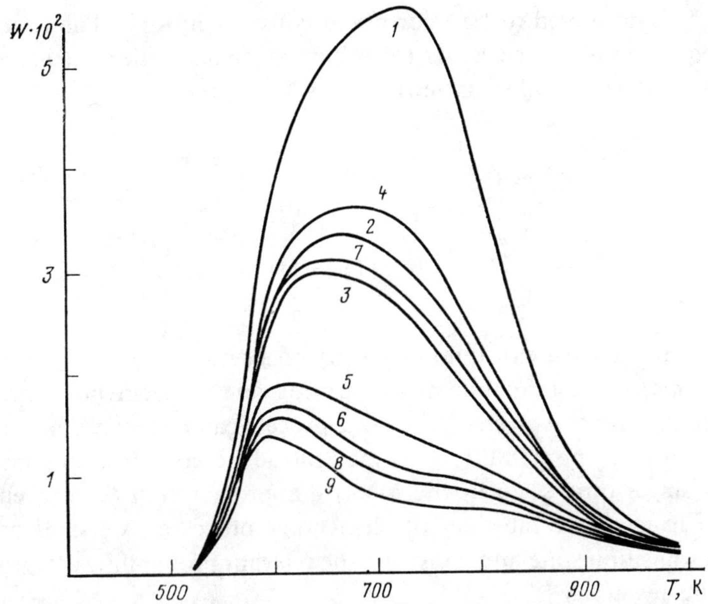

Fig. 4.4. Influence of the interaction of an activated complex with its neighboring particles on the rate of the reaction of formation of CO_2 molecules: *1-3*—$\alpha^3_{AA} = 0$; *4-6*—$\alpha^3_{AA} = 0.5$; *7-9*—$\alpha^3_{AA} = 1.0$; *1, 4, 7*—$\alpha^3_{BB} = 0$; *2, 5, 8*—$\alpha^3_{BB} = 0.5$; *3, 6, 9*—$\alpha^3_{BB} = 1.0$

prerequisites of the plurality of steady states. The influence of the coverage was found experimentally in the systems CO [220] and O_2 [221] on Pt(III). It is natural that in processes with a large number of components in a system, consideration of nonideality must also result in a large number of steady states.

4.5. Inhomogeneous Systems

For inhomogeneous systems, the probability of elementary processes depends not only on the properties of a particle and the states of occupancy of neighboring sites, but also on the types of the sites accommodating a reacting particle and its neighbors. This makes the description of inhomogeneous systems more involved, but does not alter the general principles of constructing kinetic equations. The state of a system is given by the set of quantities $\gamma^i_f \eta^q_f$, where $1 \leqslant f \leqslant N$ (see Sec. 1.3); here the set of quantities

η_f^q is considered to be known and constant in time. The fundamental kinetic equation (4.1) also holds for inhomogeneous lattices, hence expression (4.3), according to (1.16), is rewritten as follows:

$$\dot{\theta}_{f_1 \ldots f_m}^{i_1 \ldots i_m}(\tau \mid q_1 \ldots q_m) = \sum_{\{II\}} \left[\left\langle \prod_{n=1}^{m} \gamma_{f_n}^{\bar{i}_n} W_\alpha(\{II\} \to \{I\} \mid \{\eta_f^q\}) \right\rangle - \left\langle \prod_{n=1}^{m} \gamma_{f_n}^{i_n} W_\alpha(\{I\} \to \{II\} \mid \{\eta_f^q\}) \right\rangle \right] \tag{4.47}$$

where the nature of inhomogeneity of a system $\{\eta_f^q\}$ is indicated in the expressions for the probabilities of carrying out the elementary processes α. In an inhomogeneous system, the same process can be realized at different sites with a different probability. This may introduce complications into the choice of a single time scale in the Markov approximation for the entire system.

To calculate the rates of elementary processes, we shall use the model of a transition state and construct the relevant probabilities W_α for one- and two-site reactions [171].

One-Site Reaction A → B. When an activated complex of this reaction is present at site k, the total energy of the system will be:

$$H_1 = \sum_{f,\, i}{}' \nu_q^i \gamma_f^i \eta_f^q - \frac{1}{2} \sum_{f,\, g}{}' \sum_{i,\, j} \varepsilon_{qp}^{ij} \gamma_f^i \gamma_g^j \eta_f^q \eta_g^p + \nu_q^{A^*} \gamma_k^A \eta_k^q - \sum_{n,\, j} \varepsilon_{qp}^{Aj^*} \gamma_k^A \gamma_n^j \eta_k^q \eta_n^p \tag{4.48}$$

where the internal degrees of freedom of the activated complex at a site of type q are described by expression (1.54), while its interaction with a neighboring particle j at a site of type p is described by the parameter $\varepsilon_{qp}^{Aj^*}$, while n stands for the numbers of the sites adjacent to site k. The primes in the first two addends indicate the absence of contributions from an activated complex.

The energy of a system in the ground state has been determined by expression (3.1). The transition of a particle of A to the activated state changes the energy of a system by the quantity $\delta h_k^A(q\{[p]\}) = H_1 - H$ that depends on the type q of site k and on the types $p_1 \ldots p_z$ of its neighbors n (below in addition to setting the type of cluster with the aid of the notation $q\{[p]\}$, we shall use the notation $q\{[m]\}$, as indicated in Sec. 3.1):

$$\delta h_k^{\mathrm{A}}(q\{[p]\}) = \left(\nu_q^{\mathrm{A}*} - \nu_q^{\mathrm{A}} - \sum_{n,j} \delta\varepsilon_{qp}^{\mathrm{A}j}\gamma_n^j\right)\gamma_k^{\mathrm{A}}, \quad \delta\varepsilon_{qp}^{ij} = \varepsilon_{qp}^{ij*} - \varepsilon_{qp}^{ij} \quad (4.49)$$

The probability of the reaction A → B occurring at site k is the probability of a set of events consisting in a particle of A being at site k and in a change in the energy of the system because of the formation of an activated complex:

$$\begin{aligned} W([\gamma_k^{\mathrm{A}}] \to [\gamma_k^{\mathrm{B}}] \mid q\{[p]\}) &= \gamma_k^{\mathrm{A}} U_{\mathrm{A}}(k \mid q\{[p]\}) \\ U_{\mathrm{A}}(k \mid q\{[p]\}) &= \exp(-\beta\delta h_k^{\mathrm{A}}(q\{[p]\})) \end{aligned} \quad (4.50)$$

Let us see whether Eqs. (4.50) satisfy condition (4.1). For the reaction A ⇄ B, $H(\{\mathrm{I}\}) = h_k^{\mathrm{A}}(q\{[p]\}) + G_k(q\{[p]\})$ corresponds to state $\{\mathrm{I}\}$. These addends have been determined in (3.4) and (3.5), where the notation $q\{[m]\}$ has been used instead of $q\{[p]\}$. The quantity $H(\{\mathrm{II}\}) = h_k^{\mathrm{B}}(q\{[p]\}) + G_k(q\{[p]\})$ corresponds to state $\{\mathrm{II}\}$. Inserting formulas (4.50) for the forward and reverse processes into (4.1) and having in view that: (1) the parameters of an activated complex do not depend on the direction of the process, $\nu_q^{\mathrm{A}^*} = \nu_q^{\mathrm{B}^*}$ and $\varepsilon_{qp}^{\mathrm{A}j^*} = \varepsilon_{qp}^{\mathrm{B}j^*}$, and (2) in the initial and final states a particle of A ($\gamma_k^{\mathrm{A}} = 1$) and a particle of B ($\gamma_k^{\mathrm{B}} = 1$) are at site k, respectively, we obtain the identity $1 \equiv 1$.

Two-Site Reaction A + C → B + D. An activated complex occupies two neighboring sites k_1 and k_2. When it forms, the total energy is expressed as follows:

$$\begin{aligned} H_2 = {} & \sum_{f,\,i}{}' \nu_q^i \gamma_f^i \eta_f^q - \frac{1}{2}\sum_{f,\,g}{}' \sum_{i,\,j} \varepsilon_{qp}^{ij} \gamma_f^i \gamma_g^j \eta_f^q \eta_g^p + \nu_{qp}^{\mathrm{AC}^*} \gamma_{k_1}^{\mathrm{A}} \gamma_{k_2}^{\mathrm{C}} \eta_{k_1}^q \eta_{k_2}^p \\ & - \sum_{n_1,\,j} \varepsilon_{qp}^{\mathrm{A}j^*} \gamma_{k_1}^{\mathrm{A}} \gamma_{n_1}^j \eta_{k_1}^q \eta_{n_1}^p - \sum_{n_2,\,j} \varepsilon_{qp}^{\mathrm{C}j^*} \gamma_{k_2}^{\mathrm{C}} \gamma_{n_2}^j \eta_{k_2}^q \eta_{n_2}^p \end{aligned} \quad (4.51)$$

where the internal degrees of freedom of an activated complex at sites of types q and p are described by expression (1.59). The subscripts n_1 (n_2) relate to the nearest neighbors of site k_1 (k_2) except for site k_2 (k_1). The primes in the first two addends signify the absence of contributions from an activated complex.

The probability of the formation of an activated complex depends on the types of the two central sites k_1, k_2 and all their neighbors n_1, n_2. This is why we must consider clusters with two central sites (see Fig. 1.6) and for their description introduce additional distribution functions $f(qp\{[\lambda, \sigma]\})$. The latter are similar to distribution functions (3.2) and (3.3), where qp indicate the type of the central pair of sites, and $\{[\lambda]\}$ and $\{[\sigma]\}$ the type of the neighbors of sites k_1 and k_2, respectively. Depending on the set of clusters M_1 or M_2 being considered, the symbols λ and σ relate to the type of sites or to the numbers of types of sites around sites k_1 and k_2, respectively.

The probability of formation of an activated complex of the reaction A + C at sites k_1 and k_2 is the probability of a set of events consisting in the finding of a particle of A at site k_1, a particle of C at site k_2, and in the energy changing by the amount $\delta h_{k_1k_2}^{AC}(qp\{[\lambda\sigma]\}) = H_2 - H$, where H is the energy of the ground state (3.1):

$$\begin{aligned} &W([\gamma_{k_1}^{A}\gamma_{k_2}^{C}] \to [\gamma_{k_1}^{B}\gamma_{k_2}^{D}] \mid qp\{[\lambda\sigma]\}) = \gamma_{k_1}^{A}\gamma_{k_2}^{C} U_{AC}(k_1k_2 \mid qp\{[\lambda\sigma]\}) \\ &U_{AC}(k_1k_2 \mid qp\{[\lambda\sigma]\}) = \exp\left(-\beta\delta h_{k_1k_2}^{AC}(qp\{[\lambda\sigma]\})\right) \\ &\delta h_{k_1k_2}^{AC}(qp\{[\lambda\sigma]\}) = (\nu_{qp}^{AC^*} + \varepsilon_{qp}^{AC})\gamma_{k_1}^{A}\gamma_{k_2}^{C} - \nu_q^{A}\gamma_{k_1}^{A} - \nu_p^{C}\gamma_{k_2}^{C} \\ &\quad + \sum_{n_1,\, j} \delta\varepsilon_{q\lambda}^{Aj}\, \gamma_{k_1}^{A}\gamma_{n_1}^{j} + \sum_{n_2,\, j} \delta\varepsilon_{p\sigma}^{Cj}\, \gamma_{k_2}^{C}\gamma_{n_2}^{j} \end{aligned} \tag{4.52}$$

The obtained probabilities of two-site reactions (4.52) satisfy condition (4.1). The proof of this repeats word by word those for probabilities (4.28) and (4.50).

By inserting expressions (4.50) and (4.52) into equations (4.47), we obtain a system of kinetic equations describing the evolution of distribution of interacting particles among the sites of an inhomogeneous lattice. In the present chapter, we shall restrict ourselves to the case of the rapid surface mobility of the particles, therefore, as in Secs. 4.2 and 4.3, we shall construct equations only for the local concentrations of the particles, after which we shall go over to the macroscopic means. The slow mobility of particles is treated in the following chapter.

One-Site Reactions A ⇄ B. The local concentration of particles of A at site f of type q varies in time according to the following kinetic equation:

$$\dot{\theta}_f^{A}(q) = U_f^{B}(q\{[p]\}) - U_f^{A}(q\{[p]\}) \tag{4.53}$$

$$\begin{aligned} &U_f^{i}(q\{[p]\}) = \langle \gamma_f^{i} U_i(f \mid q\{[p]\})\rangle_\tau = K_q^{i} V_f^{i}(q\{[p]\}) \\ &V_f^{i}(q\{[p]\}) = \langle \gamma_f^{i} \prod_{g=1}^{z} \prod_{j=1}^{s} \gamma_g^{j} \exp \beta\delta\varepsilon_{qp}^{ij} \mid q\{[p]\}\rangle_\tau \end{aligned} \tag{4.54}$$

The mean value in (4.54) is calculated on the basis of the properties of the quantities γ_f^i in exactly the same way as has been set out in detail in Sec. 4.2 for a homogeneous lattice. A distinction consists in that the relevant molecular parameters set by the values of η_g^p correspond to each number of site g. The rate constant K_q^i has been determined in (1.57), while $V_f^i(q\{[p]\})$ is the concentration component of the rate of a one-site reaction proceeding on a cluster of type $q\{[p]\}$ with a central site f.

The right-hand side of Eq. (4.53) contains correlators of the $(z + 1)$ order. To close Eqs. (4.53), let us consider the equilibrium distribution of particles among the sites of an inhomogeneous lattice when the correlators of the first type are expressed in terms of the macroscopic means θ_q^i (1.33) by formulas of Sec. 3.6. Here the local configurations of the particles depend on the types of clusters, and not on the numbers of their sites $[V_f^i(q\{[m]\}) \to V_q^i(\{[m]\})]$. In the quasichemical approximation, the concentration component of the reaction rate on a cluster of type $q\{[m]\}$ is written as [177]:

$$V_q^i(\{[m]\} \mid \alpha) = \theta_q^i \prod_{p=1}^{t} [S_{qp}^i(\alpha)]^{m_{qp}},$$

$$S_{qp}^i(\alpha) = \sum_{j=1}^{s} t_{qp}^{ij} \exp \beta\delta\varepsilon_{qp}^{ij}(\alpha) \tag{4.55}$$

where the functions t_{qp}^{ij} are determined by solving system (3.52) and (3.54), while the index of the step α unambiguously indicates the elementary process which particle i participates in.

To obtain a kinetic equation for the macroscopic mean θ_q^i, we must average expression (4.53) over all possible types of clusters containing a central site of type q. Multiplying both sides of (4.53) by $f(q\{[m]\})$ and summing over the σ_{2q} plurality of states of the coordination sphere of the cluster $q\{[m]\}$, we obtain:

$$\dot{\theta}_q^A = \sum_{\sigma_{2q}} d(q\{[m]\})[U_q^B(\{[m]\} \mid \alpha) - U_q^A(\{[m]\} \mid \alpha)]$$

$$U_q^i(\{[m]\} \mid \alpha) = K_q^i(\alpha) V_q^i(\{[m]\} \mid \alpha) \tag{4.56}$$

where the distribution functions (3.8) have been used. Should component A change in several steps of a process, the right-hand side of (4.56) must additionally contain the sum over α.

Two-Site Reactions A + C ⇄ B + D. The change in the local concentration of component A at site f of type q is described by the following equation:

$$\dot{\theta}_f^{A}(q) = \sum_g \left[U_{fg}^{BD}(qp\{[\lambda\sigma]\}) - U_{fg}^{AC}(qp\{[\lambda\sigma]\})\right] \tag{4.57}$$

$$\begin{aligned} U_{fg}^{ij}(qp\{[\lambda\sigma]\}) &= \langle \gamma_f^i \gamma_g^j U_{ij}(fg \mid qp\{[\lambda\sigma]\})\rangle_\tau \\ &= K_{qp}^{ij} e^{-\beta\varepsilon_{qp}^{ij}} V_{fg}^{ij}(qp\{[\lambda\sigma]\}) \end{aligned} \tag{4.58}$$

$$V_{fg}^{ij}(qp\{[\lambda\sigma]\}) = \left\langle \gamma_f^i \gamma_g^j \prod_{n_1} \sum_{l=1}^{s} \gamma_{n_1}^l e^{\beta\delta\varepsilon_{q\lambda}^{il}} \prod_{n_2} \sum_{l=1}^{s} \gamma_{n_2}^l e^{\beta\delta\varepsilon_{p\sigma}^{jl}} \right\rangle_\tau$$

where the rate constant of a two-site process K_{qp}^{ij} has been determined in (1.60), while $V_{fg}^{ij}(qp\{[\lambda\sigma]\})$ is the concentration component of the rate of a two-site reaction proceeding on a cluster of type $qp\{[\lambda\sigma]\}$ with two central sites f and g. It is calculated in exactly the same way as was set out in Sec. 4.3 with account taken of the nonequivalence of the cluster points.

Let us close Eq. (4.57) by using the condition of equilibrium distribution of the particles among the sites of an inhomogeneous lattice. In this case, the correlators constructed on a cluster with two central sites can be approximately represented in a form similar to formula (4.32), after which one may use the ways of approximating correlators of the first type constructed on a cluster with one central site.

By multiplying the left-hand and right-hand sides of Eq. (4.57) by $f(qp\{[\lambda\sigma]\})$ and summing all the neighbors of the central sites of types q and p over the σ_{2qp} multitude of states and also over the types of sites p, we shall rewrite Eq. (4.57) as follows [177]:

$$\begin{aligned} \dot{\theta}_q^{A} &= z_q \sum_{p=1}^{t} d_{qp} \sum_{\sigma_{2qp}} d(qp\{[\lambda\sigma]\})[U_{qp}^{BD}(\{[\lambda\sigma]\} \mid \alpha) - U_{qp}^{AC}(\{[\lambda\sigma]\} \mid \alpha) \\ U_{qp}^{ij}(\{[\lambda\sigma]\} \mid \alpha) &= K_{qp}^{ij}(\alpha) \exp(-\beta\varepsilon_{qp}^{ij}) V_{qp}^{ij}(\{[\lambda\sigma]\} \mid \alpha) \\ V_{qp}^{ij}(\{[\lambda, \sigma]\} \mid \alpha) &= \theta_{qp}^{ij} \prod_{\xi=1}^{t} [S_{q\xi}^{i}(\alpha)]^{\lambda_{q\xi}} \prod_{\xi=1}^{t} [S_{p\xi}^{j}(\alpha)]^{\sigma_{p\xi}} \end{aligned} \tag{4.59}$$

where the functions $S_{q\xi}^i(\alpha)$ have been determined in (4.55), and the last formula in (4.59) is the concentration component of the rate of a two-site reaction on an inhomogeneous surface in a quasichemical approximation for a rapid surface mobility of the interacting particles. As previously, we have additionally introduced the step number index α in expressions (4.59). If component

A participates in several steps of the overall process, summing over α must be additionally performed on the right-hand side of Eq. (4.59).

The equations (4.59) include the conditional distribution function $d(qp\{[\lambda\sigma]\})$ characterizing the type of cluster sites surrounding central sites of types q and p:

$$f_{qp}d(qp\{[\lambda\sigma]\}) = f(qp\{[\lambda\sigma]\}) \tag{4.60}$$

To describe the kinetics of surface processes including two-site reactions in their mechanism, when considering the interaction of the nearest neighbors a knowledge of the distribution functions of various types of sites determined in Sec. 3.1. is insufficient. It is necessary to additionally set functions (4.60). Let us consider the simplest examples of constructing these functions for a point defect on a face (100) and for a stepped surface (see Sec. 3.1).

Assume that a point defect modifies the adsorption properties of its neighbors, $t = 3$: the first type of sites is the defect, the second is its neighbors, and the third — the remaining sites of the face (100). If c stands for the concentration of the defect, then, as in Sec. 3.1, we have $f_1 = c$, $f_2 = 4c$, and $f_3 = 1 - 5c$. The surface contains no pairs of nearest neighbors of type 11, 13, 22, therefore the relevant values of $f_{qp} = 0$. The sites of the first type are surrounded by ones of the second type, hence $f_{12} = c$, while each site of the second type has three neighbors of the third one, therefore $f_{23} = 3c$. The function f_{33} can be determined from the normalization condition $\sum_{q,\ p=1}^{3} f_{qp} = 1$, which yields $f_{33} = 1 - 8c$, or by direct calculations. For a crystal lattice, the total number of bonds M is expressed as $M = zN/2$. The number of bonds of type 12 and 23 on a surface is $16N_1$. The remaining bonds are ones of type 33. Consequently, $f_{33} = (M/N - 16c)/(M/N)$, where $c = N_1/N$.

Let us now consider the environment of each pair of central sites. The pair 12 has a single surrounding: $d(12[0, 3, 0; 0, 0, 3]) = 1$. The pair 23 has two types of surrounding depending on its orientation relative to pair 12. The first of them is when pairs 12 and 23 are on one straight line, and the second is when pairs 12 and 23 are at right angles to each other. Since there are altogether three ways of arranging these pairs, we have $d(23[1, 0, 2; 0, 0, 3]) = 1/3$, and $d(23[1, 0, 2; 0, 1, 2]) = 2/3$. For a central pair 33, there are three types of surroundings in which one of the sites of the third type has two, one, or zero sites of the second one. In the first case, each defect is surrounded by four sites in which two equivalent orientations of the pairs are possible, which yields $d(33[0, 2, 1; 0, 0, 3]) = 4c/f_{33}$. In the second case, there are also four sites, but in each of them the pair 33 may have one of three orientations, consequently $d(33[0, 1, 2; 0, 0, 3]) = 6c/f_{33}$. The remaining situations relate to the third case $d(33[0, 0, 3; 0, 0, 3]) = (1 - 18c)/f_{33}$.

Stepped Surface (see Fig. 3.1). There are three types of centers: (1) the row along the base of the step, (2) the row along the end of the terrace, and (3) the terrace rows. The length of the terrace equals l atoms of the surface. The given example relates to lattices in which the sites of various types have a different number of nearest neighbors. The total number of pairs for them is determined by the more general expression $M = \sum_{q=1}^{t} N_q z_q/2$.

The number of sites in all the rows is the same, therefore we shall relate the number of bonds to one site of each layer. Here $M = [3 + 3 + (l - 2) \times 4]/2 = 2l - 1$. The orientation of each central pair along or across the layers 1 and 2 is unambiguously related to the type of this pair, hence $f_{11} = f_{22} = f_{13} = f_{23} = 1/M$. And since $f_{12} = 0$, we have $f_{33} = (2l - 5)/M$. In the direct calculation of this function, it is necessary to take into account the contributions from $(l - 4)$ layers of the terrace each having four neighbors of the third type and from two layers of the terrace adjoining layers 1 and 2 and each having three neighbors of the third type. Hence, the average number of 33 pairs is $[4(l - 4) + 2 \times 3]/2$, which yields the same value of the function f_{33}.

For the pairs $qp = 11, 22, 13, 23$, the types of the surrounding neighboring sites are unambiguously related to the type of the central pair, therefore

$$d(11[1, 0, 1; 1, 0, 1]) = d(22[0, 1, 1; 0, 1, 1])$$
$$= d(13[2, 0, 0; 0, 0, 3]) = d(23[0, 2, 0; 0, 0, 3]) = 1$$

For the central pair 33 in the layers neighboring with layer 1 and 2, there are two types of surrounding depending on its orientation. In this case

$$d(33[1, 0, 2; 0, 0, 3]) = d(33[1, 0, 2; 1, 0, 2])$$
$$= d(33[0, 1, 2; 0, 0, 3]) = d(33[0, 1, 2; 0, 1, 2]) = (2l - 5)^{-1}$$

The remaining central pairs 33 are surrounded by sites of the third type, and for them $d(33[0, 0, 3; 0, 0, 3]) = (2l - 9)/(2l - 5)$.

In more involved cases too, using information on the structure of a surface and its evolution in the course of adsorbent formation, we can construct the distribution functions for clusters with two central sites in a similar way. The following approximation holds when sites of various types are distributed in equilibrium over a surface:

$$d(qp\{[\lambda\sigma]\}) = C^{\{\lambda\}}_{z_q-1} C^{\{\sigma\}}_{z_p-1} \prod_{\xi=1}^{t} (d_{q\xi})^{\lambda_{q\xi}} \prod_{\xi=1}^{t} (d_{p\xi})^{\sigma_{p\xi}},$$

$$\sum_{\xi=1}^{t} \lambda_{q\xi} = z_q - 1 \tag{4.61}$$

It is a generalization of expression (3.9). All the remarks concerning formula (3.9) hold for it.

4.6. Reactions in Inhomogeneous Systems

The kinetic equations of the surface processes on inhomogeneous surfaces for the equilibrium distribution of the particles among the sites of different types have the following general structure:

$$\begin{aligned}\dot{\theta}_{\mathrm{A}} = &\sum_{\alpha=1}^{Q_{\mathrm{A}}} \sum_{q=1}^{t} f_q \sum_{\sigma_{2q}} d(q\{[m]\})[U_q^{\mathrm{B}}(\{[m]\}\mid\alpha) - U_q^{\mathrm{A}}(\{[m]\}\mid\alpha)]\\ &+ \sum_{\alpha=1}^{Q_{\mathrm{AC}}} \sum_{q=1}^{t} f_q z_q \sum_{p=1}^{t} d_{qp} \sum_{\sigma_{2qp}} d(qp\{[\lambda\sigma]\})\\ &\times [U_{qp}^{\mathrm{BD}}(\{[\lambda\sigma]\}\mid\alpha) - U_{qp}^{\mathrm{AC}}(\{[\lambda\sigma]\}\mid\alpha)]\end{aligned} \tag{4.62}$$

As in Sec. 4.4, the particles of A are considered to participate in Q_{A} one-site reactions and in Q_{AC} two-site ones. The rates of the elementary reactions have been determined in (4.56), (4.55), and (4.59). Their calculation requires the preliminary solution of the system of equations (3.52), (3.54), and (3.58). Equation (4.62) describes the total change in the number of particles of A on an entire inhomogeneous surface according to definition (1.34). (Similar equations hold for other components of an adsorption system.) It is solved with the given initial conditions $\theta_{\mathrm{A}}(0) = \theta_{\mathrm{A}}(\tau = 0)$.

The ratios of Eqs. (3.58) taken in pairs for different $q \neq 1$ and $q = 1$, for example, enable us to exclude the partial pressures of the gas phase components in y_q^i. The number of such ratios is $t - 1$. To supplement the system of equations relative to θ_q^i to a closed form, we must employ the relation between θ_q^i and θ_i (1.34) because θ_i is known either from the initial data or from the solution of (4.62) in the preceding step (see also Appendix 1, Sec. 2). After solving the system (3.52), (3.54), and (3.58), we determine the concentration components of the rates by (4.55) and (4.53), while the rate constants have the form given by (1.57) and (1.60).

Equation (4.62) employs a detailed description of the distribution of sites of various types on a surface. If we use approximation (4.61), these equations become simplified [226]. When considering the properties of vacant sites of a surface in the explicit form, equation (4.62) will be rewritten in the form of (1.62), where the individual components are as given in (1.63) with the following concentration components of the rates of one- and two-site reactions:

$$\begin{aligned} V_q^i(\alpha) &= \theta_q^i [T_q^i(\alpha)]^{z_q}, \\ V_{qp}^{ij}(\alpha) &= \theta_{qp}^{ij} e^{-\beta\varepsilon_{qp}^{ij}} [T_q^i(\alpha)]^{z_q - 1} [T_p^j(\alpha)]^{z_p - 1} \\ T_q^i(\alpha) &= 1 + \sum_{p=1}^{t} d_{qp} \sum_{l=1}^{s-1} t_{qp}^{il} x_{qp}^{il}(\alpha), \quad x_{qp}^{il}(\alpha) = \exp \beta\delta\varepsilon_{qp}^{il}(\alpha) - 1 \end{aligned} \tag{4.63}$$

To calculate the kinetics of surface processes by Eqs. (1.62), (1.63), and (4.63), it is also necessary to solve system of equations (3.52), (3.54), and (3.58).

Let us use the oxidation of CO molecules on inhomogeneous surfaces of the platinum metals as a very simple example of constructing the kinetic equations. In this case, system of equations (4.45) and (4.46) will have the form:

$$\begin{aligned} \dot{\theta}_A &= U_v(1) - U_A(1) - U_{AB}(3) \\ \dot{\theta}_B &= U_{vv}(2) - U_{BB}(2) - U_{AB}(3) \end{aligned} \tag{4.64}$$

where the addends of system (4.64) are determined by formulas (1.63) with the concentration components (4.63). The system is solved with given initial values of $\theta_A(0)$ and $\theta_B(0)$.

It was indicated at the end of Sec. 3.6 that in the molecular-field approximation the task of describing the distribution of the particles of an adsorption system among sites of various types is simplified substantially if we adopt additional assumptions, namely, that sites of various types are distributed chaotically over a surface, and that the parameters of interaction between particles do not depend on the types of the sites accommodating them. The same assumptions also simplify the kinetic equations of the surface processes because they operate with the functions θ_i instead of with θ_q^i [193]. Such simplified equations can be useful for a qualitative description of a broad range of processes.

Adsorbed particles are also distributed inhomogeneously on homogeneous surfaces when their state of aggregation changes. Let us consider the expressions for the rates of processes of adsorption and desorption when they are ordered and stratified.

The equilibrium characteristics of the ordered phase $c(2 \times 2)$ on a face (100) were discussed in Sec. 3.5. The distribution functions of the sites for this superstructure were also indicated there. The rates of adsorption and desorption of nondissociating molecules will be written as follows:

$$U_v = (U_\alpha^v + U_\gamma^v)/2, \quad U_A = (U_\alpha^A + U_\gamma^A)/2$$
$$U_\sigma^v = K_v(1 - \theta_\sigma)(1 + t_{\sigma\lambda}^{BA} x_{BA})^z, \quad t_{\sigma\lambda}^{BA} = \theta_\lambda^A(1 - t_{\lambda\sigma}^{AA})/(1 - \theta_\sigma^A) \tag{4.65}$$
$$U_\sigma^A = K_A \theta_\sigma^A (1 + t_{\sigma\lambda}^{AA} x_{AA})^z, \quad \sigma = \gamma, \alpha; \; \lambda \neq \sigma$$

where x_{iA} were determined in (4.21), $t_{\sigma\lambda}^{AA}$, in (3.20), while the rate constants of adsorption and desorption were determined in (1.44) and (1.43). U_σ^v and U_σ^A are the rates of adsorption and desorption at sites of type σ (see also Sec. 7.2). In a similar way, formulas (4.62) or (1.62), (1.63), and (4.63) can yield expressions for the rates of adsorption processes when taking into account their ordered distribution on inhomogeneous surfaces. As for an equilibrium state, the task consists in setting the relevant distribution functions for sites of various types and solving the system of equations relative to the equilibrium relation between θ_A and θ_q^A.

In stratifying adsorption systems, adsorption and desorption processes in both phases occur independently at each instant, although the density of the

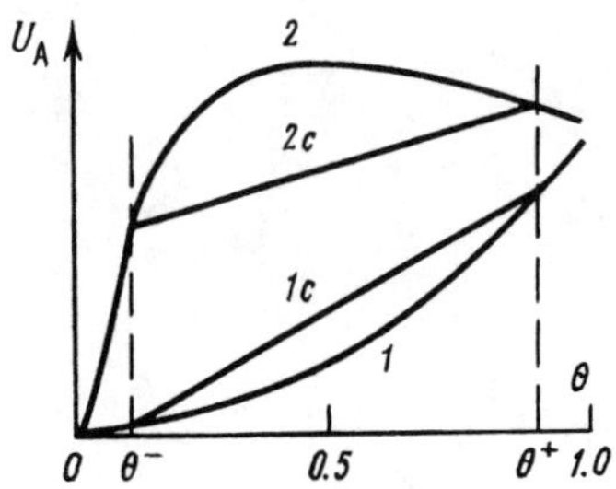

Fig. 4.5. Concentration dependences of rate of desorption for attracted particles: *1*—$\varepsilon_{AA}^* > \varepsilon_{AA}$; *2*—$\varepsilon_{AA}^* < \varepsilon_{AA}$; the symbol *c* relates to curves for a two-phase adsorption system

substance in each phase and the fraction of the surface falling to the share of each phase are related by the conditions of their coexistence (Sec. 2.3). The rates of the adsorption and desorption processes are given in the form of (2.31): θ_{AA} $(\theta^{\pm})$ must be replaced with $U_V(\theta^{\pm})$ or $U_A(\theta^{\pm})$. These elementary processes are determined by formulas (4.21) for each of the phases, respectively.

Figure 4.5 shows the rates of desorption of particles that attract one another with and without account of their two-dimensional condensation. When $\varepsilon^*_{AA} > \varepsilon_{AA}$, the rate of desorption grows monotonically with increasing coverage, and the effective activation energy of desorption drops with increasing θ. When $\varepsilon^*_{AA} < \varepsilon_{AA}$, the rate of desorption passes through a maximum because at high coverages the attraction of neighboring particles prevents desorption. In the two-phase region, the rate of desorption varies linearly with increasing θ. In the first case, consideration of two-dimensional condensation of the particles increases the rate of desorption because of the growth in the contribution of the flow from the denser phase, while in the second case, consideration of two-dimensional condensation lowers the contribution because the latter is lower for the dense phase than with moderate coverages. The rates of adsorption with explicit consideration of stratification of the adsorption system also change appreciably. This is why the phase state of the adsorption system must be taken into consideration for a quantitative description of the surface processes.

5 Surface Processes. Slow Mobility of Particles

5.1. Kinetic Equations of Processes

The evolution of the state of a condensed phase is described by system of equations (4.3) [or (4.47)]. The first equations of this system relate to the evolution of the local concentrations of the phase components at various sites of a lattice. On the right-hand sides of Eqs. (4.7), (4.29), (4.53), and (4.57), unknown correlators with a dimension greater than unity appear that were approximated on the basis of equilibrium relations in terms of local concentrations. Such closing of the kinetic equations corresponds to the traditional region of chemical kinetics, when the steps of chemical transformation—the kinetic conditions of a process, are the limiting steps. In the present chapter, we deal with the more general case when the particles have a limited migration velocity and the equilibrium relations of Chaps. 2 and 3 employed in Chap. 4 are not observed.

The first unknown correlator appearing on the right-hand side of Eqs. (4.7) and (4.53) is the correlator of the dimension $(z + 1)$ (4.13), (4.54). The number of such correlators is Ns^z, where N is the number of lattice sites. Even for two species of particles on a lattice $z = 4$, the dimension of the system will be so large that we shall actually have to limit ourselves to small values of N. This necessitates the use of the approximation of correlators (4.13) in terms of correlators with a smaller dimension. Approximation (1.17) is the simplest one. Insertion of this expression into formulas (4.13), (4.31), (4.54), and (4.63) closes the kinetic equations relative to θ_f^i. We obtain expressions for the reaction rates in the chaotic approximation, while with an additional limiting process $\beta\delta\varepsilon_{qp}^{ij}(\alpha) \to 0$ — in the molecular-field approximation. In the last case, concentration components have the form:

$$V_f^i(q\{[p]\}|\alpha) = \theta_f^i(q) \exp \beta \sum_{g=1}^{z_q} \sum_{j=1}^{s} \delta\varepsilon_{qp}^{ij}(\alpha)\theta_g^j(p)$$

$$V_{fg}^{ij}(qp\{[\lambda\sigma]\}|\alpha) = \theta_f^i(q)\theta_g^j(p) \exp \beta \left[\sum_{h=1}^{z_q-1} \sum_{l=1}^{s} \delta\varepsilon_{q\lambda}^{il}(\alpha)\theta_h^l(\lambda) + \sum_{h=1}^{z_p-1} \sum_{l=1}^{s} \delta\varepsilon_{p\sigma}^{jl}(\alpha)\theta_h^l(\sigma) \right] \qquad (5.1)$$

But the absence of correlation effects in approximation (1.17) may lead to qualitative distinctions in the dynamic characteristics of a system, as occurred when calculating the TDS (Fig. 4.2). This is why more accurate approximations are needed.

In the quasichemical approximation, the concentration components of the reaction rates are written as follows:

$$V_f^i(q\{[p]\}|\alpha) = \theta_f^i(q) \prod_{g=1}^{z_q} S_{fg}^i(qp|\alpha)$$

$$S_{fg}^i(qp|\alpha) = \sum_{j=1}^{s} t_{fg}^{ij}(qp) \exp \beta\delta\varepsilon_{qp}^{ij}(\alpha) \tag{5.2}$$

$$V_{fg}^{ij}(qp\{[\lambda\sigma]\}|\alpha) = \theta_{fg}^{ij}(qp) \prod_{h=1}^{z_q-1} S_{fh}^i(q\lambda|\alpha) \prod_{h=1}^{z_p-1} S_{gh}^i(p\sigma|\alpha)$$

where $t_{fg}^{ij}(qp) = \theta_{fg}^{ij}(qp)/\theta_f^i(q)$. In other words, the right-hand sides of the kinetic equations for the local concentrations contain the unknown functions $\theta_{fg}^{ij}(qp)$; their calculation requires the construction of the relevant system of kinetic equations. Let us consider the contributions introduced separately by the one- and two-site reactions into the right-hand sides of the equations for the two-body correlators. We shall first limit ourselves to a homogeneous lattice.

The **one-site reaction** A ⇄ B proceeding at site f changes the function θ_{fg}^{ij} if only the species of the particles i or j (or both) coincide with those of A or B particles, otherwise $\dot{\theta}_{fg}^{ij} = 0$. In accordance with Eq. (4.3) for $m = 2$ and expression (4.6) for the probability of realizing an elementary process, we have

$$\begin{aligned} \dot{\theta}_{fg}^{AB} &= U_{fg}^{\bar{B}\bar{B}}(\alpha) - U_{fg}^{\bar{A}B}(\alpha) + U_{fg}^{A\bar{A}}(\alpha) - U_{fg}^{A\bar{B}}(\alpha) \\ \dot{\theta}_{fg}^{Ai} &= U_{fg}^{\bar{B}i}(\alpha) - U_{fg}^{\bar{A}i}(\alpha), \quad i \neq A,\ B \end{aligned} \tag{5.3}$$

where the rates of the elementary one-site processes which a particle indicated by a bar capping its symbol participates in is expressed as follows:

$$\begin{aligned} U_{fg}^{\bar{i}j}(\alpha) &= K_i(\alpha)\left\langle \gamma_f^i\gamma_g^j \prod_h \sum_{l=1}^{s} \gamma_h^l \exp \beta\delta\varepsilon_{il}(\alpha)\right\rangle_T \\ &= U_f^i(\alpha)\psi_{fg}^{ij}(\alpha) \\ \psi_{fg}^{ij}(\alpha) &= t_{fg}^{ij} \exp \beta\delta\varepsilon_{ij}(\alpha)/S_{fg}^i(\alpha) \end{aligned} \tag{5.4}$$

When calculating the mean value in (5.4), we used the projection properties of the quantities γ_f^i that were repeatedly employed previously; the subscript h corresponds to all the neighboring sites f accommodating a reacting particle. A neighboring particle j does not participate in a reaction, but its presence affects the height of the activation barrier. It should be noted that the dimension of correlators (5.4) does not exceed that of the correlators in Eq. (4.7). In the final form, the addends of Eqs. (5.3) are represented in terms of the product of the rate of the one-site reactions (4.13), (5.2) and of the perturbation introduced by a neighboring particle j.

Similar equations hold if in the first of equations (5.3) we consider the pairs AA, BA, and BB instead of the pair AB and in the second of equations (5.3) the pairs iA, Bi, and iB instead of the pair Ai. The obtained equations (5.3) and (5.4) contain no new unknown functions, i.e. are closed.

The **two-site reaction** A + C ⇄ B + D proceeding at neighboring sites of a lattice changes the magnitude of the function θ_{fg}^{ij} if the given reaction occurs directly at sites f and g, or one of the particles i or j can enter into a reaction with a neighboring particle at site h. Inserting expression (4.27) into Eq. (4.3) for $m = 2$, we obtain three forms of the right-hand sides of the equations for two-body correlators depending on the species of the particles at sites f and g, namely, (1) A and C or B and D; (2) any pairs of particles A, B, C, and D instead of AC and BD; and (3) the pairs I, J, where I = A, B, C, or D, and J ≠ I:

$$\begin{aligned}
\dot{\theta}_{fg}^{AC} &= U_{fg}^{BD}(\alpha) - U_{fg}^{AC}(\alpha) - \sum_h [U_{hfg}^{\overline{CA}C}(\alpha) - U_{hfg}^{\overline{DB}C}(\alpha)] \\
&\quad - \sum_h [U_{fgh}^{A\overline{CA}}(\alpha) - U_{fgh}^{A\overline{DB}}(\alpha)] \\
\dot{\theta}_{fg}^{AB} &= \sum_h [U_{hfg}^{\overline{DB}B}(\alpha) - U_{hfg}^{\overline{CA}B}(\alpha)] + \sum_h [U_{fgh}^{A\overline{AC}}(\alpha) - U_{fgh}^{A\overline{BD}}(\alpha)] \\
\dot{\theta}_{fg}^{BJ} &= \sum_h [U_{hfg}^{\overline{CA}J}(\alpha) - U_{hfg}^{\overline{DB}J}(\alpha)]
\end{aligned} \tag{5.5}$$

where the subscript h to the left of f relates to its neighboring sites, and to the right of g relates to the latter's neighboring sites. A bar capping a pair signifies a pair of particles participating in an elementary reaction α. The third particle (not capped) does not participate in the reaction, it affects the magnitude of the activation barrier. The first addends of the equation for pair AC are determined by expressions (4.31) and (5.2), while the remaining addends are:

$$U_{hfg}^{\overline{ijk}}(\alpha) = K_{ij}(\alpha)e^{-\beta\varepsilon_{ij}} \left\langle \gamma_h^i \gamma_f^j \gamma_g^k \prod_{\eta} \sum_{l=1}^{s} \gamma_\eta^l e^{\beta\delta\varepsilon_{il}(\alpha)} \times \prod_{\xi} \sum_{l=1}^{s} \gamma_\xi^l e^{\beta\delta\varepsilon_{jl}(\alpha)} \right\rangle_\tau = U_{hf}^{ij}(\alpha)\psi_{fg}^{jk}(\alpha) \tag{5.6}$$

where η numbers the sites neighboring with site h, but $\eta \neq f$, and ξ numbers the sites neighboring with site f, but $\xi \neq h$. The dimension of the correlators (5.6) does not exceed that in formula (4.31). As above, the addends of Eqs. (5.5) are the rates of two-site reactions (4.31) and (5.2) modified as a result of the perturbation introduced by a neighboring particle k, while Eqs. (5.5) and (5.6) are closed.

The constructed equations are naturally generalized for multistep processes. Assume that a particle of A participates in Q_A steps of the one-site reactions $A \rightleftarrows B_\alpha$ and Q_{AC} steps of the two-site reactions $A + B_\alpha \rightleftarrows C_\alpha + D_\alpha$, while the process itself proceeds on an inhomogeneous lattice. The kinetic equation for a pair of particles AB_1, for example, at the pair of sites fg will be written as

$$\begin{aligned} \dot{\theta}_{fg}^{AB_1}(qp) = & \sum_{\alpha=1}^{Q_A} [U_{fg}^{\overline{B_\alpha} B_1}(q\{[p]\}|\alpha) - U_{fg}^{\overline{A}B_1}(q\{[p]\}\alpha)] \\ & + [U_{fg}^{C_1D_1}(qp\{[\lambda\sigma]\}|1) - U_{fg}^{AB_1}(qp\{[\lambda\sigma]\}|1)] \\ & + \sum_{\alpha=1}^{Q_{AC}} \sum_{h} [U_{hfg}^{\overline{C_\alpha D_\alpha} B_1}(\xi q\{[\lambda\sigma]\}|\alpha) - U_{hfg}^{\overline{B_\alpha A} B_1}(\xi q\{[\lambda\sigma]\}|\alpha) \\ & + \sum_{h} [U_{fgh}^{A\overline{D_1 C_1}}(p\xi\{[\lambda\sigma]\}|1) - U_{fgh}^{A\overline{B_1 A}}(p\xi\{[\lambda\sigma]\}|1) \end{aligned} \tag{5.7}$$

The first addend describes the participation of a particle of A in one-site reactions at the central site f of a type q $[p]$ cluster in the presence of a neighboring particle B_1, it being assumed that the B_1 particle does not participate in the other steps. The remaining addends describe two-site reactions. The second addend relates to the reactions of particles at the central sites f and g of a cluster $qp\ \{[\lambda\sigma]\}$ (5.2). The third one relates to the reaction of a particle of A with particles B_α at the central sites h and f of a cluster $\xi q\{[\lambda\sigma]\}$, where the sites h of type ξ are the nearest neighbors of site f, but $h \neq g$. The last addend relates to a reaction of a particle of B_1 with a particle of A at the central sites g and h of a cluster $p\xi\{[\lambda\sigma]\}$, but $h \neq f$. The type of cluster is set unambiguously by the numbers of the central sites by means of the quantities η_f^q, therefore the sets of indices $\{[\lambda\sigma]\}$ for different central sites differ. The addends of Eqs. (5.7) have the form

$$U_{fg}^{\overline{ij}}\ (q\{[p]\}|\alpha) = U_f^i(q\{[p]\}|\alpha)\psi_{fg}^{ij}\ (qp_g|\alpha)$$

$$U_{hfg}^{\overline{ij}k}\ (qp\{[\lambda\sigma]\}|\alpha) = U_{hf}^{ij}\ (qp\{[\lambda\sigma]\}|\alpha)\psi_{fg}^{jk}\ (p\xi_g|\alpha) \tag{5.8}$$

$$\psi_{fg}^{ij}\ (qp_g|\alpha) = t_{fg}^{ij}\ (qp)\ \exp\ \beta\delta\varepsilon_{qp}^{ij}\,(\alpha)/S_{fg}^i(qp|\alpha)$$

where p_g is the type of a site numbered g. It should be noted that the function describing the influence of a neighboring particle on the rate of a two-site reaction coincides with the similar function for a one-site reaction because in the model being considered only the direct correlations of the nearest interacting particles are taken into account. Equations (5.7) must be supplemented with the relevant equations for the local concentrations (4.53) and (4.57) in which the participation of the reactants in various steps is also considered.

In the following sections of this chapter, we shall give examples of constructing closed systems of equations for specific multistep processes with account taken on the slow mobility of the particles.

5.2. Local Redistribution of Particles

The models of adsorption processes at relatively high densities ($\theta \geqslant 0.05$) do not generally include the step of migration between neighboring centers. It is customarily assumed that the particles cover distances of the order of magnitude of several surface lattice constants quite rapidly, i.e. the assumption of their local equilibrium distribution is used. This assumption is justified if the rate of migration of the particles substantially exceeds the rate of their adsorption or desorption. If the rate of migration and the rate of adsorption or desorption are commensurable, explicit consideration of the migration step is needed when studying adsorption processes. At present, the mobility of particles is taken into account in problems involving their macrotransfer over large distances [49-52] or when calculating the rates of recombination of molecules at low densities of the particles in diffusion-controlled reactions [58].

At the same time, a number of experimental facts point to the dependence of the kinetics of adsorption processes on the mobility of particles at large coverages. These are, in particular, experiments involving the desorption of CO molecules adsorbed at various temperatures on the nonreorganizing face (111) of Pd [227], and of oxygen molecules when the same number of oxygen atoms on the surface of Pt was obtained as a result of different preparation procedures [228, 229]. In both systems, the shape of the TDS at $\theta_A(0)$ = const depended on the conditions of formation of the initial coverage. Here the average spacing of the particles changed from one to several lattice constants. In such a situation, macroflows of the particles are absent in the system, and their migration results in processes of local redistribution.

Let us study the influence of the local mobility of the particles in the least "profitable" case—the adsorption of undissociating molecules. To establish the role of particle mobility, it is essential that the resultant flow of the particles between any pairs of sites be other than zero. On a homogeneous surface in an ideal model, all the resultant flows equal zero and the kinetic equation of adsorption has the same form for any particle mobility [230]. When the interaction between the particles is considered, the rates of the elementary processes depend not only on the average concentration of the particles, but also on their local configurations characterized by the number of neighboring particles. The probabilities of local configurations were approximated above in terms of the average numbers of both occupied sites θ and occupied pairs of neighboring sites θ_{AA}. With rapid mobility of the particles, the entire spectrum of local configurations of the particles is realized around any site during the characteristic times of the adsorption or desorption processes. The probabilities of the configurations obey an equilibrium distribution (see Sec. 2.2), and the process is described only by Eq. (4.21) relative to θ.

With a limited mobility of the particles, the characteristic times of the adsorption-desorption and migration processes are commensurable, and the local configurations change in time at the expense of the indicated competing processes, which, according to Sec. 5.1, leads to the relevant equations for θ and θ_{AA}. The migration of particles in the forward and reverse directions over all the pairs of neighboring sites of a homogeneous surface leads to a zero resultant contribution to the kinetic equation for θ. But in the kinetic equation for θ_{AA} a nonzero resultant flow appears because of the predominating transition of the atoms to energetically more profitable configurations. For instance, for a pair of sites fg (see the top right-hand corner of Fig. 5.1) having three neighboring particles around site f and one particle around site g, the resultant flow will be directed from f to g if the particles repel one another and in the opposite direction if they attract one another.

The system of kinetic equations for the adsorption of nondissociating

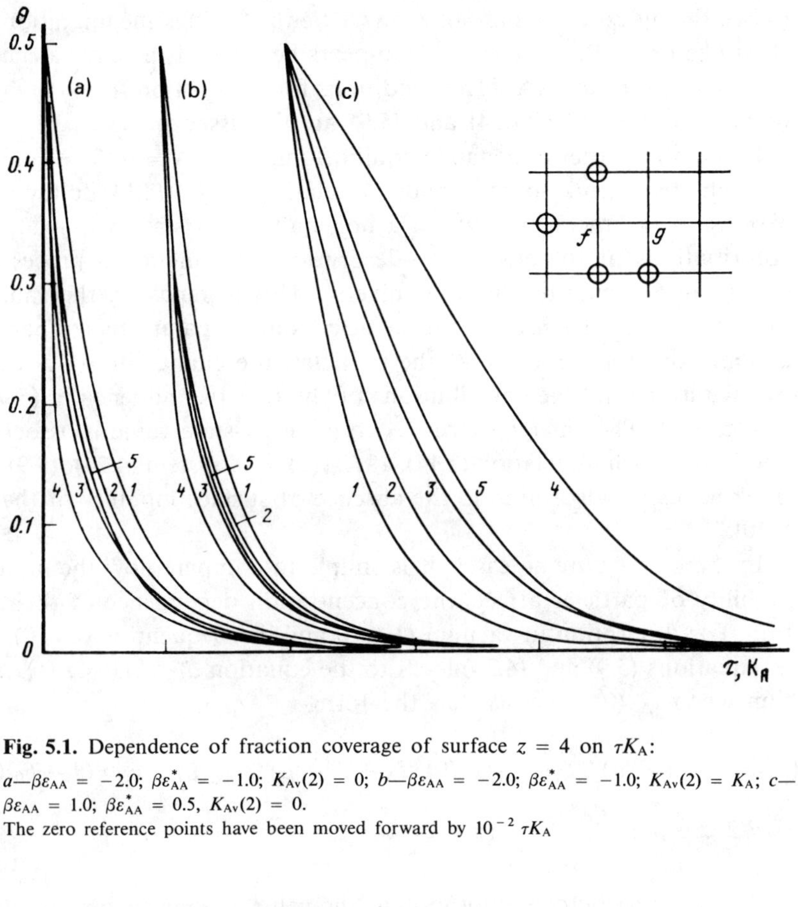

Fig. 5.1. Dependence of fraction coverage of surface $z = 4$ on τK_A:
a—$\beta\varepsilon_{AA} = -2.0$; $\beta\varepsilon^*_{AA} = -1.0$; $K_{Av}(2) = 0$; b—$\beta\varepsilon_{AA} = -2.0$; $\beta\varepsilon^*_{AA} = -1.0$; $K_{Av}(2) = K_A$; c—$\beta\varepsilon_{AA} = 1.0$; $\beta\varepsilon^*_{AA} = 0.5$, $K_{Av}(2) = 0$.
The zero reference points have been moved forward by $10^{-2}\ \tau K_A$

molecules with consideration of the restricted mobility of the particles in the quasichemical approximation consists of an equation for θ (4.24) and the following equation for θ_{AA} [231]:

$$\begin{aligned}
1/2\dot{\theta}_{AA} = {} & K_v(\theta - \theta_{AA})e^{\beta\varepsilon^*_{AA}}(1 + t_{vA}x_{vA})^{z-1} - K_A\theta_{AA}e^{\beta\delta\varepsilon_{AA}} \\
& \times (1 + t_{AA}x_{AA})^{z-1} - (z-1)K_{Av}(2)(\theta - \theta_{AA}) \\
& \times [(1 + t_{vA}x_{vA}(2))(1 + t_{AA}x_{AA}(2))]^{z-2} \\
& \times \{t_{AA}e^{\beta\delta\varepsilon_{AA}(2)}(1 + t_{vA}x_{vA}(2)) - t_{vA}e^{\beta\varepsilon^*_{AA}(2)}(1 + t_{AA}x_{AA}(2))\} \\
x_{ij}(\alpha) = {} & \exp \beta\delta\varepsilon_{ij}(\alpha) - 1
\end{aligned} \tag{5.9}$$

where the first two addends describe the change in θ_{AA} at the expense of adsorption and desorption [the step $\alpha = 1$; for the notation to coincide with

(4.24), the index $\alpha = 1$ has not been written]; $K_{Av}(2)$ is the migration rate constant (step $\alpha = 2$); the factor 1/2 appears because of the equivalence of both particles of the pair AA. [The conditions for a transition from the expressions of Sec. 5.1 to system (4.24) and (5.9) are discussed in Sec. 5.3.]

In a state of thermodynamic equilibrium, $\dot{\theta} = \dot{\theta}_{AA} = 0$. The contributions of each step shrink to zero, and the relation $\theta_{AA}(\theta)$ (2.14) follows from the expression in braces in (5.9). In a nonequilibrium state, the relation of the contributions of the adsorption-desorption and migration processes to the change in θ_{AA} may be the most diverse. This depends on the values of the rate constants, the density of the particles, and the parameters of particle interaction. For rapid mobility of the particles, the change in θ_{AA} is determined by migration, and the contributions of the first two addends in (5.9) may be disregarded. The emergence of θ_{AA} to a steady-state value is associated with $\theta_{AA}(\theta)$ approaching relation (2.14). If $K_{Av}(2) = 0$, system (4.24), (5.9) describes the kinetics of adsorption in the absence of surface mobility of the particles [230].

In these two extreme cases, it is simple to compare how the nature of the mobility of particles affects the concentration dependence of their distribution. For desorption in vacuum ($P = 0$ and, consequently, $K_v = 0$), the ratio of equations (5.9) and (4.24) leads to the equation $dt_{AA}/d\theta = F(t_{AA})$; its solution when $t_{AA}(0) = t_{AA}(\theta_0)$ has the form

$$(t_{AA}x_{AA} + 1 - \alpha)^{\alpha}/t_{AA} - (t_{AA}(0)x_{AA} + 1 - \alpha)^{\alpha}/t_{AA}(0) = \exp\,(\theta - \theta_0)(1 - \alpha),$$

$$\alpha = 2\exp\beta\delta\varepsilon_{AA}, \quad \theta_0 = \theta_A(\tau = 0) \qquad (5.10)$$

When the particles are not mobile, the value of the function t_{AA} monotonically decreases with their decreasing concentration. But unlike (2.14), the relation $t_{AA}(\theta)$ is determined by the value of $\delta\varepsilon_{AA}$ and also by the initial distribution of the particles $t_{AA}(0)$ and θ_0. From a similar ratio of equations (5.9) and (4.24) for $K_{Av}(2) = K_A = 0$, it follows that $t_{AA}(\theta)$ depends on ε^*_{AA}, $t_{AA}(0)$, and θ_0. In other words, the quantity t_{AA} is determined by the prehistory of particle distribution and the magnitudes of the changes in the activation barriers at the expense of the local surroundings, and not only by the concentration of the particles and their interaction in the ground state as for rapid mobility. (Other examples of the influence rendered by the absence of mobility of the particles have been given in [230].) It follows that in systems with a limited particle mobility, the initial distribution of the particles is of major significance.

Figure 5.1 presents isothermal curves of the desorption of particles for their following four initial distributions at $\theta_0 \leqslant 0.5$: (1) $\theta_{AA}(0) = 0$ for isolated particles, (2) $\theta_{AA}(0) = \theta_0/z$ for isolated pairs of particles, (3) $\theta_{AA}(0) = \theta_0^2$ for chaotic distribution, and (4) $\theta_{AA}(0) = \theta_0$ for macrodomains (Fig. 5.2). Curves *5* in Fig. 5.1 correspond to the equilibrium distribution of the particles. For chemisorption (*a*, *b*), curves *4* change the most in the initial period of time because of repulsion in the macrodomains. Curves *1* change the slowest—isolated particles do not affect one another. The differences in the initial distributions result in the same value of θ being reached during different intervals of time. In the initial periods, these intervals may vary up to three-five times. In the region of low coverages, each particle behaves independently, and all the curves coincide. With an increase in the mobility of the particles (see Fig. 5.1*b*), the influence of the initial distributions diminishes—curves *1* and *4* approach curve *5*. In physical adsorption, the attraction of the particles retains them on the surface, hence curve *1* changes the most rapidly and curve *4* the slowest in time.

The influence of the initial particle distribution tells still more in the TDS (Fig. 5.2). When the fraction of neighboring pairs of particles is larger, desorp-

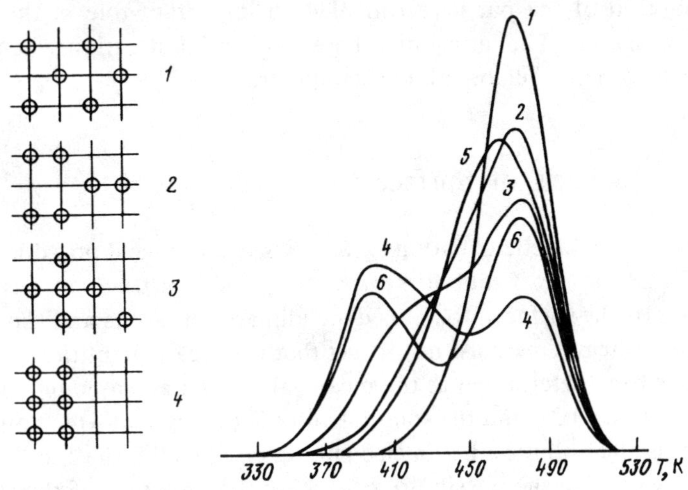

Fig. 5.2. Dependence of the thermodesorption spectra on the initial distribution of the particles:

$\hat{K}_A = 10^{12}\ s^{-1}$, $E_A = 63$, $\varepsilon_{AA} = -8.4$, $\varepsilon^*_{AA} = -4.2$ kJ/mol, $\hat{K}_{Av}(2) = \hat{K}_A$, $\theta_A(0) = 0.5$, $b = 50$ K/s. For curve *6*, $\theta_{AA}(0) = \theta_A(0)$, $K_{Av}(2) = K_A$, $E_{Av}(2) = 54.6$ kJ/mol. The numbers of curves *1*—*4* coincide with the numbers of the initial distributions

tion begins at a lower temperature. Splitting of the TDS at $\theta_{AA}(0) = \theta_0$ for any values of θ_0 is a new effect. Recall that with rapid mobility splitting of the TDS occurs at $\theta_0 > 0.6$-0.7, while a low-temperature shoulder appears in the region $0.5 < \theta < 0.6$-0.7 [145]. With restricted mobility of the particles, a low-temperature shoulder is present at any values of θ_0 for chaotic particle distribution. Curves *4* and *6* show that the shape of the TDS is sensitive to small changes in the flows of particles over the surface and in the exchange with the gas phase. Moreover, an increase in the speed of migration of 20 times does not yet allow us to speak of rapid migration. Otherwise, all the laws governing the influence of the model parameters on the shape of the TDS considered by the author in a different publication [145] remain in force. These features of the behavior of the TDS enable one to explain the TDS in systems qualitatively [227-229].

The local redistribution of particles in the absence of interaction on inhomogeneous surfaces have been considered by the author elsewhere [231]. For these surfaces, the nonzero resultant flows of particle redistribution are associated with the conditions of the equilibrium occupancy of sites of various types being disturbed. As previously, this appreciably changes the curves of the TDS. The above examples show that in a number of systems the local redistribution of the particles can play an important role in the kinetics of surface processes. The latter must be given special attention when describing the kinetics in the region of low temperatures.

5.3. Models of Surface Processes

The kinetic equations given in Sec. 5.1 make it possible to consider the different nature of mobility of particles and construct various systems of equations of the surface processes depending on the space and time scale used to consider them. One must not forget that the mean quantities in these equations have been determined as the mean values over an ensemble of equivalent lattices (see Sec. 1.3), and the smallest volume of a system which one can operate with is the volume of a lattice site. Consequently, the kinetic equations of Sec. 5.1 provide the possibility of tracing the evolution of the states of occupancy for each site of a system in detail.

We shall consider the catalytic process of oxidation of CO molecules on a homogeneous face (100) of Pt as an example of constructing systems of kinetic equations. The process consists of three steps, namely, adsorption-desorption of CO and O_2 molecules and a reaction between the adsorbed CO and O. We shall use the notation of Sec. 4.4 for the species of the particles and numbers of the steps. The steps of migration of the adsorbed particles will be written as $Z_fCO + vZ_g \rightleftarrows vZ_f + Z_gCO$ ($\alpha = 4$) and $Z_fO + vZ_g \rightleftarrows vZ_f + OZ_g$ ($\alpha = 5$), where g is a neighboring site of site f.

The most detailed description of a process is given by a distributed model. The kinetic equations for local concentrations of particles, as follows from (4.7) and (4.29), have the form

$$\begin{aligned} \dot{\theta}_f^{A} &= U_f^{v}(1) - U_f^{A}(1) - \sum_g [U_{fg}^{AB}(3) + U_{fg}^{Av}(4) - U_{fg}^{vA}(4)] \\ \dot{\theta}_f^{B} &= \sum_g [U_{fg}^{vv}(2) - U_{fg}^{BB}(2) - U_{fg}^{BA}(3) - U_{fg}^{Bv}(5) + U_{fg}^{vB}(5)] \end{aligned} \tag{5.11}$$

The local concentrations satisfy normalization condition (1.11). The addends on the right-hand side are the rates of elementary processes whose form is determined by formulas (4.13), (4.31), and (5.2) [For a homogeneous surface, the subscripts indicating the type of the sites should be discarded in formulas (5.2).] These rates will be written as follows:

for the adsorption of Co and O_2 molecules

$$\begin{aligned} U_f^{v}(1) &= K_v(1)\theta_f^{v} \prod_g S_{fg}^{v}(1) \\ U_{fg}^{vv}(2) &= K_{vv}(2)\theta_{fg}^{vv} \prod_h S_{fh}^{v}(2) \prod_h S_{gh}^{v}(2) \end{aligned} \tag{5.12}$$

for the desorption of CO and O_2 molecules

$$\begin{aligned} U_f^{A}(1) &= K_A(1)\theta_f^{A} \prod_g S_{fg}^{A}(1) \\ U_{fg}^{BB}(2) &= K_{BB}(2)e^{-\beta\varepsilon_{BB}}\theta_{fg}^{BB} \prod_h S_{fh}^{B}(2) \prod_h S_{gh}^{B}(2) \end{aligned} \tag{5.13}$$

for the reaction of formation of CO_2 molecules

$$U_{fg}^{AB}(3) = K_{AB}(3)e^{-\beta\varepsilon_{AB}}\theta_{fg}^{AB} \prod_h S_{fh}^{A}(3) \prod_h S_{gh}^{B}(3) \tag{5.14}$$

for the migration of the adsorbed particles (i = A when $\alpha = 4$, and i = B when $\alpha = 5$)

$$U^{i\mathrm{v}}_{fg}(\alpha) = K_{i\mathrm{v}}(\alpha)\theta^{i\mathrm{v}}_{fg} \prod_h S^{i}_{fh}(\alpha) \prod_h S^{\mathrm{v}}_{gh}(\alpha) \tag{5.15}$$

where the functions $S^{i}_{fh}(\alpha)$ have been determined in (5.2) and do not take into account the zero values of the parameters of interactions with vacant sites i = v in the explicit form; here $\varepsilon^{*}_{\mathrm{v}j}(\alpha) = \varepsilon^{*}_{ij}(\alpha)$, where i = A for $\alpha = 1$ and 4, and i = B for $\alpha = 2$ and 5; the constants $K_{\mathrm{v}}(1)$ and $K_{\mathrm{vv}}(2)$ include the partial pressures P(CO) and P(O_2) in (1.44) and (1.52). Sites h are the nearest neighbors of a site f, when $h \neq g$, or of g, when $h \neq f$.

We shall supplement kinetic equations (5.11) with equations for the functions θ^{ij}_{fg} . Having in view that the two-body correlators obey normalizing conditions (1.13), we shall select the pairs ij = AA, BB, AB, and BA as the linearly independent variables.

The equations for the pairs AA are:

$$\begin{aligned}\dot{\theta}^{\mathrm{AA}}_{fg} = {} & U^{\bar{\mathrm{v}}\mathrm{A}}_{fg}(1) - U^{\bar{\mathrm{A}}\mathrm{A}}_{fg}(1) + U^{\mathrm{A}\bar{\mathrm{v}}}_{fg}(1) - U^{\mathrm{A}\bar{\mathrm{A}}}_{fg}(1) - \sum_h U^{\overline{\mathrm{BA}}\mathrm{A}}_{hfg}(3) \\ & - \sum_h U^{\mathrm{A}\overline{\mathrm{AB}}}_{fgh}(3) - D^{\bar{\mathrm{A}}\mathrm{A}}_{fg}(4) - D^{\mathrm{A}\bar{\mathrm{A}}}_{fg}(4)\end{aligned} \tag{5.16}$$

where the first four addends describe the processes of adsorption-desorption of CO molecules at sites f and g, the next two addends describe the consumption of the CO molecules in the oxidation reaction, and the last two addends describe the migration of the CO molecules. The bar indicates the species of a migrating particle that is at site f or g, respectively

$$D^{\bar{i}j}_{fg}(\alpha) = \sum_h [U^{\bar{\mathrm{v}}ij}_{hfg}(\alpha) - U^{\bar{i}\mathrm{v}j}_{hfg}(\alpha)] \tag{5.17}$$

In our example, the particle species i = v corresponds to a vacancy.
The equations for the pairs BB are:

$$\begin{aligned}\dot{\theta}^{\mathrm{BB}}_{fg} = {} & U^{\mathrm{vv}}_{fg}(2) - U^{\mathrm{BB}}_{fg}(2) + \sum_h [U^{\overline{\mathrm{vv}}\mathrm{B}}_{hfg}(2) - U^{\overline{\mathrm{BB}}\mathrm{B}}_{hfg}(2)] \\ & + \sum_h [U^{\mathrm{B}\overline{\mathrm{vv}}}_{fgh} - U^{\mathrm{B}\overline{\mathrm{BB}}}_{fgh}(2)] - \sum_h U^{\overline{\mathrm{AB}}\mathrm{B}}_{hfg}(3) - \sum_h U^{\mathrm{B}\overline{\mathrm{BA}}}_{fgh}(3) \\ & - D^{\bar{\mathrm{B}}\mathrm{B}}_{fg}(5) - D^{\mathrm{B}\bar{\mathrm{B}}}_{fg}(5)\end{aligned} \tag{5.18}$$

The addends with the index $\alpha = 2$ describe the adsorption-desorption of O_2 molecules at the central pair of sites fg and at one central site f or g with their neighbors h ($h \neq g$ or $h \neq f$), the addends with $\alpha = 3$ describe the consumption of oxygen atoms in the oxidation reaction, while the last two addends describe the migration of oxygen atoms.

The equations for the pairs AB are:

$$\dot{\theta}_{fg}^{AB} = U_{fg}^{\bar{v}B}(1) - U_{fg}^{\bar{A}B}(1) + \sum_{n} [U_{fgh}^{A\overline{vv}}(2) - U_{fgh}^{A\overline{BB}}(2)]$$

$$- U_{fg}^{AB}(3) - \sum_{h} U_{hfg}^{\overline{BA}B}(3) - \sum_{h} U_{fgh}^{A\overline{BA}}(3) - D_{fg}^{\bar{A}B}(4) - D_{fg}^{A\bar{B}}(5) \qquad (5.19)$$

Here the adsorption and desorption of the O_2 molecules occur at site g and its neighbor site h, while oxidation occurs both at the central sites f and g and at one of them and its neighbor site h. In the equations for the pair BA, the places of the indices A and B should be changed.

In formulas (5.16)-(5.19), the reacting particles are indicated by a bar over them. The contributions of the elementary processes are determined in (5.8). For visualization, we shall rewrite them as applied to a homogeneous lattice:

$$U_{fg}^{\bar{i}j}(\alpha) = U_{f}^{i}(\alpha)\psi_{fg}^{ij}(\alpha),$$

$$U_{hfg}^{\overline{ij}k}(\alpha) = U_{hf}^{ij}(\alpha)\psi_{fg}^{jk}(\alpha)$$

$$\psi_{fg}^{ij}(\alpha) = t_{fg}^{ij} \exp \beta\delta\varepsilon_{ij}(\alpha)/S_{fg}^{i}(\alpha), \qquad (5.20)$$

$$S_{fg}^{i}(\alpha) = \sum_{j=1}^{s} t_{fg}^{ij} \exp \beta\delta\varepsilon_{ij}(\alpha)$$

where the rates of one- and two-site reactions are expressed by formulas (5.12) and (5.15).

The constructed closed system of equations is solved with the preset initial conditions $\theta_f^i(\tau = 0)$ and $\theta_{fg}^{ij}(\tau = 0)$. Its dimension is large, and for practical realization it can be employed for a small number of sites of a two-dimensional lattice. (The use of algorithms for solving the "rigorous" systems of equations [232, 233], for example, even with economical utilization of the memory of a computer with compaction of rarefied matrices [234], will evidently not make it possible to consider lattices with a square side exceeding $L \approx 8$ in the nearest future.)

The question naturally arises on the possibility of reducing the dimension of the system of equations. The latter is associated with coarsening of the detailed description of particle distribution, and a hierarchy of mathematical models of a process may be obtained that differ in the accuracy of describing the distribution of the particles. We shall begin their discussion with a transition from two-dimensional models to one-dimensional ones.

We can go over to a one-dimensional model if we assume that a square lattice is represented in the form of a set of layers of sites k (k is the number of the layer) in which the distribution of the particles i is characterized by mean values of the concentrations of the particles θ_k^i and the fractions θ_{kq}^{ij} ($q = k,\ k + 1$) of pairs of neighboring particles ij. These functions are obtained from the distribution functions for a two-dimensional model by averaging over the sites of each layer. Since $f = (k, m)$, where m is the number of a site in layer k, we have

$$\theta_k^i = \sum_{m=1}^{L} \theta_{km}^i / L, \quad \theta_{kq}^{ij} = \sum_{m=1}^{L} \theta_{km,qn}^{ij} / L \tag{5.21}$$

Here $n = m + 1$ if $q = k$, and $n = m$ if $q = k + 1$.

Each coarsening procedure is related to the relevant averaging; here loss of part of the information is possible. For instance, in a formal transition to one-dimensional models by averaging (5.21), the possibility is excluded of describing the ordered distribution of the particles inside a layer. To consider ordering of the particles, one must reflect the difference in occupancies of sites relating to different sublattices. For unambiguity, we shall consider the superstructure $c(2 \times 2)$ [see Sec. 3.5 and Eqs. (4.65)], in which each site of sublattice α is surrounded by sites of sublattice γ, and vice versa. In this case, the transition to a one-dimensional model is achieved by the following averaging:

$$\theta_{k\alpha}^i = 2 \sum_{m=0}^{L/2} \theta_{k,2m}^i / L, \quad \theta_{k\gamma}^i = 2 \sum_{m=0}^{L/2} \theta_{k,2m+1}^i / L \tag{5.22}$$

where $\theta_{k\sigma}^i$ is the local concentration of particles i at sites of types σ ($\sigma = \alpha$, $\tilde{\gamma}$) of layer k. The two-body distribution functions $\theta_{k\sigma,q\lambda}^{ij}$, where $q = k,\ k \pm 1$, and $\lambda \neq \sigma$, are determined similarly. Their determinations, like those of the local concentrations of the particles, depend on the type of superstructure.

The next step in coarsening the description of particle distribution is the transition from distributed models to point ones in which averaging is performed over all the lattice sites. The concentrations of the particles and the fraction of neighboring pairs of particles are characterized by macroscopic means determined as follows:

$$\theta_\sigma^i = \sum_{k=1}^{L} \theta_{k\sigma}^i/L, \quad \theta_{\sigma\lambda}^{ij} = \sum_{k=1}^{L} \sum_{q} \theta_{k\sigma,q\lambda}^{ij}/(z_{\sigma\lambda}L) \tag{5.23}$$

where $z_{\sigma\lambda}$ is the number of sites of type λ that are neighbors of sites of type σ. Distributions (5.23) take into account the ordered arrangement of the particles in accordance with the superstructure $c(2 \times 2)$. For a disordered system, expressions (5.23) will be rewritten in the form

$$\theta_i = \sum_{k=1}^{L} \theta_k^i/L, \quad \theta_{ij} = \sum_{k=1}^{L} \sum_{q} \theta_{kq}^{ij}/(zL) \tag{5.24}$$

Unlike distributed models, the point models do not reflect the processes of transport of a substance, but consider only its local redistribution. A very simple example of such a model has been treated in Sec. 5.2.

A typical time scale corresponds to each typical dimension of a region L for which a definite coarsening of the description of particle distribution (space scale) is used so that the description being considered would correspond to a real process. Accordingly, various expressions are obtained for the right-hand sides of the kinetic equations (their specific expressions are considered in Appendix 3). But the form of the kinetic equations is also affected by the nature of mobility of the reactants. If the mobility of the reactants differs sharply, then when we are considering the evolution of the mobile particles during the times $\Delta\tau$, the "slow" particles remain virtually stationary. If we are considering the evolution of the slow particles during the times $\Delta\tau$, the distribution of the mobile particles virtually coincides with the equilibrium one. In the latter case, the dimension of the system of kinetic equations lowers. This can occur both for distributed (one- and two-dimensional) and point models (see Appendix 3).

All the models relating to different hierarchial levels operate with a single set of parameters of interactions and rate constants of reactions. Without changing the values of the parameters of a model, we can employ a more general model to obtain the same results that are obtained in particular models, i.e. models constructed on the basis of the system of equations of Secs. 4.2, 4.3, and 5.1 consecutively "fit" into one another. This enables us, by comparing the results of calculations of the models relating to adjacent heirarchial levels, to find the optimal conditions of solving a problem so that, on the one hand, the dimension of the system of differential equations would be minimal, and on the other, to take the basic properties of a real experiment into consideration.

In concluding this section, we shall write out the closed equations of a point model of the oxidation of CO with a view to the limited mobility of the reactants [determination of the mean values by (5.24)]. The kinetic equations for the concentrations of the adsorbed particles coincide with equations (4.45) in which the addends have the form of (4.46). But for evaluating the functions $S_i(\alpha)$ the solution of the system of kinetic equations relative to the pair correlators θ_{ij}, where ij = AA, BB, and AB (owing to the symmetry $\theta_{ij} = \theta_{ji}$, an equation is not required for the pair BA).

$$\begin{aligned}
1/2\dot{\theta}_{AA} &= U^{\bar{v}A}(1) - U^{\bar{A}A}(1) - (z-1)[U^{\overline{BA}A}(3) \\
&\quad + U^{\bar{v}AA}(4) - U^{\overline{Av}A}(4)] \\
\dot{\theta}_{BB} &= U^{vv}(2) - U^{BB}(2) + 2(z-1)[U^{\overline{vv}B}(2) - U^{\overline{BB}B}(2) \\
&\quad - U^{\overline{AB}B}(3) - U^{\bar{v}BB}(5) + U^{\bar{B}vB}(5)] \\
\dot{\theta}_{AB} &= U^{\bar{v}B}(1) - U^{\bar{A}B}(1) - U^{AB}(3) + (z-1)[U^{A\overline{vv}}(2) \\
&\quad - U^{A\overline{BB}}(2) - U^{\overline{BA}B}(3) - U^{A\overline{BA}}(3) - U^{\bar{v}AB}(4) \\
&\quad + U^{\bar{A}vB}(4) - U^{A\bar{B}v}(5) + U^{Av\bar{B}}(5)]
\end{aligned} \qquad (5.25)$$

The addends of system (5.25) are expressed in terms of correlators (5.24), therefore the indices of the sites have been omitted in them. The addends have the following form, where the rates of the elementary processes have been determined in (4.46):

$$\begin{aligned}
&U^{\bar{i}j}(\alpha) = U_i(\alpha)\psi^{ij}(\alpha), \quad U^{\bar{i}jk}(\alpha) = U^{ij}(\alpha)\psi^{jk}(\alpha) \\
&U^{vi}(\alpha) = K_{vi}(\alpha)\theta_{iv}[S_v(\alpha)S_i(\alpha)]^{z-1}, \ \psi^{ij}(\alpha) = t_{ij}e^{\beta\delta\varepsilon_{ij}(\alpha)}/S_i(\alpha)
\end{aligned} \qquad (5.26)$$

where $S_i(\alpha)$ has been determined in (4.18); in the expressions for $U^{vi}(\alpha)$, i = A for $\alpha = 4$ and i = B for $\alpha = 5$. Kinetic equations (4.45), (4.46), (5.25), and (5.26) are solved with the preset initial values of θ_i ($\tau = 0$) and θ_{ij} ($\tau = 0$), where i, j = A, B.

The influence of the reactant mobility on the kinetic characteristics of the desorption of CO molecules and the reaction of formation of CO_2 molecules on the face (111)Pt is shown in Fig. 5.3. In the calculations, the parameters of a model with rapid surface mobility of the reactants were employed (Sec. 4.4). Diminishing of the reactant mobility shifts the process of CO oxidation to the region of higher temperatures and increases the fraction of desorbed CO molecules in comparison with their participation in the oxida-

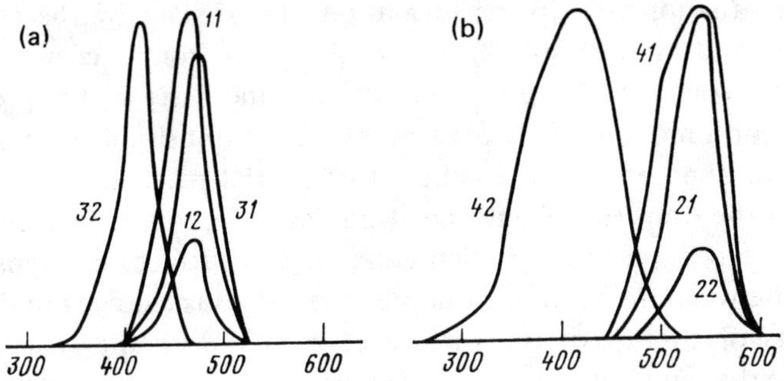

Fig. 5.3. Rates of desorption of CO (11, 21, 31, 41) and reactions of formation of CO_2 (12, 22, 32, 42).

The first digit in the curve number stands for the number of the variant. The following rate constants and parameters of particle interaction have been used [216] (all the energies are in kJ/mol):
a—$E_{AB}(3) = 98.5$, $K_{Av}(4) = 2 \times 10^{13}$, $K_{Bv}(5) = 5 \times 10^{13}\ s^{-1}$, $\theta_A(0) = 0.37$, $\theta_B(0) = 0.22$, $\theta_{AA}(0) = 0.165$, $\theta_{AB}(0) = 0.01$, $\theta_{BB}(0) = 0.064$. Variant *1*—$E_{Av}(4) = 119$, $E_{Bv}(5) = 210$; variant *3*—$E_{Av}(4) = 98.5$, $E_{Bv}(5) = 173$;
b—$E_{AB}(3) = 90$, $K_{Av}(4) = 10^{12}$, $K_{Bv}(5) = 10^{12}\ s^{-1}$, $\theta_A(0) = 0.37$, $\theta_B(0) = 0.18$, $\theta_{AA}(0) = 0.12$, $\theta_{AB}(0) = 0$, $\theta_{BB}(0) = 0.006$; variant *4*—$E_{Av}(4) = 107$, $E_{Bv}(5) = 188$; variant *4*—$E_{Av}(4) = 61$, $E_{Bv}(5) = 142$ (curves *a* and *b* are not normalized with each other)

tion reaction. Curves *3* and *4* are consistent qualitatively with data of Gland and Kollin [223]. As for the kinetics of desorption, the initial arrangement of the reactants is of major significance. With a regular alternation of the CO and O particles at the initial instant, the rate of the reaction shifts greatly to the region of low temperatures. With an increase in the mobility of the reactants, the model with limited mobility gives the same solution as the model of Sec. 4.4 with rapid mobility of the reactants. An exception here is the small initial interval during which the functions θ_{ij} "go over" to their equilibrium values (2.45), while the right-hand sides of Eqs. (5.25) shrink to zero.

5.4. Evolution of the Concentration Profile in the Subsurface Region of a Solid

The results of Secs. 3.8 and 3.9 indicate the possibility of a major difference in the distribution of the components of a solid in the subsurface region in comparison with their distribution in the bulk. The kinetic equations of Sec. 5.1 enable us to describe the evolution of these distributions in the following situations: (1) in the formation of a surface by the cleavage off a part of a solid in vacuum; (2) in the transition from a clean surface to an equilibrium coverage of the surface whose value is set by the pressure of the adsorbate or a mixture of adsorbates; (3) in the transition from one equilibrium state to another one when the set of partial pressure values is changed; and (4) in the transition from one steady state of the surface concentrations of the adsorbed particles participating in the surface reactions to another steady state when the set of partial pressure values is changed and under other conditions of contact of the surface of a solid with the gas phase.

Assume that an s-component solid solution is one of the substitutional type; the considered ranges of the change in the concentrations of the alloy components correspond to the region of homogeneity (Secs. 3.8 and 3.9). Let us construct the kinetic equations of a process describing the change in the state of the concentration profile of the subsurface region of the alloy and the adsorbed particles. These equations as a result of interaction between the particles and their mobility are mutually related to one another and are a single system of equations [210].

Let us consider a process occurring on the surface of an alloy. The three-step mechanism of the process coincides with that of the reaction of oxidation of CO (Sec. 4.4). In addition to the three indicated steps, $\alpha = 4$ corresponds to the step of migration of the particles to the neighboring vacant sites: $jZ_k + vZ_q \rightleftarrows vZ_k + jZ_q$, where vZ_k is a vacant point of layer k, and jZ_k is a particle j at a site of layer k (a vacancy mechanism). For simplicity, we shall consider in the present section that the atoms of the solid do not emerge beyond a flat face on the surface (they are in the layers $1 \leqslant k \leqslant \varkappa$) and that the adsorbed molecules (the layer $k = 0$) do not dissolve in the solid (a rough surface of an alloy has been considered in Sec. 5.5, and the dissolution of adsorbed particles in the interstices of a solid will be considered in Sec. 5.6).

A description of the distribution of the particles in the subsurface region by layers (see Fig. 3.19) corresponds to a one-dimensional sequence of sites of various types (the type of a site is characterized by the number of the layer). This is why all the functions of the conditional distributions for clusters with one and two central sites equal unity, while the correlators θ_k^i and θ_{kq}^{ij} are determined by expressions (5.21), in which L is the number of sites in a two-dimensional layer (in Sec. 5.3, the system is two-dimensional and a layer is one-dimensional).

Alloy-Vacuum Interface. The state of the subsurface region of an alloy changes with a change in the temperature or in a transition from a freshly cleaved surface characterized by the bulk distributions of the particles to an equilibrium state. As a result of the migration of the components of the alloy, its concentration profile changes in accordance with the following system of equations:

$$
\begin{aligned}
\dot{\theta}_k^i &= \sum_{q=k\pm 1} z_{kq}[U_{kq}^{vi}(4) - U_{kq}^{iv}(4)], \quad i \neq v \\
\dot{\theta}_{kq}^{ij} &= -D_{kq}^{\bar{i}j} - D_{kq}^{i\bar{j}} \\
D_{kq}^{\bar{i}j} &= \sum_{h=k+1}^{k-1} (z_{kh} - \Delta_{qh})[U_{hkq}^{\overline{vi}j}(4) - U_{hkq}^{\overline{iv}j}(4)]
\end{aligned}
\tag{5.27}
$$

where the addends of the equations have the form

$$
\begin{aligned}
U_{kq}^{ij}(\alpha) &= \hat{K}_{ij}(\alpha) e^{-\beta\varepsilon_{ij}\theta_{kq}^{ij}} \prod_{p=k+1}^{k-1} [S_{kp}^i(\alpha)]^{z_{kp} - \Delta_{qp}} \\
&\quad \times \prod_{\xi=q+1}^{q-1} [S_{q\xi}^j(\alpha)]^{z_{q\xi} - \Delta_{k\xi}}, \\
U_{hkq}^{\overline{ij}n}(\alpha) &= U_{hk}^{ij}(\alpha)\psi_{kq}^{jn}(\alpha) \\
\psi_{kq}^{ij}(\alpha) &= t_{kq}^{ij} \exp \beta\delta\varepsilon_{kq}(\alpha)/S_{kq}(\alpha), \\
S_{kq}^i(\alpha) &= \sum_{j=1}^{s_q} t_{kq}^{ij} \exp \beta\delta\varepsilon_{kq}^{ij}(\alpha)
\end{aligned}
\tag{5.28}
$$

Formulas (5.27) and (5.28) differ from the similar expressions in Secs. 5.1 and 5.3 in that the numbers of the sites are replaced by those of the layers, which simultaneously characterize the type of the sites. As in formula (5.17), the neighboring sites h do not include the site in layer q containing a particle j (this is taken into account by the Kroneker symbol). The equation for the concentration of the particles contains no contributions from migration inside a layer, which are compensated. In vacuum for layer $k = 1$. It should be remembered that $z_{10} = 0$, and the subscripts q and h are not equal to zero. In formulas (5.28), as in formula (3.68), in the general case, the species of the particles and their number may depend on the type of the site. All the interactions of the particles with their surroundings are taken into account by the functions $S_{kq}^i(\alpha)$, hence formula (5.28) for the rate of a two-site reaction contains the preexponential factor of the rate constant. System (5.27) is solved with the preset initial conditions θ_k^i ($\tau = 0$) and θ_{kq}^{ij} ($\tau = 0$), where $1 \leqslant k \leqslant \varkappa$ and $q = k, k+1$.

Alloy-Reactive Gas Phase Interface. The molecules of the gas phase are adsorbed and fill the layer $k = 0$; their number depends on the partial pressures and on the rate of the reaction between them. The adsorbed particles affect the state of the subsurface region of the alloy through the layer $k = 1$, hence in the given model the kinetic equations (5.27) and (5.28) for the layers $k \geqslant 2$ remain unchanged. For the layer $k = 1$, the equation for $\dot{\theta}_1^i$ also contains no addends with z_{10} because it is forbidden for the atoms to emerge into the autoadsorbed state, but in the expressions for the addends of (5.28) the multipliers raised to the power z_{10} do not equal zero: the particles of the first layer interact with the adsorbed particles of the zero layer, and the latter affect the rate of migration of the alloy components.

Two cases must be distinguished for the functions $S_{kq}^i(4)$: (1) $q \geqslant 1$, when the sum over j in the expression for $S_{kq}^i(4)$ over s components of the alloy, and (2) $q = 0$, when the sum over j is taken over the components of the adsorption system: A, B, v. In exactly the same way in the equations for $\dot{\theta}_{lq}^{ij}$, $q = 1$, 2, addends are absent corresponding to the transition of the alloy atoms into the layer $k = 0$, but the multipliers $S_{10}^i(4)$ affect the rate of the elementary jumps.

The adsorption layer contains three species of particles, namely AZ, BZ, and vZ. The kinetic equations for their concentration are written as:

$$\dot{\theta}_0^{\mathrm{A}} = U_0^{\mathrm{v}}(1) - U_0^{\mathrm{A}}(1) - z_{00}U_{00}^{\mathrm{AB}}(3) \tag{5.29}$$

$$\dot{\theta}_0^{\mathrm{B}} = z_{00}[U_{00}^{\mathrm{vv}}(2) - U_{00}^{\mathrm{BB}}(2) - U_{00}^{\mathrm{AB}}(3)], \quad \sum_{i=\mathrm{A}}^{\mathrm{v}} \theta_0^i = 1$$

The surface concentrations change only at the expense of adsorption-desorption and a reaction. The diffusion addends are absent—for the zero layer this is a point model with consideration of the limited mobility of the particles that describes the effects of their local redistribution (5.24). System of equations (5.29) coincides in its appearance with system (4.45), but its addends are expressed as follows:

$$U_0^i(\alpha) = \hat{K}_i(\alpha)\theta_0^i \prod_{q=1}^{0} [S_{0q}^i(\alpha)]^{z_{0q}}$$

$$U_{00}^{ij}(\alpha) = \hat{K}_{ij}(\alpha)\theta_{00}^{ij} \exp(-\beta\varepsilon_{ij}) \prod_{q=1}^{0} [S_{0q}^i(\alpha)S_{0q}^j(\alpha)]^{z_{0q}} \tag{5.30}$$

The number z in formulas (5.46) corresponds to z_{00} in the employed notation. The number z_{01} reflects the number of bonds of an adsorbed particle with the atoms of the alloy. The quantities $\delta\varepsilon_{01}^{ij}(\alpha)$ contained in $S_{01}^{i}(\alpha)$ are the energies of the bond of an adsorbed particle i to the atoms j of the alloy in the activated state. On a homogeneous surface, the quantity $z_{01}\delta\varepsilon_{01}^{ij}(\alpha)$ is the activation energy of the reaction and is included in the reaction rate constant. This is why formula (5.30) includes the preexponential factors of the rate constants of one- and two-site reactions, which, if i = v, includes the partial pressures of the gas phase molecules (See. 1.5).

We shall supplement the kinetic equations for the concentrations of the particles in the layers $k = 0$ and 1 with equations for the pairs θ_{0q}^{ij}, where $q = 0$, 1. With a view to the normalizing relations, we shall select the functions for the pairs AA, AB, BB ($q = 0$) and Aj and Bj, where $1 \leqslant j \leqslant s$, but $j \neq$ v ($q = 1$) as the linearly independent functions.

$$1/2\dot{\theta}_{00}^{AA} = U_{00}^{\bar{v}A}(1) - U_{00}^{\bar{A}A}(1) - (z_{00} - 1)U_{000}^{\overline{BA}A}(3) - D_{00}^{\bar{A}A}$$

$$\dot{\theta}_{00}^{BB} = U_{00}^{vv}(2) - U_{00}^{BB}(2) + 2(z_{00} - 1)[U_{000}^{\overline{vv}B}(2) - U_{000}^{\overline{BB}B}(2) - U_{000}^{\overline{AB}B}(3)] - 2D_{00}^{\bar{B}B}. \qquad (5.31)$$

$$\dot{\theta}_{00}^{AB} = U_{00}^{\bar{v}B}(1) - U_{00}^{\bar{A}B}(1) - U_{00}^{AB}(3) - (z_{00} - 1)[U_{000}^{\overline{BA}B}(3) + U_{000}^{A\overline{BA}}(3) + U_{000}^{A\overline{BB}}(2) - U_{000}^{A\overline{vv}}(2)] - D_{00}^{\bar{A}B} - D_{00}^{A\bar{B}}$$

where the subscript h in formula (5.27) for $D_{00}^{\bar{i}j}$ (layer $k = 0$) takes on the value $h = 0$. The addends of two-site reactions have been determined in (5.28) and (5.30), while $U_{00}^{\bar{i}j}(1) = U_{0}^{i}(1) \times \psi_{00}^{ij}(1)$. Equations (5.31) coincide in their appearance with Eqs. (5.25), but their addends differ in exactly the same way as formulas (5.30) differ from (4.46).

For the other pairs, the kinetic equations have the form

$$\left.\begin{aligned} \dot{\theta}_{01}^{Ai} &= U_{01}^{\bar{v}i}(1) - U_{01}^{\bar{A}i}(1) - z_{00}U_{001}^{\overline{BA}i}(3) - D_{01}^{\bar{A}i} - D_{01}^{A\bar{i}} \\ \dot{\theta}_{01}^{Bi} &= z_{00}[U_{001}^{\overline{vv}i}(2) - U_{001}^{\overline{BB}i}(2) - U_{001}^{\overline{AB}i}(3)] - D_{01}^{\bar{B}i} - D_{01}^{\bar{B}i} \end{aligned}\right\} \qquad (5.32)$$

where for $D_{01}^{\bar{n}i}$, the subscript h in formula (5.27) is zero, while for $D_{01}^{n\bar{i}}$ it is 1 and 2.

The constructed equations are a closed system describing the process of the change in the subsurface region of a solid when it is in contact with the gas phase. It is solved with the preset initial conditions θ_k^i $(\tau = 0)$ and θ_{kq}^{ij} $(\tau = 0)$. The boundary conditions are set by the state of the gas phase and that of the solid in the volume of $k = \varkappa + 1$ (constancy of the values of the bulk concentrations of the solid's atoms or equality of the flows of its atoms between the layers $\varkappa$ and $\varkappa + 1$).

The constructed kinetic curves describe the atomic and molecular processes in the subsurface region of an alloy in detail and make it possible to determine the relations between the characterisitic times (the relaxation times) of the processes on the surface and in the subsurface region, and also to determine which relations between the parameters correspond (1) to the practical absence of changes in the distributions in the subsurface region notwithstanding the processes in the surface layer, or (2) to the quasistationary proceeding of the surface processes in the course of evolution of the subsurface region of an alloy. In other words, an analysis of these equations enables us to determine the region of applicability of the indicated assumptions actively employed in the theory of surface processes.

5.5. Rough Gas-Solid Interface

In many real situations, the surface layer of a solid is not ideally flat. From the viewpoint of a multilayer subsurface region (see Secs. 3.8 and 3.9), the surface atoms are in different layers thereof (see Fig. 3.19). Such a surface is said to be rough [196]. It is chiefly formed as a result of interaction of the solid with a reactive gas phase or in the course of growth (sublimation) of a crystal from the gas phase. A definite role may possibly also be played by local overheating because of the proceeding of surface reactions, although the contributions of the thermal fluctuations to the creation of a rough surface are apparently small.

The evolution of the states of a rough surface makes it possible to consider both the change in the concentration profile of the subsurface region and the motion of the phase interface. At present, the processes of motion of a phase interface (the crystallization or sublimation of solids, catalytic processes which complications are introduced into by poisoning of the catalysts, topochemical reactions, etc. [22, 42-44, 175]) are described at a macroscopic level [22, 43, 44] or on the basis of models [196-199, 208] excluding the presence of vacancies in the solid. This does not enable one to describe the formation of defective crystals and the migration of particles into them. These restrictions are re-

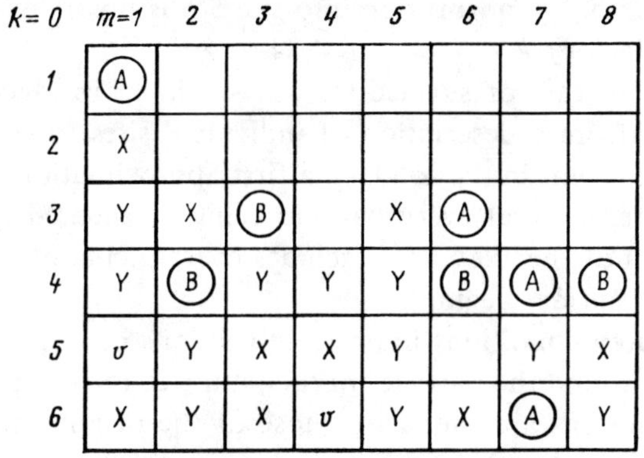

Fig. 5.4. Schematic views of a rough surface of the alloy $X_{\delta}Y_{1-\delta}$ contacting with the gas phase:
A, B = adsorbed particles, v = vacancy in the bulk of the alloy, $\varkappa = 6$; $k - m$ = number of a site; k = layer number; and m = column number

moved when employing the equations of Sec. 5.1 for the model of Sec. 3.8 [209].

Let s_a and s_g stand for the numbers of the alloy components and adsorbed gas particles, respectively. Each site of layer k has s states of occupancy $s = s_a + s_g + 1$, the number 1 signifying vacancies. The layer $k = 0$ corresponds to the bulk of the gas phase, and the layer $k = \varkappa + 1$ to the bulk of the alloy. One layer may contain atoms of the alloy and adsorbed particles; it is assumed here that dissolution proceeds into the alloy lattice sites, and not into the interstices (see the following section 5.6). Surface reactions proceed only on the surface atoms of the alloy, and we shall therefore introduce the following feature for them. A given atom of layer k is a surface one if there is a "phase" of vacancies above it, i.e. if there are vacancies in layers $k - 1$, $k - 2$, and so on up to $k = 1$. This definition signifies the absence of hanging of particles over an atom in the layer k if we consider the sites above it in one column (Fig. 5.4). For example, particles are forbidden to occupy site 5-6 when particles 4-6 and 4-7 are absent because particle 3-6 is over this site. Particles may get into this vacancy as a result of migration from sites 5-5, 6-6, 5-7, 4-6, or from the gas phase after a reaction between particles of A (3-6) and B (4-6). If we consider the higher sites in different columns, the overhanging effects will be taken into account. For instance, after the reaction of particles 4-6 and 4-7, adsorption into site 5-6 is possible if a particle travels along the trajectory $k - 7$ (k = 1-4), 4-6, 5-6.

In addition to the feature of surface atoms, we shall introduce a second condition to go over from a description of individual sites to consideration of the bonding of the atoms in the solid. In a first approximation, this condition can be formulated so that every particle being considered has at least one neighboring atom of the alloy (this excludes overhanging of the particles in vacuum in the subsurface region).

Each of the indicated conditions imposes restrictions on the states of occupancy of the sites around the particles participating in the elementary reactions. Their manifestations are similar to those of the neighboring particles in the addends of Eqs. (5.4) and (5.6). The second condition signifies that any atoms of an alloy can be neighbors, i.e. expressions of type (5.4) and (5.6) must contain a sum over their species. The first condition signifies that we are considering a sequence of neighboring sites from the given layer k to $k = 0$ that should be vacant (we shall limit ourselves below to sites in one column).

As in Sec. 5.4, let us assume that a three-step catalytic process is occurring ($s_g = 2$), and all the particles migrate by a vacancy mechanism. The state of occupancy of any site of the subsurface region change at the expense of the processes of migration of particles between layers and within the confines of each layer, and at the expense of adsorption-desorption processes and a reaction. The system of kinetic equations will be written as follows:

$$\begin{aligned} &\dot{\theta}_k^n = (\dot{\theta}_k^n)_m + (\dot{\theta}_k^n)_r, \quad \dot{\theta}_k^i = (\dot{\theta}_k^i)_m, \quad \dot{\theta}_{kq}^{ij} = (\dot{\theta}_{kq}^{ij})_m \\ &\dot{\theta}_{kq}^{nl} = (\dot{\theta}_{kq}^{nl})_m + (\dot{\theta}_{kq}^{nl})_r, \quad \dot{\theta}_{kq}^{ni} = (\dot{\theta}_{kq}^{ni})_m + (\dot{\theta}_{kq}^{ni})_r \end{aligned} \tag{5.33}$$

where n, l = A, B, $1 \leqslant i, j \leqslant s_a$, the subscript "m" signifies migrational contributions ($\alpha = 4$), and the subscript "r"—reaction contributions (α = 1-3). The migration of all the particles is described by equations of the type of (5.27) in which the quantities $U_{kq}^{vi}(4)$ and $U_{hkq}^{\overline{vij}}(4)$ have been replaced by the quantities $W_{kq}^{vi}(4)$ and $W_{hkq}^{\overline{vij}}(4)$. The difference between them is due to the second condition, because in Sec. 5.4 the coherence condition was satisfied by the initial formulation of the problem. Here this condition changes the expressions for the addends of equations of type (5.27):

$$W_{kq}^{iv}(4) = U_{kq}^{iv}(4) \times \begin{cases} \psi_{kk+1}^{i}(4)\psi_{kk+1}^{v}(4), & q = k \\ \psi_{kk}^{i}(4)\psi_{qq}^{v}(4), & q = k \pm 1 \end{cases} \tag{5.34}$$

$$W_{hkq}^{ivj}(4) = W_{hk}^{iv}(4)\psi_{kq}^{vj}(4)\Lambda_{hkq}^{i}(4), \quad \psi_{kq}^{i}(\alpha) = \sum_{j=1}^{s_a} \psi_{kq}^{ij}(\alpha)$$

$$\Lambda_{hkq}^{i}(4) = [\psi_{kq}^{i}(4)]^{-1}, \quad q = k + 1,\ h = k;\ q = k,$$

$$h = k \pm 1;\ q = k - 1,\ h = k - 1$$

In the remaining cases, $\Lambda_{hkq}^{i}(4) = 1$. As a result, the addends of (5.34) are determined by modified formulas (5.28). The functions $\psi_{kq}^{i}(\alpha)$ reflect the fact that alloy atoms must be present next to a reacting particle i. The adopted way of taking into account that a pair iv belongs to the alloy corresponds to the real inhomogeneity of particle distribution in the subsurface region, namely, the density of the vacancies diminishes with an increasing layer number. If $q = k$, the alloy atoms must be present in the layer $k + 1$, which is an analogue of migration on a flat surface; if $q = k + 1$, the alloy atoms must be present in the same layers k and q—the analogue of migration along a vertical wall.

The reaction contributions to the kinetic equations for the local concentrations of the adsorbed particles have the form

$$\begin{aligned} \dot{\theta}_k^{A} &= W_k^{v}(1) - W_k^{A}(1) - \sum_{q=k+1}^{k-1} z_{kq} W_{kq}^{AB}(3) \\ \dot{\theta}_k^{B} &= \sum_{q=k+1}^{k-1} z_{kq}[W_{kq}^{vv}(2) - W_{kq}^{BB}(2) - W_{kq}^{BA}(3)] \end{aligned} \tag{5.35}$$

where the two-site reaction addends of Eqs. (5.35) are expressed in terms of formulas (5.28), which determine the rate in the absence of additional conditions,

$$W_{kq}^{nl}(\alpha) = U_{kq}^{nl}(\alpha)\begin{cases} \psi_{kk+1}^{n}(\alpha)\psi_{kk+1}^{l}(\alpha)\Phi_{k(q)}^{n(l)}(\alpha), & q = k \\ \psi_{kk}^{n}(\alpha)\psi_{qq}^{l}(\alpha)\Phi_{k(q)}^{n(l)}(\alpha), & q = k \pm 1 \end{cases} \tag{5.36}$$

while the addends of the one-site reactions are expressed as

$$\begin{aligned} W_k^{n}(\alpha) &= \hat{K}_n(\alpha)\theta_k^{n} \prod_{q=k+1}^{k-1} [S_{kq}^{n}(\alpha)]^{z_{kq}} \psi_{kk+1}^{n}(\alpha)\Phi_k^{n}(\alpha) \\ \Phi_k^{n}(\alpha) &= \frac{\theta_{kk-1}^{nv}}{\theta_k^{n} S_{kk-1}^{n}(\alpha)} \prod_{l=1}^{k-1} (\theta_{k-l,k-l-1}^{vv}/\theta_{k-l}^{v}) \end{aligned} \tag{5.37}$$

where $n = \mathrm{v}$ corresponds to adsorption of particles of A, $n = \mathrm{A}$ corresponds to their desorption, $\hat{K}_n(\alpha)$ is the preexponential factor of the rate constant; the functions $\psi^n_{kk+1}(\alpha)$ take into account that a particle n has alloy atoms under it, while the function $\Phi^n_k(\alpha)$ takes into account that there is a vacancy phase over a particle n in the layer k; the index n or l in the function $\Phi^{n(l)}_{k(q)}$ in (5.36) corresponds to the smaller number of the layer k or q (here orientation of the molecule B_2 is disregarded, and it is considered that to fly up to a surface atom of the alloy it is sufficient to "pass" through one column of vacant sites).

The reaction contributions to the kinetic equations for two-body functions (5.33) will be written as follows:

$$(\dot{\theta}^{nl}_{kq})_p = W^{ij}_{kq}(\alpha) - W^{nl}_{kq}(\alpha) + W^{n(l)}_{kq} + W^{l(n)}_{qk}$$

$$W^{n(l)}_{kq} = \begin{cases} W^{\overline{\mathrm{v}}l}_{kq}(1) - W^{\overline{\mathrm{A}}l}_{kq}(1) - \sum\limits_{h=k+1}^{k-1} (z_{kh} - \Delta_{qh}) W^{\overline{\mathrm{AB}}l}_{hkq}(3), & n = \mathrm{A} \\ \sum\limits_{h=k+1}^{k-1} (z_{kh} - \Delta_{qh})[W^{\overline{\mathrm{vv}}l}_{hkq}(2) - W^{\overline{\mathrm{BB}}l}_{hkq}(2) - W^{\overline{\mathrm{AB}}l}_{hkq}(3)], & n = \mathrm{B} \end{cases} \tag{5.38}$$

where the first addends correspond to the contribution of the reaction between the central particles, namely, $n = \mathrm{A}$, $l = \mathrm{B}$ or vice versa for $\alpha = 3$ (the pair ij is absent); $n = l = \mathrm{B}$ and $i = j = \mathrm{v}$ for $\alpha = 2$. The quantity $W^{n(l)}_{kq}$ corresponds to the contribution of the reactions of particle n at site k in the presence of a neighboring particle q at site q. If l corresponds to the components of the alloy, then the contribution $W^{l(n)}_{qk} = 0$. The contributions of the individual steps are expressed as follows:

$$W^{\overline{n}j}_{kq}(\alpha) = W^n_k(\alpha) \times \begin{cases} \psi^{nj}_{kq}(\alpha) & q = k-1,\ z_{kk-1} > 1;\ q = k \\ \psi^{nj}_{kq}(\alpha)/\psi^n_{kq}(\alpha), & q = k+1 \end{cases}$$

$$W^{\overline{ln}j}_{hkq}(\alpha) = W^{ln}_{hk}(\alpha) \tag{5.39}$$

$$\times \begin{cases} \psi^{nj}_{kq}(\alpha), & h = q = k,\ k-1;\ h = k-1,\ q = k+1 \\ \psi^{nj}_{kq}(\alpha)/\psi^n_{kq}(\alpha), & q = k,\ h = k \pm 1;\ q = k+1,\ h = k,\ k+1 \end{cases}$$

If $q = k - 1$ and $z_{kk-1} = 1$, the addends of (5.39) equal zero, which signifies prohibition of a reaction because of overhanging of the particles.

As a result, system of equations (5.33) describes the behavior of the concentration profile for a rough surface of a multicomponent alloy when the catalyt-

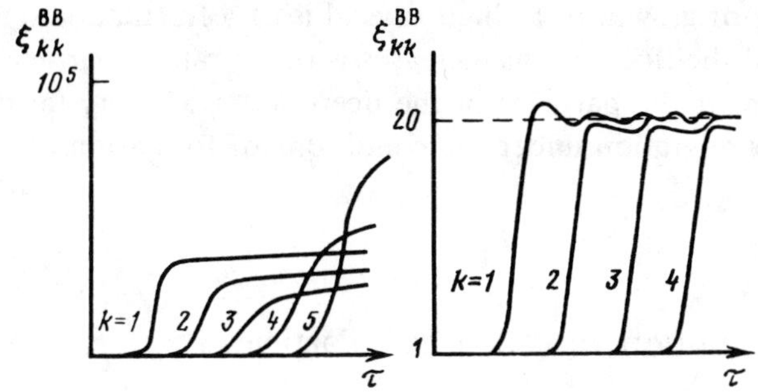

Fig. 5.5. Time dependences of the functions ξ^{vv}_{kk} in the absence of migration of the atoms in a crystal (*a*) and when they are rapidly stirred (*b*):

$\beta\varepsilon_{AA} = 3.0$, $\beta\varepsilon^{*}_{AA} = 1.5$, $\hat{K}_{v} = 1$, $K_{A} = 0.05$

ic reaction $A + \frac{1}{2} B_2 \rightarrow C$ proceeds. The constructed system is solved with the given initial and boundary conditions for the states of the bulk phases of the gas and solid or for the flows through the boundary layers of the subsurface region. The solution of the system yields the values of the functions θ^{i}_{k} and θ^{ij}_{kq} in terms of which the temporal evolution of the surface roughness is expressed because its determination [195] [see also (3.67)] remains in force in the dynamics of the process.

On the basis of the above approach to a description of the evolution of a rough surface Tovbin and Filimonova [235] have considered the process of growth of a single crystal from the gas phase with a view to its defectiveness. Figure 5.5 shows the change in time of the fraction of bivacancies in the initial layers of a growing crystal for the cases of the absence of migration and the rapid mobility of the particles. In both cases, the function $\zeta^{vv}_{kk} = \theta^{vv}_{kk}/(\theta^{v}_{k})^2$ increases for each layer to a value depending on all the parameters of the model. For rapid stirring, they tend to one value, while in the absence of mobility these quantities are different for each layer.

Under the conditions being considered, there is a trend to the formation of bivacancies, which can characterize the trend of pore formation. In the absence of mobility of the particles, this trend is greater than for rapid stirring. The growth in the interaction of particles in the transition state relative to interaction in the ground state sharply increases the linear rate of crystal growth and in the absence of migration increases the width of the transition region. The rate of growth of a single crystal is of a fluctuational nature, and the amplitude of the fluctuations increases with the rate of particle migration. The distribution of the particles in the deep layers of a crystal depends on the rate of their migration under the conditions of formation of a transition region.

5.6. Dissolution of Adsorbed Particles

Many systems, e.g. hydrogen-metal (Pd, Ni), hydrogen-alloy (Pd-Ag), nitrogen-iron, oxygen-metal (Mo, Nb) [45, 236], and oxygen-oxides [237-239] are characterized by the adsorption of a large amount of the relevant gas, which substantially alters the surface concentration of the reactants. The change in the reactant concentration, in turn, changes the rate of the surface reactions affecting the rate of adsorption of the component being dissolved. This is why the relations between the processes of particle dissolution in the bulk and the chemical transformation of the particles on the surface play a major role in processes of the mass transfer of substances.

Let us construct the kinetic equations of the dissolution of atoms of dissociating molecules B_2 that react on the surface with adsorbed particles of A (a three-step mechanism similar to that in Sec. 4.4). Assume that the particles of B dissolve in the interstices of a one-component crystal with a flat interface. As above, the value of $\alpha = 4$ corresponds to the step of migration along the interstices. We shall have in view that the surface boundary changes the properties of the crystal atoms in its subsurface region. This, in turn, changes the properties of the absorption centers, in particular the binding energy of a particle.

We shall consider the adsorbed particles to be in the zero layer, the interstices of the first and second layers of the crystal atoms to form the first layer of adsorption sites, and so on. For visualization, we shall limit ourselves to a simple cubic lattice. Let the number of absorption centers in the subsurface region be $\varkappa$, outside the critical region this number is usually not large. All the absorption sites in each layer are considered to be equivalent. We shall assume that the gas being sorbed does not cause rearrangement of the structure of the initial crystal. (The absorption in the interstices of a rearranging multicomponent alloy is treated in Sec. 7.3)

The kinetic equations for the surface layer $k = 0$ include equations for the concentrations of particles and their pairs. For a nondissolving component A, the equations for $\dot{\theta}_0^{A}$ and $\dot{\theta}_{00}^{AA}$ have the form of (5.29) and (5.31) in which the addends are determined by formulas of the type of (5.30). Their distinction from formulas (5.30) consists in that the preexponential factors of the rate constants have been replaced by the rate constants $K_0^i(\alpha)$ and $K_{00}^{ij}(\alpha)$ including the activation energies of the processes with zero coverages of the surface layer sites—these are the binding energies of the particles to the crystal atoms, and when $q = 1$ the sum of the species of the particles in the functions $S_{01}^i(\alpha)$ is taken over the vacant and occupied sites of the absorption layer $k = 1$. The equations for the component B additionally contain the addends of its migration between the layers $k = 0$ and $k = 1$:

$$
\begin{aligned}
\dot{\theta}_0^{B} &= z_{00}[U_{00}^{vv}(2) - U_{00}^{BB}(2) - U_{00}^{AB}(3)] - z_{01}[U_{01}^{Bv}(4) - U_{01}^{vB}(4)] \\
\dot{\theta}_{00}^{BB} &= U_{00}^{vv}(2) - U_{00}^{BB}(2) - 2(z_{00} - 1)[U_{000}^{\overline{vv}B}(2) - U_{000}^{\overline{BB}B}(2) \\
&\quad - U_{000}^{\bar{A}BB}(3) - 2D_{00}^{\bar{B}B} - 2z_{01}[U_{100}^{\bar{v}BB}(4) - U_{100}^{\overline{Bv}B}(4)] \\
\dot{\theta}_{00}^{AB} &= U_{00}^{\bar{v}B}(1) - U_{00}^{\bar{A}B}(1) - U_{00}^{AB}(3) - (z_{00} - 1)[U_{000}^{\overline{BA}B}(3) \\
&\quad - U_{000}^{A\overline{vv}}(2) + U_{000}^{A\overline{BB}}(2) + U_{000}^{A\overline{B}A}(3)] - D_{00}^{\bar{A}B} - D_{00}^{A\bar{B}} \\
&\quad - z_{01}[U_{100}^{\overline{vB}A}(4) - U_{100}^{\overline{Bv}A}(4)]
\end{aligned}
\tag{5.40}
$$

where all the addends have been determined in Sec. 5.4 with the differences indicated above, namely, the rate constants are taken instead of their preexponential factors and the sum over the species of the particles is taken over the species A, B, and v in the layer $k = 0$ and over the species B and v in the layers $k \geqslant 1$. The same differences concern the expressions for the pairs θ_{01}^{AB} and θ_{01}^{BB} in Eqs. (5.32), where the numbers h also change in the sums $D_{01}^{i\bar{B}}$ ($h = 0, 1, 2$), $i =$ A, B, and $D_{01}^{\bar{B}B}$ ($h = 0$ and 1), and the expressions (5.27) for the particle of B in the layers $k \geqslant 1$. In the given case for the layer $k = 1$, the indices q and h can take on a zero value (the emergence of B onto the surface).

It should be noted that in the general case $K_{kq}^{Bv}(4)$, the rate constants of migration of B particles depend on the numbers of the layers between which the jumps occur. When $k > \varkappa$ (the bulk phase), these constants and parameters of interaction between the particles of B remain constant, and the numbers of the layers may be discarded. Consequently, equations of type (5.27) are equations of the absorption of a dissolving component, while the equations for the layer $k = 0$ describe surface reactions that are excluded in the absence of component A [240].

The system of equations is solved with the given initial values of the concentrations of all the components and their pairs in all the layers of the region of the solid being considered.

With longer times ($K_{Bv}\tau \gg 1$), we may replace the discrete system of equations when describing the migration of particles over large distances from the phase interface with the diffusion equation where a is the lattice constant.

$$\frac{\partial\theta}{\partial\tau} = \frac{\partial}{\partial x}\left[D(\theta)\frac{\partial\theta}{\partial x}\right], \quad x = (k - \varkappa)a \tag{5.41}$$

For Eq. (5.41), the discrete system of equations for the layers of the subsurface region $0 \leqslant k \leqslant \varkappa$ is a boundary condition relating the bulk value of the concentration θ to the concentration of the particles of B in the adsorption layer. At the diffusion limit, $\dot{\theta}_{\varkappa}^{B} = 0$ [241], which allows us to write the equation for the layer $k = \varkappa$ as follows:

$$D(\theta)\frac{d\theta}{dx} + z_{\varkappa\varkappa-1}[U_{\varkappa\varkappa-1}^{vB}(4) - U_{\varkappa\varkappa-1}^{Bv}(4)] = 0 \tag{5.42}$$

where the functions $t_{\varkappa\varkappa+1}^{ij}$ in the multipliers $S_{\varkappa\varkappa+1}^{i}(4)$ in expressions of the type (5.28) for the rates of the jumps of B particles are related to the local concentration of the particles at the bulk point $k = \varkappa + 1$ by the equilibrium expressions of Sec. 2.2. The boundary condition for the diffusion equation far from the surface depending on the situation may be of the first or second kind, or be similar equations for the other side of the specimen if it is limited in volume.

The form of the other equations for $\dot{\theta}_{k}^{B}$ and $\dot{\theta}_{kq}^{BB}$, where $0 \leqslant k \leqslant \varkappa$, depends on the relation between K_{Bv} and K_{kq}^{Bv} [240]. If $K_{kq}^{Bv}\,\tau \leqslant 1$, it is necessary to employ the initial system of differential equations, otherwise for all the transition layers $\dot{\theta}_{k}^{B} = 0$, which leads to an algebraic relation between θ_{B} ($x = 0$) and the concentrations of the molecules in the gas phase. When $K_{kq}^{Bv} < K_{Bv}$, this relation is determined by the steady-state expressions $\dot{\theta}_{k}^{B} = 0$, while when $K_{kq}^{Bv} \gg K_{Bv}$ it is determined by the equilibrium expressions. The condition of a steady state for the kinetic equations relative to the functions θ_{kq}^{Bv} in the bulk of the solid also corresponds to the transition to a diffusion approximation. For qualitative appraisals, we can use the condition of the equilibrium distribution of the particles instead of the steady-state condition (see the following section). For the subsurface layers, the equations for the functions θ_{kq}^{Bv} remain differential if $K_{kq}^{Bv}\tau \ll 1$, or transform into the steady-state equations $\dot{\theta}_{kq}^{Bv} = 0$ when $K_{kq}^{Bv} \leqslant K_{Bv}$ or into the equilibrium expressions of Sec. 3.2 when $K_{kq}^{Bv} \gg K_{Bv}$.

5.7. Diffusivities

For longer times, the macroscopic flows of particles are described in the diffusion approximation. The initial discrete system of conventional differential equations of Sec. 5.1 is replaced by a system of equations in partial derivatives for the local (on a macroscopic scale) concentrations of the particles. The diffusivities (or diffusion coefficients) are the basic characteristic of the transfer processes described by the diffusion equations. The diffusivities determine the relation between the flow of particles and the concentration gradients. For an isotropic lattice, the one-dimensional macroscopic flows of the particles in the diffusion approximation are as follows:

$$\frac{\partial \theta_i}{\partial \tau} = \frac{\partial}{\partial x}\left(\sum_{j=1}^{s-1} D_{ij}\frac{\partial \theta_j}{\partial x}\right) \tag{5.43}$$

where D_{ij} is the diffusivity of component i at the expense of the gradient of concentration of component j (the particles of species s are vacancies), and x is a space coordinate. The diffusivity, which is a parameter of the phenomenological theory, can be evaluated on the basis of notions on the migration mechanism and on the parameters of particle interactions. Such calculations are based on the model of an elementary jump of a particle between two localization positions. The characteristic distances in Eq. (5.43) are much larger than the lattice constant. Consequently, within the limits of a small elementary volume a state of local equilibrium is realized. Let us consider an elementary jump as a "very simple" chemical reaction whose rate is expressed by relations (4.16), (4.31), and (4.33) and construct expressions for the self-diffusivity (label transfer) and diffusivity (mass transfer).

Self-Diffusivity (Self-Diffusion Coefficient). Let us denote the fraction of the labeled component i by θ_i^*. The rate of exchange of particles i with vacancies does not depend on whether a given particle is labeled or not. The flow of labeled particles J along the x-axis will be represented as the difference between the numbers of exchanges in unit time between two neighboring planes 1 and 2 ($U_{12}^{iv} - U_{12}^{vi}$) [49-51, 134]. Under conditions of the uniform distribution of the particles (the quasichemical approximation), this flow is expressed as [212, 242]

$$J = K_{iv}(\theta_{i1}^* - \theta_{i2}^*)t_{iv}(S_iS_v)^{z-1} \tag{5.44}$$

where θ_{ik}^* is the fraction of labeled particles i in layer k. By comparing this formula with the expression for Fick's first law $J = -D_i^* d\theta_i^*/dx$, provided that there is only a small change in the concentration of the labeled particles between neighboring layers $\theta_{i1}^* = \theta_{i2}^* - ad\theta_i^*/dx$, we obtain

$$D_i^* = a^2 K_{iv} t_{iv} (S_i S_v)^{z-1} \tag{5.45}$$

The self-diffusivity is obtained similarly in other approximations and for other migration mechanisms. The self-diffusivity of particles in a multicomponent system can be introduced ambiguously [49], but in each specific case it can be constructed in terms of the rate of exchanges of neighboring particles, although in general matters are more involved than what we have set out here in the style of the traditional approach [134]. Complications are introduced by the appearance in a number of mechanisms of correlation between consecutive jumps of particles. This leads to the appearance of a correlation multiplier in formula (5.45) [49, 243]. Its appraisals for a binary system in the absence of interaction reveal that it changes slightly from 1 to 0.65 when $0 \leqslant \theta \leqslant 1$ [244]. The concentration dependences of the self-diffusivity (5.45) in a binary system are shown in Fig. 5.6. The behavior of the curves is determined by the relation between the parameters of particle interaction.

Diffusivity. When there is a gradient of concentration of substances, every plane or line is characterized for surface diffusion by its own values of the concentrations $\theta_k^i = \theta_{k+1}^i - ad\theta_i/dx$, where θ_k^i is the concentration of particles i in layer k. It is easy to obtain a qualitative appraisal of the concentration dependence of the diffusivity by considering the flows of particles between two planes if we ignore the change in the functional relation between θ_i and θ_{ij} at the expense of the concentration gradient [245]. Let us single out the local contributions of neighboring planes in the expression for the rate of particle exchange:

$$\begin{aligned} U_{kk+1}^{iv} &= K_{iv} \psi_k^i \Phi_{k+1}^v, \\ \psi_k^i &= \theta_i (S_i)^{z-1}, \\ \Phi_{k+1}^v &= t_{iv} (S_v)^{z-1} \end{aligned} \tag{5.46}$$

Now, as previously, by writing the expression for the flow J and comparing it with the expression for Fick's law, provided that the concentration gradient is small, we obtain:

$$D_{ij} = D_i^* \left[\Delta_{ij} - \theta_i \frac{\partial \ln t_{iv}}{\partial \theta_j} + \theta_i (z-1) \left\{ \frac{\partial \ln S_i}{\partial \theta_j} - \frac{\partial \ln S_v}{\partial \theta_j} \right\} \right] \tag{5.47}$$

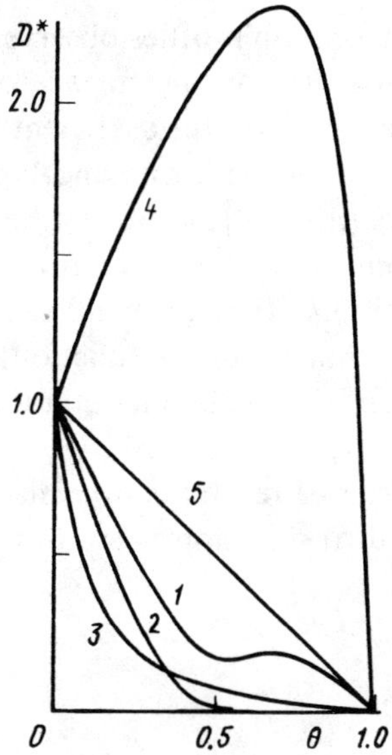

Fig. 5.6. Concentration dependences of the self-diffusivity:

$\beta\varepsilon_{AA} = -3.0$ (*1* and *2*); 1.2 (*3* and *4*); and 0 (*5*); $\beta\varepsilon^*_{AA} = -1.5$ (*1*); -3.0 (*2*); 1.2 (*3*); and 0 (*4* and *5*)

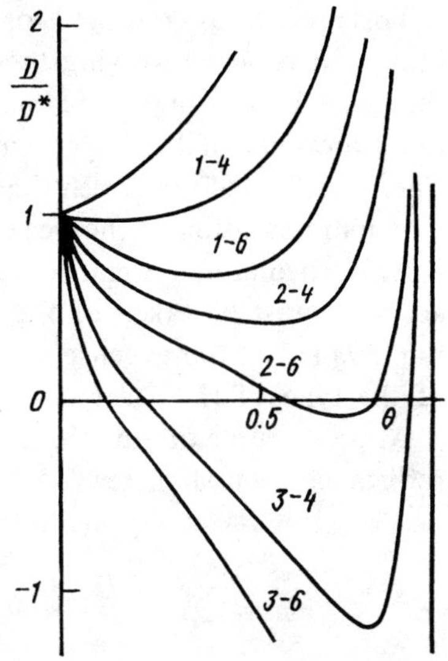

Fig. 5.7. Concentration dependences of the diffusivity, $\beta\varepsilon^*_{AA} = 0.5\beta\varepsilon_{AA}$.

In the curve numbers $n - z$, n stands for the variant number, and z for the number of nearest neighbors (*4* and *6*). $\beta\varepsilon = 0.6$ (*1*); 1.5 (*2*); and 2.0 (*3*)

where D_i^* have been determined in (5.45), the functions S_i — in (4.16), while the derivatives of the functions θ_{ij} are determined from differentiation of systems of equations in Sec. 2.4. In the particular case of a binary system, we have [245]

$$D_{AA} = D_A^* \left[1 + \theta \frac{\partial t_{AA}}{\partial \theta} \{(1 - t_{AA})^{-1} + (z - 1)(1 - u)/(t_{AA} + (1 - t_{AA})u)\}\right] \tag{5.48}$$

where $u = \exp \beta\varepsilon$, while the function t_{AA} is expressed by formula (2.14). The nature of the curves is determined by the relation between the parameters ε_{AA} and ε^*_{AA}. When $\beta > \beta_c$, where β_c is a quantity corresponding to the critical temperature, the diffusivity may be negative—condensation of the diffusing component occurs (Fig. 5.7).

This model was employed for describing the concentration profiles observed in an experiment involving the diffusion of oxygen atoms on the surface of the face (110) of tungsten [246]. The found parameters point to the sufficiently rapid mobility of the oxygen atoms and their weak bond with one another. This is well consistent with the notions set out by Nartikoev [247], according to whom migration of the weakly bound oxygen atoms over a layer of strongly chemisorbed atoms is realized under the given conditions. In the general case, account must be taken of the contribution of the change in the functional bond θ_{ij} (θ_i) at the expense of the gradient of concentrations by an equation of the type of (5.5) (see also [84]).

As an example of equations considering the transfer of reactants over macrodistances, we shall rewrite system (4.45) in the diffusion approximation:

$$\begin{aligned} \frac{\partial \theta_{\mathrm{A}}(x)}{\partial \tau} &= I_{\mathrm{A}}(x) + \frac{\partial}{\partial x}\left(\sum_{j=\mathrm{A}}^{\mathrm{B}} D_{\mathrm{A}j}\frac{\partial \theta_j}{\partial x}\right) \\ \frac{\partial \theta_{\mathrm{B}}(x)}{\partial \tau} &= I_{\mathrm{B}}(x) + \frac{\partial}{\partial x}\left(\sum_{j=\mathrm{A}}^{\mathrm{B}} D_{\mathrm{B}j}\frac{\partial \theta_j}{\partial x}\right) \end{aligned} \tag{5.49}$$

where $I_{\mathrm{A}}(x)$ and $I_{\mathrm{B}}(x)$ are the right-hand sides of Eqs. (4.45) with addends (4.46) describing the adsorption-desorption processes and the reaction in a local region with the coordinate x; here $0 \leqslant x \leqslant L$, L is the length of a section of the surface. Equations (5.49) are solved with the preset initial and boundary conditions whose form depends on the specific experimental situation.

6 Equilibrium of a Gas-Solid System

6.1. Two-Body Interactions of a Finite Number of Neighbors

In the preceding chapters, we constructed a theory of surface processes with account taken of the interaction of the nearest neighbors. This enabled us to give a self-consistent description of their kinetics and equilibrium within broad ranges of the values of the concentrations of particles, temperature of the system, and the mobilities of the reaction system particles. In the following chapters, we shall again revert to the same processes and see how the mathematical models of the processes will change when more realistic potentials of particle interaction are employed (two-body interaction of a finite number of neighbors and multiparticle potentials) and what new opportunities are associated with their use. The initial formulations of the problems remain unchanged, and they will not be repeated.

Let us consider an s-component system of particles on a homogeneous lattice with a two-body potential of interaction between them. Let $\varepsilon_{ij}(r)$ be the energy parameters corresponding to the value of the two-body potential of interaction at a distance r, where $1 \leqslant r \leqslant R$; here R is the interaction radius—any preset number. The distances will be measured in numbers of the coordination spheres.

The energy of a system will be written as follows [248]:

$$H = \sum_{f,i} \nu_i \gamma_f^i - \frac{1}{2} \sum_{r=1}^{R} \sum_{f,g_r} \sum_{i,j} \varepsilon_{ij}(r) \gamma_f^i \gamma_{g_r}^j \qquad (6.1)$$

where ν_i has been determined in (1.23). The index f runs through all the sites of a lattice, while the sum over g_r is taken over $z(r)$ sites in the r-th coordination sphere of site f.

To describe the equilibrium distribution of the particles, we shall limit ourselves to clusters with one central site. The initial lattice is imaged on a set of clusters consisting of a central site and its R coordination spheres. The energy of the central particles is determined by formula (2.2) whose addends have the form

$$h_f^i = \left(\nu_i - \sum_{r=1}^{R} \sum_{g_r} \sum_{j=1}^{s} \varepsilon_{ij}(r)\gamma_{g_r}^j\right)\gamma_f^i \tag{6.2}$$

Let us determine the two types of cluster correlators that characterize the probabilities of configurations of particles on clusters of the indicated size. The correlators of the first type can be represented as follows [248]:

$$\theta_{f(g_1\ldots g_z)R}^{i(j_1\ldots j_z)R} = \left\langle \gamma_f^i \prod_{r=1}^{R} \prod_{g_r} \gamma_{g_r}^j \right\rangle = \left\langle \gamma_f^i \prod_{r=1}^{R} \prod_{j=1}^{s} (\gamma_{g_r}^j)^{n_r^j} \right\rangle$$

$$= \theta_{\{[n]\}R}^i = \frac{\text{const}\,(\{[n]\}R)}{Q} \exp\beta\left(-\nu_i + \sum_{r=1}^{R} \sum_{j=1}^{s} \varepsilon_{ij}(r)n_r^j\right) \tag{6.3}$$

Expression (6.3) has been written on the basis of the same reasoning as the derivation of formula (2.5) from (2.3). The energy bonds of a central particle with the neighboring particles j at a distance of r are set by the numbers n_r^j. The square brackets signify that each coordination sphere contains exactly n_r^j particles of species j, where $1 \leqslant j \leqslant s$. For the unambiguous setting of the state of occupancy of the sphere sites, it is sufficient to set $(s - 1)$ values of n_r^j because

$$n_r^s = z(r) - \sum_{j=1}^{s-1} n_r^j, \quad 0 \leqslant n_r^j \leqslant z(r) \tag{6.4}$$

The braces in (6.3) signify the completeness of the sets of all the n_r^j's relating to different species of particles and to different numbers of the coordination spheres, $r \leqslant R$.

The correlators of the second type are determined similarly to (2.6):

$$\theta_{\{n\}R}^i = \left\langle \gamma_f^i \prod_{r=1}^{R} \prod_{j=1}^{s-1} (\gamma_{g_r}^j)^{n_r^j} \right\rangle \tag{6.5}$$

They characterize the probability of the following configurations: a particle i is at the center, while the r-th coordination sphere of a cluster contains at least n_r^j particles j, where $1 \leqslant j \leqslant s - 1$. The equations (6.4) are not observed for correlators (6.5), hence the square brackets are absent in the notation.

The correlators of the first and second types are related by linear expressions generalizing expressions (2.40). Since the sites of each sphere are occupied independently, these relations are written in the following compact form:

$$\begin{aligned}\theta^i_{\{n\}R} &= \prod_{r=1}^{R} \left\{ \sum_{k_1=0}^{\tilde{z}} \cdots \sum_{k_{s-1}=0}^{\tilde{z}-l} C_{\tilde{z}}^{\{k\}} \right\}_r \theta^i_{\{[n+k]\}R} \\ \theta^i_{\{[n]\}R} &= \prod_{r=1}^{R} \left\{ \sum_{k_1=0}^{\tilde{z}} \cdots \sum_{k_{s-1}=0}^{\tilde{z}-l} (-1)^{\tilde{k}} C_{\tilde{z}}^{\{k\}} \right\}_r \theta^i_{\{n+k\}R}\end{aligned} \tag{6.6}$$

where the quantities $\tilde{z}$, l, $\tilde{k}$, and $C_{\tilde{z}}^{\{\tilde{k}\}}$ have been determined in (2.39) for each coordination sphere r, but for simplicity of the notation, the index r is discarded. The abbreviated notation $\{n + k\}R$ signifies the complete set of values $n_r^1 + k_r^1, \ldots, n_r^{s-1} + k_r^{s-1}$ for all the values of $r \leqslant R$.

Systems of equations are obtained for correlators of the second type if we exclude the unknown constants and the partition function by considering the relation between correlators (6.3) with an identical state of occupancy of the sites in the coordination spheres, but differing in the state of occupancy of the central site

$$\theta^i_{\{[n]\}R} = \theta^j_{\{[n]\}R} M^{ij}_{\{[n]\}R} \tag{6.7}$$

where $M^{ij}_{\{[n]\}R}$ is the ratio of the exponents of (6.3), and express the correlators of the first type in (6.7) in terms of the correlators of the second type (6.6). By performing these linear transformations and regrouping the addends with account taken of the explicit form of $M^{ij}_{\{[n]\}R}$, we obtain

$$\begin{aligned}\theta^i_{\{[n]\}R} &= \exp\left[-\beta\left(\nu_{ij} + \sum_{r=1}^{R}\sum_{m=1}^{s-1} \varepsilon_{ijm}(r) n_r^m\right)\right] \\ &\times \prod_{r=1}^{R} \left\{ \sum_{k_1=0}^{\tilde{z}} \cdots \sum_{k_{s-1}=0}^{\tilde{z}-l} C_{\tilde{z}}^{\{k\}} \prod_{m=1}^{s-1} (x^r_{ijm})^{k_r^m} \right\}_r \theta^j_{\{n+k\}R}\end{aligned} \tag{6.8}$$

where the following notation has been used:

$$\begin{aligned}\nu_{ij} &= \nu_i - \nu_j + \sum_{r=1}^{R} z(r)[\varepsilon_{js}(r) - \varepsilon_{is}(r)] \\ x^r_{ijm} &= \exp(-\beta\varepsilon_{ijm}(r)) - 1; \\ \varepsilon_{ijm}(r) &= \varepsilon_{jm}(r) - \varepsilon_{im}(r) - \varepsilon_{js}(r) + \varepsilon_{is}(r)\end{aligned} \tag{6.9}$$

Let us consider very simple ways of closing system of equations (6.8) [249].

Consideration of Direct Correlations. We postulate decoupling of the multiparticle correlators of the second type in terms of the functions θ_i and θ_{ij}:

$$\theta^i_{\{[n]\}R} = \theta_i \prod_{r=1}^{R} \prod_{j=1}^{s-1} [t_{ij}(r)]^{n_r^j}, \; t_{ij}(r) = \theta_{ij}(r)/\theta_i \tag{6.10}$$

where $\theta_{ij} = \langle \gamma_f^i \gamma_{g_r}^j \rangle$ is the probability of particle i and j being at a distance r, and $t_{ij}(r)$ is the conditional probability of finding a particle j at a distance r from a particle i. The uncoupling of (6.10) is the generalization of quasichemical approximation (2.41) over arbitrary distances. The closed system of equations relative to θ_i and $\theta_{ij}(r)$ has the following form:

$$\begin{aligned} &\theta_s = \theta_i \exp \beta(\nu_i - \nu_s) \prod_{r=1}^{R} [S_i(r|s)]^{z(r)}, \; S_i(r|s) = \sum_{j=1}^{s} t_{ij}(r) e^{\beta(\varepsilon_{sj}(r) - \varepsilon_{ij}(r))} \\ &\theta_{ii}(r)\theta_{jj}(r) = \theta_{ij}^2(r) \exp \beta(\varepsilon_{ij}(r) + \varepsilon_{jj}(r) - 2\varepsilon_{ij}(r)), \; \sum_{j=1}^{s} \theta_{ij}(r) = \theta_i \end{aligned} \tag{6.11}$$

Its solution enables us to calculate all the correlators of the second type (6.10) and the correlators of the first type:

$$\theta^i_{\{[n]\}R} = \theta_i \prod_{r=1}^{R} \prod_{j=1}^{s} [t_{ij}(r)]^{n_r^j} \tag{6.12}$$

Absence of Correlations. By expressing the multiparticle correlators in terms of θ_i, we obtain

$$\theta^i_{\{[n]\}R} = \theta_i \prod_{j=1}^{s} (\theta_j)^{\sigma_j}, \quad \theta^i_{\{n\}R} = \theta_i \prod_{j=1}^{s-1} (\theta_j)^{\sigma_j}, \; \sigma_j = \sum_{r=1}^{R} n_r^j \tag{6.13}$$

where the functions θ_i in the molecular-field approximation are determined by solving the system of equations

$$\begin{aligned} &\theta_i = \tilde{y}_i \Big/ \sum_{i=1}^{s} \tilde{y}_i, \\ &\tilde{y}_i = \exp \beta \left(-\nu_i + \sum_{r=1}^{R} \sum_{j=1}^{s} z(r)\varepsilon_{ij}(r)\theta_j \right) \end{aligned} \tag{6.14}$$

If we introduce the quantities $\varepsilon'_{ij} = \sum_r \varepsilon_{ij}(r)z(r)/z$, then provided that $|\varepsilon'_{ij}| < \infty$, system (6.14) reduces to a system similar to (2.47).

Equations (6.8) can be used to construct "intermediate" approximations considering the direct correlations at distances of $R_1 < R$. In this case

$$\theta_s = \theta_i \exp \beta \left[\nu_i - \nu_s - z \sum_{j=1}^{s} (\tilde{\varepsilon}_{ij} - \tilde{\varepsilon}_{sj})\theta_j \right] \prod_{r=1}^{R_1} [S_i(r|s)]^{z(r)} \tag{6.15}$$

where $\tilde{\varepsilon}_{ij} = \sum_{r=R_1+1}^{R} z(r)\varepsilon_{ij}(r)/z$, for $r \leqslant R_1$ the functions $\theta_{ij}(r)$ are found from the equations for two-body correlators (6.11).

Consideration of Indirect Correlations. As for the interaction of nearest neighbors, the consideration of indirect correlations is associated with the difference in the probabilities of the configurations of particles arranged differently in the coordination spheres. Here the binomial coefficients determining the weights of the configurations with an identical energy of the central particle in (6.8) break up into separate groups of weights corresponding to different configurations. If we employ the two-body approximation when considering indirect correlations, all the higher correlators become decoupled in terms of $\theta_{ij}(r)$, as occurs in the superposition approximation for three particles [76, 77], and the number of equations of the system remains the same as in the quasichemical approximation. In this case, the system of equations will be written as follows:

$$\theta_s = \theta_i \exp(-\beta\nu_{si}) \prod_{r=1}^{R} \left\{ \sum_{k_1=0}^{z} \cdots \sum_{k_{s-1}=0}^{z-l} C_z^{\{k\}} \prod_{j=1}^{s-1} (t_{ij}(r)x_{sij}^{r})^{k_r^j} \right\} \Phi_{\{k_1\}\ldots\{k_R\}}$$

$$\begin{aligned} \theta_{sj}(r) = {} & \theta_{ij}(r) \exp \beta(\varepsilon_{sij}(r) - \nu_{si}) \\ & \times {\prod_{\varrho=1}^{R}}' \left\{ \sum_{k_1=0}^{z} \cdots \sum_{k_{s-1}=0}^{z-l} C_z^{\{k\}} \prod_{m=1}^{s-1} (t_{im}(\varrho)x_{sim}^{\varrho})^{k_\varrho^m} \right\}_\varrho \\ & \times \sum_{k_r^1=0}^{\tilde{z}(r)} \cdots \sum_{k_r^{s-1}=0}^{\tilde{z}(r)-l_r} C_{\tilde{z}_r}^{\{k_r\}} \prod_{m=1}^{s-1} (t_{im}(r)x_{sim}^{r})^{k_r^m} \Phi_{\{k_1\}\ldots\{k_r\}'\ldots\{k_R\}} \end{aligned} \tag{6.16}$$

where the prime in the product over ϱ signifies the absence of the multiplier $\varrho = r$ for which summation is conducted over $\tilde{z}(r) = z(r) - 1$ sites because a particle j has been registered at one of the sites of the r-th coordination sphere. The functions $\Phi_{\{k_1\}\ldots\{k_R\}}$ takes into account the indirect correlations between the particles in R coordination spheres of a cluster. These functions with respect to their construction are expressed only in terms of $\xi_{ij}(r) = \theta_{ij}(r)/\theta_i\theta_j$. The principle underlying the construction of functions of indirect correlators is set out in Appendix 1 (Sec. 3). The abbreviated notation $\{n_r\}'$, stands for $n_r^1 \ldots n_r^j + 1 \ldots n_r^{s-1}$.

At the relevant values of the parameters ν_i, $\varepsilon_{ij}(r)$, and $z(r)$, the obtained equations describe the local distributions of the particles in problems of adsorption, absorption, and solid solutions.

6.2. Equilibrium Characteristics of Sorption

Models of ad- and absorption, or simply sorption, were formulated in Sec. 2.1, while their equilibrium characteristics with consideration of the interaction of nearest neighbors were discussed in Secs. 2.3 and 2.5. Let us see how the interaction of the first and following neighbors affects the equilibrium characteristics of sorption of gas-phase molecules. We shall first deal with the simplest case of considering the effects of correlation—direct correlations (6.11), and then go over to consideration of the contributions of indirect correlations (6.16) (in the molecular-field approximation the concentration dependences of the equilibrium characteristics do not depend on the kind of potential of two-body interaction—only the value of the effective parameter of interaction of the nearest neighbors changes). As in Sec. 2.3, we shall omit the indices A for the particles being sorbed, and shall designate the quantities $\varepsilon_{AA}(r)$, $t_{AA}(r)$, and $z(r)$ relating to different distances by ε_r, t_r, and z_r. Having in view that the molecular parameters of the vacant sites ν_v and $\varepsilon_{iv}(r)$ equal zero, in the quasichemical approximation the distribution of the particles is described by the expressions $\theta_{AA}(r) = \theta t_r$ [250], where

$$t_r = 1 - 2(1 - \theta)/(1 + \delta_r),$$
$$\delta_r = \{1 + 4\theta(1 - \theta)(\exp \beta\varepsilon_r - 1)\}^{1/2} \tag{6.17}$$

The nature of particle distribution does not depend on the specific structure of the lattice set by the numbers z_r.

In this approximation, the sorption isotherm is written as follows:

$$y = \frac{\theta}{1 - \theta} (S_1)^{z_1}(S_2)^{z_2}, \quad S_r = 1 + t_r x_r, \quad x_r = e^{-\beta\varepsilon_r} - 1 \tag{6.18}$$

In the region of low coverages with an accuracy to the terms θ^2, the isotherm will be rewritten as $y = \theta(1 + \varkappa\theta)$, where $\varkappa = 1 + \sum_r z_r(1 - \exp \beta\varepsilon_r)$. To achieve a fixed coverage when $\varkappa > 1$, the pressure of the gas must be greater

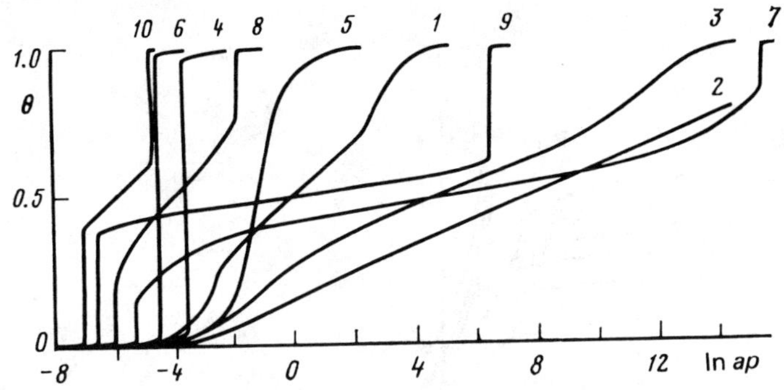

Fig. 6.1. Adsorption and absorption isotherms:
The values of the parameters z_1, z_2, $\beta\varepsilon_1$, and $\beta\varepsilon_2$ for curves *1-10* are, respectively: *1*—4, 4, 2, −2; *2*—8, 6, −1, −1.25; *3*—8, 6, −2, 1; *4*—6, 12, −0.2, 0.7; *5*—8, 6, 0.05, 0.35; *6*—6, 12, 2, −0.2; *7*—8, 6, −3.5, 3; *8*—4, 4, −2, 4; *9*—6, 12, −4.5, 2.3; *10*—8, 6, 3, −2

than for noninteracting particles ($\varkappa = 1$) and smaller when $\varkappa < 1$. In the repulsion of the particles at both distances, we have $\varkappa > 1$, and its limit value is $1 + z_1 + z_2$. If both parameters are positive, then $\varkappa < 1$, including $\varkappa < 0$. If the parameters have different signs ($\varepsilon_1\varepsilon_2 < 0$), then depending on their relation, cases are possible when $\varkappa > 1$, $0 \leqslant \varkappa \leqslant 1$, or $\varkappa < 0$. Negative values of $\varkappa$ are obtained, for example, when the first neighbors are repelled and the second ones are attracted, if $\beta\varepsilon_2 > \ln[(1 + z_1 + z_2(1 - \exp \beta\varepsilon_1))/z_2]$. Negative values of $\varkappa$ have a meaning when $dy/d\theta = 1 + 2\varkappa\theta > 0$, otherwise we are in the region of a phase transition, and it is necessary to use the complete expression (6.18). In the region of large coverages, the sorption isotherm has the form $y \approx \exp(\beta\Delta Q)/(1 - \theta)$, where $\Delta Q = -m\sum_r z_r\varepsilon_r$ is the total difference of the heat of sorption from $\theta = 0$ to $\theta = 1$ (2.34).

Sorption isotherms of various shapes in the entire region of coverages are shown in Fig. 6.1. The isotherms are featured by the presence of one or two phase transitions (the positions of the secants were determined with the aid of Maxwell's rule). The curves of the isotherms are antisymmetrical relative to a point with the coordinates [$\theta = 0.5$, $y(\theta = 0.5)$]. In the region of moderate coverages in the absence of phase transitions or when there are two phase transitions, an isotherm is approximated by relation (2.28) whose constants depend on the interaction parameters and the coordination numbers.

The critical values of $\beta\varepsilon_r$ for a number of surface and bulk structures are

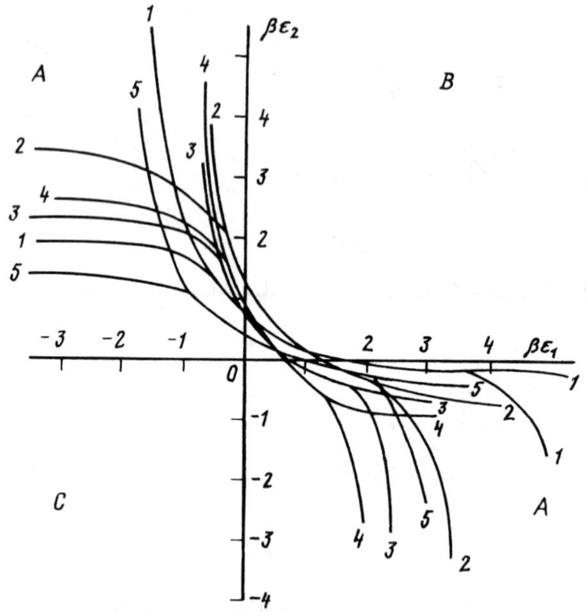

Fig. 6.2. Critical values of $\beta\varepsilon_1$ and $\beta\varepsilon_2$ for structures having the following numbers of first (z_1 = 3, 4, 6, 8, 6) and second (z_2 = 6, 4, 6, 6, 12) neighbors, respectively

shown in Fig. 6.2. For each lattice there are three regions of parameters corresponding to the realization of L phase transitions, $L = 0, 1, 2$. In region A (isotherms of the type of *7-10*), which is in the second and fourth quadrants, $L = 2$. With an increase in the repulsion of the first or second neighbors with constant values of z_r and the parameter corresponding to the attraction of particles, the distance along the pressure axis between two transitions increases, while with an increase in attraction (the constants z_r and $\varepsilon_r < 0$), the density jump between the coexisting phases grows. In region B (isotherms of type *4*, *6*), $L = 1$, the density jump corresponds to almost the entire region of coverages. In region C, $L = 0$. For lattices with $z_1 = z_2$, the curves are symmetrical relative to the bisectrix, $\varepsilon_1 = \varepsilon_2$. When $\varepsilon_2 = 0$, only the regions B and C are realized.

The origin of coordinates corresponds to $T \to \infty$. A drop in the temperature signifies motion from the origin of coordinates along a ray. There are numerous experimental observations of one phase transition in monolayer physical

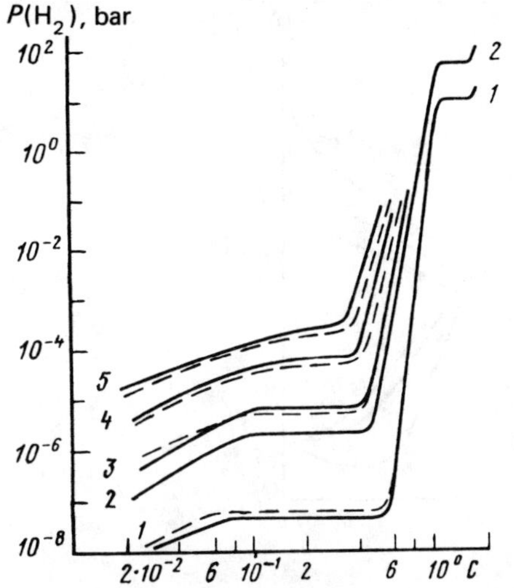

Fig. 6.3. Experimental (solid lines) and theoretical (dashed lines) of adsorption isotherms for hydrogen and niobium at 352 (1), 406 (*2*), 420 (*3*), 473 (*4*) and 503 K (*5*), c = H/Nb

adsorption [182, 251, 252]. Consideration of the interaction of the first and second neighbors points to the possibility of changing the numbers of phase transitions with lowering of the temperature. This case has been found experimentally by Gerot and Duval [253]. The same model [250] enables one to describe the adsorption isotherm of H/Nb [254] in which two phase transitions are observed (Fig. 6.3), and also the absorption of H/V [255] with one phase transition.

The heat of adsorption by homogeneous sorbents, according to definition (2.31), is written as follows [256]:

$$Q(\theta) = \varepsilon_{\mathrm{A}} + m \sum_{r=1}^{R} Q_r,$$

$$Q_r = \frac{z_r \varepsilon_r}{S_r \delta_r} [t_r \delta_r (x_r + 1) - \theta(1 - t_r)^2 (1 - \exp \beta \varepsilon_r)] \qquad (6.19)$$

In the region of small coverages, $Q(\theta) = \varepsilon_{\mathrm{A}} + \varkappa\theta m$, where $\varkappa = \sum_r z_r \varepsilon_r \exp(\beta\varepsilon_r)$. The quantity $\varkappa$ can be seen to be determined practically

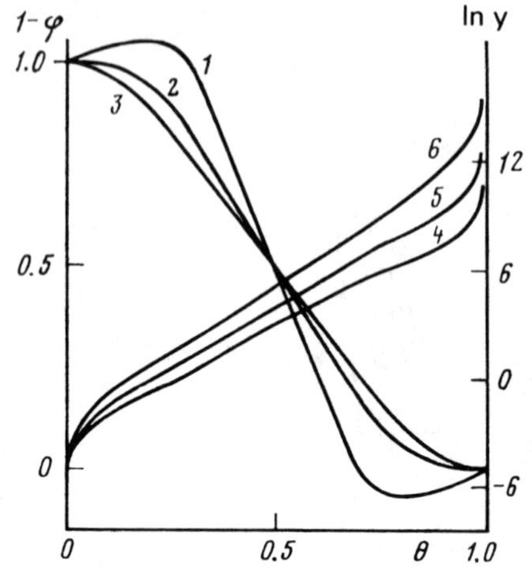

Fig. 6.4. Dependence of the heat $\varphi = [Q(\theta) - Q(0)]/\Delta Q$ (curves *1-3*) and isotherms (curves *4-6*) of adsorption on θ at $z_1 = z_2 = 4$, $\beta\varepsilon_1 = -2.0$, $\beta\varepsilon_2 = 0.5$ (*1*, *4*), 0(*2*, *5*), and -0.5 (*3*, *6*)

completely by attraction of the particles ($\varepsilon_r > 0$), their repulsion at low coverages gives virtually no contribution. The dependence of $\varkappa$ on the kind of potential was discussed by the author in greater detail in a different publication [256]. Figure 6.4 shows the concentration dependences of the heat of adsorption for various contributions of the second neighbors and the adsorption isotherms corresponding to them.

Consequently, on homogeneous surfaces, the heat of adsorption at low coverages should either grow or remain practically constant. It is just such behavior of the heat of adsorption that is observed for many systems in physical [182] and chemical adsorption, e.g. O_2 on Mo, W, and Au [257-259], and H_2 and CO on W, Fe, and Ni [260-262]. The experimentally observed decrease in Q with a growth in θ at low coverages can only be a result of surface inhomogeneity (see Sec. 3.4, the role of impurities has been studied in [263]). The failure to consider the correlation effects {for instance, the molecular-field or chaotic approximations (sec. 2.3) and the model of a two-dimensional electron gas [161, 264-267]} result in a linear decrease in Q vs. θ in the region of coverages being treated. This circumstance, consequently, implicitly reflects the surface inhomogeneity. A similar reduction in $Q(\theta)$, but smaller in magni-

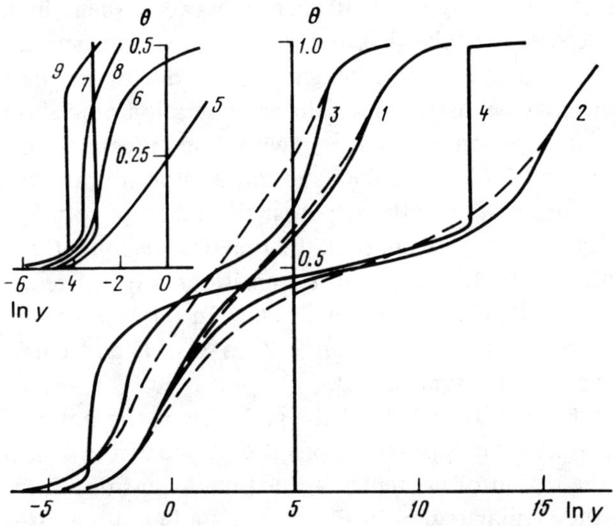

Fig. 6.5. Adsorption isotherms $z_1 = z_2 = 4$ [comparison of consideration of indirect correlations (solid curves *1-4*) with the quasichemical approximation (dashed curves)]:

1—$\beta\varepsilon_1 = -2.0$; *2*—$-4.0$; *3*—$-2.0$; *4*—$-4.0$; *5*—$-1.0$; *6*—$-3.0$; *7*—$-1.0$; *8*—$-1.0$; *9*—$\beta\varepsilon_1 = -1.0$; *1*—$\beta\varepsilon_2 = 0.1$; *2*—0.1; *3*—1.0; *4*—2.0; *5*—-0.5; *6*—1.5; *7*—0.5; *8*—2.0; *9*—$\beta\varepsilon_2 = 2.5$

tude, is given by the contributions of the far neighbors in "intermediate" approximation (6.15), which in terms of a two-dimensional electron gas was employed in ref. [158] to describe the equilibrium characteristics of the adsorption of potassium and barium on tungsten (see also [161]).

A closed system of sorption equations with account taken of the contributions of indirect correlations in a two-body approximation is given in Appendix 1 (Sec. 3). The calculations were performed for a two- ($z_1 = z_2 = 4$) and a three-dimensional ($z_1 = 6$, $z_2 = 12$) lattices with consideration of the interaction of the first two neighbors [268]. The adsorption isotherms for a square lattice are shown in Fig. 6.5. We can conditionally single out two regions of parameter values. The first one is realized at $|\beta\varepsilon_1| < 2\text{-}3$, $|\beta\varepsilon_2| < |\beta\varepsilon_1|/2$, and the second one at the opposite signs of the above relations.

In the first region, the isotherms are practically antisymmetric relative to a point with the coordinates [$\theta = 0.5$; $y(\theta = 0.5)$] and are sufficiently close to isotherms (6.18), though the difference between them increases with a growth in the repulsion of the first and, especially, the attraction of the second neighbors. For repulsion of both neighbors, the distinction between the two approximations may be disregarded. The consideration of indirect correlations causes θ to change more sharply with the pressure in the regions $0.15 < \theta < 0.30$ and $0.70 < \theta < 0.85$ and more smoothly for $0.40 < \theta < 0.60$. The extreme values of the isotherms at small and large coverages do not depend on the type of approximation. In the first region of the parameters, isotherms with $L = 0$, 1, and 2 phase transitions are realized. Their critical values are close to those indicated in Fig. 6.2. In the insert in Fig. 6.5, curves *5*, *6*, and *8* correspond to $L = 0$, curve *7*, to $L = 1$, and curve *9*, to $L = 2$ (at $\theta > 0.5$ they are antisymmetric).

In the second region of values of the parameters, the point of intersection of the isotherms in both approximations is displaced from $\theta = 1/2$ to smaller values. Here the region of existence of one phase transition increases in comparison with the similar region in Fig. 6.2. A similar situation is realized for three-dimensional lattices for which $|\beta\varepsilon_1| < 0.8\text{-}0.9$ and $|\beta\varepsilon_r| < |\beta\varepsilon_1|/4$ cor-

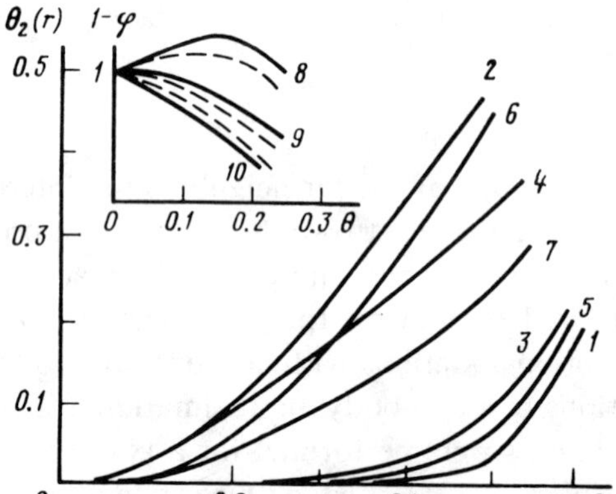

Fig. 6.6. Second correlation functions of a square lattice ($z = 4$) at $\beta\varepsilon_1 = -3.0$:

Curves *1*, *3*, *5*—first neighbors; *2*, *4*, *6*—second neighbors. Curves *3* and *4* have been calculated in a quasichemical approximation:
1-4—$\beta\varepsilon_2 = 1.5$; *5-6*—$\beta\varepsilon_2 = 0$; *8-10*—$\beta\varepsilon_1 < 0$; *8*—$\beta\varepsilon_2 > 0$; *9*—$\beta\varepsilon_2 = 0$; *10*—$\beta\varepsilon_2 < 0$

respond to the first region. Such a distinction is associated with a considerable increase in the number of first and second neighbors. It should be noted that the parameters relating to the second region correspond to the conditions of realization in the system of phase transitions that change the nature of uniform distribution of the particles employed when obtaining the expressions of Sec. 6.1.

Outside of the critical region, consideration of the indirect correlations substantially refines the description of particle distribution. Figure 6.6 shows the concentration dependences $\theta_{AA}(r)$. At $\theta > 0.85$, all the curves transform to the relation $\theta_{AA}(r) = \theta^2$, and curve *7* corresponds to this chaotic distribution. Consideration of the indirect correlations increases $\theta_{AA}(2)$ in comparison with (6.17) at any values of ε_1 and ε_2, while $\theta_{AA}(1)$ is greater than (6.17) if ε_1 and ε_2 are of the same sign, and is smaller if they are of opposite signs. The insert in Fig. 6.6 shows the change in the heat of adsorption (the quantity φ was defined in Sec. 2.3) in the region of small coverages in comparison with (6.19) (the dashed lines). As previously, at $\varepsilon_1 < 0$ the nature of the relations $Q(\theta)$ is determined by the sign of ε_2. Consideration of the indirect correlations changes the shape of the curves differently for positive and negative values of ε_2.

6.3. Equilibrium State of Inhomogeneous Systems

The results of Sec. 6.1 and Chap. 3 enable us to easily describe the distribution of the particles of a multicomponent system among the sites of an inhomogeneous lattice with consideration of the finite radius R of interaction between the particles. The total energy of such a system in a grand canonical ensemble will be written as follows:

$$H = \sum_{f,i} \nu_q^i \gamma_f^i \eta_f^q - \frac{1}{2} \sum_{r=1}^{R} \sum_{f,g_r} \sum_{ij} \varepsilon_{qp}^{ij}(r) \gamma_f^i \gamma_{g_r}^j \eta_f^q \eta_{g_r}^p \tag{6.20}$$

where $\varepsilon_{qp}^{ij}(r)$ is the parameter of interaction of particles i and j at sites of types q and p at a distance r; here $1 \leqslant r \leqslant R$. To use expression (6.20), one must know the distribution functions of clusters containing central sites and their coordination spheres. Let us denote these functions for one central site by $f(q\{[m]\}R)$ [the set M_2 (Sec. 3.1)], where the argument of the function is a list of the values of $m_{qp}(r)$ indicating how many sites of type p are at a distance r from the central site q. Expression (3.8) is generalized for these functions:

$$f(q\{[m]\}R) = f_q d(q\{[m]\}R), \quad \sum_{\sigma_{2q}} d(q\{[m]\}R) = 1 \tag{6.21}$$

where the sum is taken over all the types R of the coordination spheres.

The distribution of particles on the cluster $YR = q\{[m]\}R$ is described by correlators of two types. The correlators of the first type are determined as follows:

$$\theta_{YR}^{i\{[n]\}R} = \frac{\text{const }(\{[n]\}R)}{Q} \exp \beta \left[-\nu_q^i + \sum_{r=1}^{R} \sum_{p=1}^{t} \sum_{j=1}^{s} \varepsilon_{qp}^{ij}(r) n_{qp}^j(r) \right] \tag{6.22}$$

where $n_{qp}^j(r)$ is the number of particles j at $m_{qp}(r)$ sites of type p at a distance of r; here $m_{qp}(r) = \sum_{j=1}^{s} n_{qp}^j(r)$, while the correlators of the second type $\theta_{YR}^{i\{n\}R}$ contain at least $n_{qp}^j(r)$ particles j at $m_{qp}(r)$ sites of type p at a distance r from the center of a cluster of type q containing a particle i. The correlators of the two types are related by a system of linear coupling:

$$\begin{aligned} \theta_{YR}^{i\{[n]\}R} &= \prod_{r=1}^{R} \prod_{p=1}^{t} \left\{ \sum_{k_{qp}^1=0}^{\tilde{m}_{qp}} \cdots \sum_{k_{qp}^{s-1}=0}^{\tilde{m}_{qp}-\varkappa_{qp}} (-1)^{\tilde{k}_{qp}} C_{\tilde{m}_{qp}}^{\{k_{qp}\}} \right\}_r \theta_{YR}^{i\{n+k\}R} \\ \theta_{YR}^{i\{n\}R} &= \prod_{r=1}^{R} \prod_{p=1}^{t} \left\{ \sum_{k_{qp}^1=0}^{\tilde{m}_{qp}} \cdots \sum_{k_{qp}^{s-1}=0}^{\tilde{m}_{qp}-\varkappa_{qp}} C_{\tilde{m}_{qp}}^{\{k_{qp}\}} \right\}_r \theta_{YR}^{i\{[n+k]\}R} \end{aligned} \tag{6.23}$$

where the quantities $\tilde{m}_{qp}$, $\varkappa_{qp}$, etc. are determined for each coordination sphere in (3.47), but to simplify the notation, the subscript r is omitted. The abbreviated notation $\{n + k\}R$ signifies a complete set of values of $n_{qp}^j(r) + k_{qp}^j(r)$ for all the types of sites p in the coordination sphere r at all values of $r \leqslant R$.

A system of equations for correlators of the second type on the cluster YR is obtained in the standard way:

$$\begin{aligned} \theta_{YR}^{l\{n\}R} &= \exp \left[-\beta \left(\nu_q^{li}(YR) + \sum_{r=1}^{R} \sum_{p=1}^{t} \sum_{j=1}^{s-1} \varepsilon_{qp}^{lij}(r) n_{qp}^j(r) \right) \right] \\ &\times \prod_{r=1}^{R} \prod_{p=1}^{t} \left\{ \sum_{k_{qp}^1=0}^{\tilde{m}_{qp}} \cdots \sum_{k_{qp}^{s-1}=0}^{\tilde{m}_{qp}-\varkappa_{qp}} \prod_{j=1}^{s-1} (x_{qp}^{lij}(r))^{k_{qp}^j(r)} C_{\tilde{m}_{qp}}^{\{k_{qp}\}} \right\}_r \theta_{YR}^{i\{n+k\}R} \end{aligned} \tag{6.24}$$

$$\nu_q^{li}(YR) = \nu_q^l - \nu_q^i + \sum_{r=1}^{R} \sum_{p=1}^{t} m_{qp}(r)(\varepsilon_{qp}^{is}(r) - \varepsilon_{qp}^{ls}(r))$$

$$\varepsilon_{qp}^{lij}(r) = \varepsilon_{qp}^{ij}(r) + \varepsilon_{qp}^{ls}(r) - \varepsilon_{qp}^{lj}(r) - \varepsilon_{qp}^{is}(r) \tag{6.25}$$

$$x_{qp}^{lij}(r) = \exp(-\beta\varepsilon_{qp}^{lij}(r)) - 1$$

System of equations (6.24) generalizes Eqs. (3.49) for two-body interaction with a radius of R and Eqs. (6.8) for an inhomogeneous lattice. We shall consider only direct correlations. In this case, the correlators of the second and first types are expressed in terms of the functions θ_q^i and $\theta_{qp}^{ij}(r)$ in the form:

$$\theta_{YR}^{i\{n\}R} = \theta_q^i \prod_{r=1}^{R} \prod_{p=1}^{t} \prod_{j=1}^{s-1} (t_{qp}^{ij}(r))^{n_{qp}^j(r)}, \quad t_{qp}^{ij}(r) = \theta_{qp}^{ij}(r)/\theta_q^i$$

$$\theta_{YR}^{i\{[n]\}R} = \theta_q^i \prod_{r=1}^{R} \prod_{p=1}^{t} \prod_{j=1}^{s} (t_{qp}^{ij}(r))^{n_{qp}^j(r)}, \quad n_{qp}^s(r) = m_{qp}(r) - \sum_{j=1}^{s-1} n_{qp}^j(r) \tag{6.26}$$

While the functions themselves are determined by solving the following system of equations:

$$\theta_q^s = \theta_q^i \exp\beta(\nu_q^i - \nu_q^s) \sum_{\sigma_{2q}} d(q\{[m]\}R) \prod_{r=1}^{R} \prod_{p=1}^{t} [S_{qp}^i(r|s)]^{m_{qp}(r)}$$

$$S_{qp}^i(r|s_q) = \sum_{j=1}^{s_p} t_{qp}^{ij}(r) \exp\beta[\varepsilon_{qp}^{sj}(r) - \varepsilon_{qp}^{ij}(r)] \tag{6.27}$$

$$\theta_{qp}^{ij}(r)\theta_{qp}^{lk}(r) = \theta_{qp}^{ik}(r)\theta_{qp}^{lj}(r) \exp\beta[\varepsilon_{qp}^{ij}(r) + \varepsilon_{qp}^{lk}(r) - \varepsilon_{qp}^{ik}(r) - \varepsilon_{qp}^{lj}(r)]$$

The expression for $S_{qp}^i(r|s_q)$ takes into account that in the general case the number of species of particles at sites of different types may vary, as indicated in (3.68). We shall encounter situations of this kind in Sec. 6.5 below. For sorption problems considering the properties of the vacant sites, system (6.27) will be rewritten as follows:

$$y_q^i\theta_q^v = \theta_q^i \sum_{\sigma_{2q}} d(q\{[m]\}R) \prod_{r=1}^{R} \prod_{p=1}^{t} [S_{qp}^i(r|v)]^{m_{qp}(r)}$$

$$S_{qp}^i(r|v) = 1 + \sum_{j=1}^{s-1} t_{qp}^{ij}(r)x_{qp}^{ij}(r), \quad x_{qp}^{ij}(r) = \exp(-\beta\varepsilon_{qp}^{ij}(r)) - 1 \tag{6.28}$$

where the functions $t^{ij}_{qp}(r)$ are also determined from the solution of system of equations (6.27) for two-body functions, while y^i_q has been determined in (1.29). If the function of cluster distribution can be represented as follows [generalization (3.9)]:

$$d(q\{[m]\}R) = \prod_{r=1}^{R} C^{\{m_{qp}\}_r}_{z_q(r)} \prod_{p=1}^{t} (d_{qp}(r))^{m_{qp}(r)}, \quad \sum_{p=1}^{t} m_{qp}(r) = z_q(r) \tag{6.29}$$

where $z_q(r)$ is the number of sites in the coordination sphere r of a site of type q, then expression (6.28) become simplified:

$$y^i_q \theta^{\text{v}}_q = \theta^i_q \prod_{r=1}^{R} \left\{1 + \sum_{p=1}^{t} d_{qp}(r) \sum_{j=1}^{s-1} t^{ij}_{qp}(r) x^{ij}_{qp}(r)\right\} \tag{6.30}$$

In the particular case of two species of particles, the system for two-body probabilities (6.27) becomes simplified and is reduced to an explicit relation [generalization (3.20)]:

$$t^{\text{AA}}_{qp}(r) = 2\theta^{\text{A}}_p/(b_{qp}(r) + \delta_{qp}(r)), \quad \delta_{qp}(r) = 1 + x^{\text{AA}}_{qp}(r)(1 - \theta^{\text{A}}_q - \theta^{\text{A}}_p)$$
$$b_{qp}(r) = [(\delta_{qp}(r))^2 + 4x^{\text{AA}}_{qp}(r)\theta^{\text{A}}_q\theta^{\text{A}}_p]^{1/2} \tag{6.31}$$

In the molecular-field approximation, consideration of the second and following neighbors does not change the concentration dependence of the equations for θ^i_q (3.59) because system (6.25) reduces to system (3.59) if the effective parameters of interaction of the nearest neighbors are introduced:

$$\bar{\varepsilon}^{ij}_{qp} = \sum_{r=1}^{R} z_q(r) d_{qp}(r) \varepsilon^{ij}_{qp}(r)/(d_{qp}(1) z_q(1)) \tag{6.32}$$

To use the obtained adsorption isotherms, one must know the functions of distribution of sites of various types. A larger interaction radius increases the dimension of the required distribution functions. Let us consider for illustration the distribution functions of a point impurity and a stepped surface for $R = 2$.

A point impurity on a face (100) changes the properties of its nearest neighbors $t = 3$: $f_1 = c$, $f_2 = zc$, and $f_3 = 1 - (z + 1)c$, where c is the concentration of the impurity. The complete notation for the conditional probabilities (6.21) is written as $d(q[m_{q1}(1), m_{q2}(1), m_{q3}(1); m_{q1}(2), m_{q2}(2), m_{q3}(2)])$. The surface will be depicted on the following clusters:

$$d(1[0,\ 4,\ 0;\ 0,\ 0,\ 4]) = d(2[1,\ 0,\ 3;\ 0,\ 2,\ 2]) = 1$$

$$d(3[0,\ 2,\ 2;\ 1,\ 2,\ 1]) = d(3[0,\ 1,\ 3;\ 0,\ 0,\ 4]) = d(3[0,\ 0,\ 4;$$

$$0,\ 1,\ 3]) = 4c/f_3,\ d(3[0,\ 0,\ 4;\ 0,\ 0,\ 4]) = 1 - (12c/f_3)$$

Stepped Surface (Sec. 3.1), $t = 3$. The surface is imaged on five types of clusters as for consideration of the interaction of the nearest neighbors because the first and second neighbors are in the same adjacent row of sites (the distances for them are a and $a\sqrt{2}$, respectively);

$$d(1[2,\ 0,\ 1;\ 0,\ 1,\ 2]) = d(2[0,\ 2,\ 1;\ 1,\ 0,\ 2]) = 1,$$

$$d(3[1,\ 0,\ 3;\ 2,\ 0,\ 2]) = d(3[0,\ 1,\ 3;\ 0,\ 2,\ 2]) = (l - 2)^{-1},$$

$$d(3[0,\ 0,\ 4;\ 0,\ 0,\ 4) = (l - 4)/(l - 2)$$

To simplify the description of a surface, the function $d_{qp}(r)$ is employed. It characterizes the probability of finding a site of type p at a distance r from a site of type q. With their aid, the same inhomogeneous surfaces are set as follows [135].

A point impurity on a face (100): $z_q(r) = 4$, $r = 1$ and 2, $q = 1, 2, 3$.

$$d_{12}(1) = 1,\ d_{21}(1) = 1/4,\ d_{23}(1) = 3/4,\ d_{32}(1) = 3c/f_3,$$

$$d_{33}(1) = 1 - (3c/f_3),\ d_{22}(2) = 1/2,\ d_{23}(2) = 1/2,\ d_{13}(2) = 1,$$

$$d_{31}(2) = c/f_3,\ d_{32}(2) = c/2f_3,\ d_{33} = 1 - (3c/2f_3)$$

The remaining values of $d_{qp}(r)$ equal zero.

Stepped Surface. $z_1(r) = z_2(r) = 3$, $z_3(r) = 4$, $r = 1$ and 2. The functions $d_{qp}(1)$ are indicated in Sec. 3.1, therefore we shall write the nonzero distribution functions for the second neighbors:

$$d_{12}(2) = d_{21}(2) = 1/3,\ d_{13}(2) = d_{23}(2) = 2/3$$

$$d_{31}(2) = d_{32}(2) = [2(l - 2)]^{-1},\ d_{33}(2) = (l - 3)/(l - 2)$$

The distribution functions for other inhomogeneous surfaces are constructed similarly.

6.4. The Ordering of Particles in Sorption Problems

Consideration of interaction with the second and following neighbors enables us to describe the real distribution of the particles more accurately, in particular to describe the processes of their ordering. The use of a more accurate interaction potential expands the possibilities of describing a larger number of superstructure types. In the present section, we shall limit ourselves to the interaction of the first two neighbors and consider the sorption isotherms for a two- ($z = 4$) and a three-dimensional lattice ($z = 6$) for the formation of the superstructure $c(2 \times 2)$, and also construct equations for the flat superstructures $p(2 \times 1)(z = 4)$ and $c(2 \times 2)(z = 6)$.

In the first case, the lattice is divided into two sublattices consisting of α and γ sites. Every site α is surrounded by sites γ, and vice versa (Sec. 3.5). The lattice is depicted on two types of clusters: $d(\alpha[0, z_1; z_2, 0])$ and $d(\gamma[z_1, 0; 0, z_2])$, where z_1 and z_2 are the numbers of first and second neighbors. This allows us to write the adsorption isotherm as in (3.46), but the occupancies of the sites are determined by equations (6.27). Having in view that the sorption coefficient and the interaction parameters do not depend on the type of sublattice, we have [206, 269]:

$$y = \frac{\theta_\alpha}{1 - \theta_\alpha} (S_{\alpha\gamma})^{z_1} (S_{\alpha\alpha})^{z_2}, \quad y = \frac{\theta_\gamma}{1 - \theta_\gamma} (S_{\gamma\alpha})^{z_1} (S_{\gamma\gamma})^{z_2}$$

$$S_{\sigma\lambda} = 1 + t_{\sigma\lambda} x_r, \quad r = \begin{cases} 1, & \sigma \neq \lambda, \ \lambda = \alpha, \gamma \\ 2, & \sigma = \lambda = \alpha, \gamma \end{cases} \tag{6.33}$$

where the functions $t_{\sigma\lambda}$ are determined by formulas (6.31), while the setting of the subscripts σ and λ unambiguously determines r.

Figure 6.7 shows the occupancies of both sublattices and the sorption isotherms for the lattices $z_1 = z_2 = 4$ (curves *1-3*) and $z_1 = 6$, $z_2 = 12$ (curve *4*). The shape of the curves is similar to that of the curves in Fig. 3.12. The width of the ordered region depends on the relation of the parameters. When the repulsion of the first neighbors is larger, the ordered region is wider with respect to the pressure (curves *1* and *3*). This is facilitated by the increase in the attraction of the second neighbors (compare curves *1* and *2*).

Figure 6.8 shows the region of parameters corresponding to the existence of the superstructure $c(2 \times 2)$ (a fragment of Fig. 6.2). Ordering is absent below curve *CA* and is present in region *DCA*. An increase in the number of

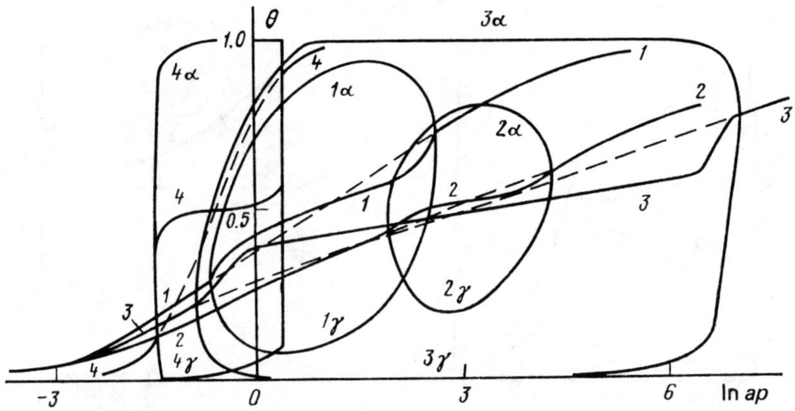

Fig. 6.7. Isotherms with (solid lines) and without (dashed lines) consideration of ordering of the particles and occupancy of the sublattices α and γ:

1—$\beta\varepsilon_1$ = −1.0; *2*— −1.5; *3*— −2.0; *4*—$\beta\varepsilon_1$ = −1.0; *1*—$\beta\varepsilon_2$ = 0.5; *2*—0; *3*— −0.5; *4*—$\beta\varepsilon_2$ = −1.5

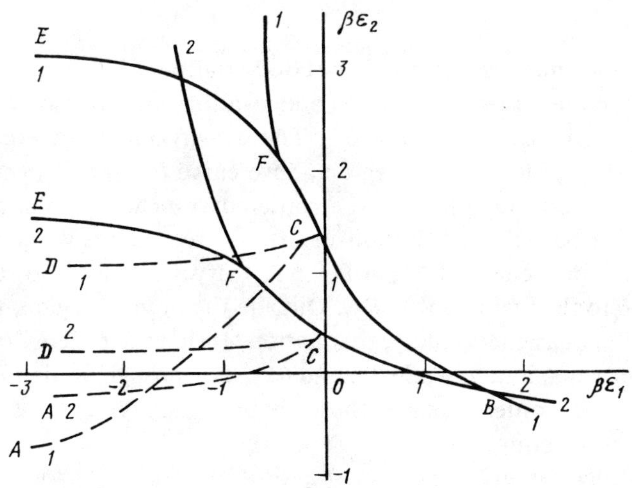

Fig. 6.8. Regions of the interaction parameters corresponding to different phase states of the particles:

1—$z_1 = z_2 = 4$; *2*—$z_1 = 6$, $z_2 = 12$

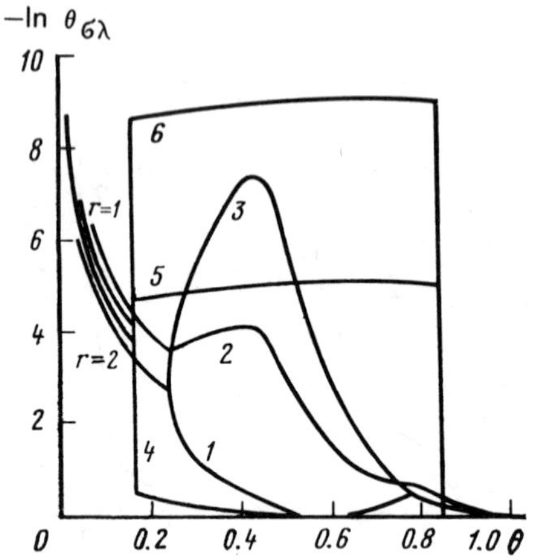

Fig. 6.9. Second correlation functions in a system that can be ordered, $z_1 = 6$, $z_2 = 12$: *1-3*—$\beta\varepsilon_1 = -1.0$; $\beta\varepsilon_2 = 0.1$; *4-6*—$\beta\varepsilon_1 = -0.5$; $\beta\varepsilon_2 = 0.4$; *1, 4*—$\theta_{\alpha\alpha}$; *2, 5*—$\theta_{\alpha\gamma}$; *3, 6*—$\theta_{\gamma\gamma}$

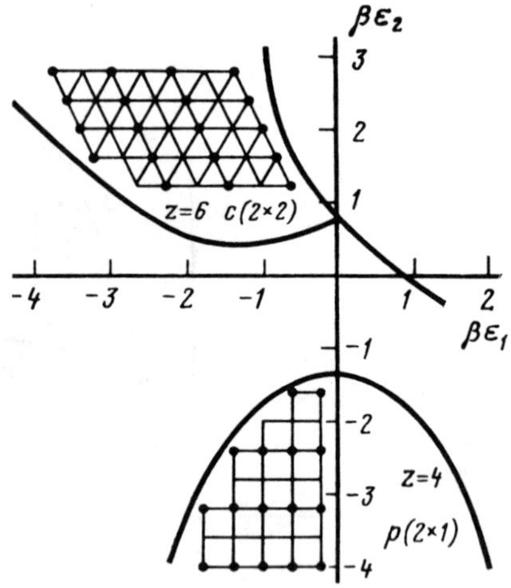

Fig. 6.10. Region of existence of ordered phases $p(2 \times 1)$ ($z = 4$) and $c(2 \times 2)$ ($z = 6$)

second neighbors may be attended by condensation of the ordered phase. Curve *DC* is the line of ordered phase condensation, and the isotherms above it have the shape of curve *4* in Fig. 6.7. The density jumps were determined by Maxwell's rule. Finally, in the region above curve *EF*, jumps in the density are observed both on the ordered and disordered branches of solutions (6.33).

The nature of particle distribution in the ordered region is shown by the concentration dependences of $\theta_{\sigma\lambda}$ in Fig. 6.9. Curves *1-3* relate to region *AC*, and curves *4-6* to the region above *DC*. Outside the ordered region, the curves behave as in the quasichemical approximation. Inside the ordered region, $\theta_{\alpha\alpha}$ passes through a maximum, while $\theta_{\alpha\gamma}$ and $\theta_{\gamma\gamma}$ pass through a minimum. With stratification of the ordered phase, these changes mainly occur at the order-disorder transition points.

The superstructures $p(2 \times 1)$ ($z = 4$) and $c(2 \times 2)$ ($z = 6$) for flat lattices are presented in Fig. 6.10. In both cases, the lattice sites are divided into two sublattices α and γ forming one-dimensional rows of sites of one type. As previously, $f_\alpha = f_\gamma = 1/2$ and $\theta = (\theta_\alpha + \theta_\gamma)/2$. The surfaces are imaged on two types of clusters, namely, $z = 4$, $d(\alpha[2, 2; 0,4]) = d(\gamma[2, 2; 4,0]) = 1$ and $z = 6$, $d(\alpha[2, 4; 2, 4]) = d(\gamma[4, 2; 4, 2]) = 1$. By equations (6.27), the occupancies of sites of various types are related by the following equations that are obtained when the quantity y is eliminated:

$$\frac{(1-\theta_\alpha)\theta_\gamma}{(1-\theta_\gamma)\theta_\alpha} = \left[\frac{S_{\alpha\gamma}(1)S_{\alpha\alpha}(1)[S_{\alpha\gamma}(2)]^2}{S_{\gamma\alpha}(1)S_{\gamma\gamma}(1)[S_{\gamma\alpha}(2)]^2}\right]^2$$
$$\frac{(1-\theta_\alpha)\theta_\gamma}{(1-\theta_\gamma)\theta_\alpha} = \left[\frac{S_{\alpha\alpha}(1)[S_{\alpha\gamma}(1)S_{\alpha\gamma}(2)]^2 S_{\alpha\alpha}(2)}{S_{\gamma\gamma}(1)[S_{\gamma\alpha}(1)S_{\gamma\alpha}(2)]^2 S_{\gamma\gamma}(2)}\right]^2 \qquad (6.34)$$

where the functions $S_{\sigma\lambda}(r)$ have been defined in (6.27), while the quantities $t_{\sigma\lambda}(r)$ in them are determined by formulas (6.31). As previously, the only solutions of Eqs. (6.34) correspond to a disordered system, while a plurality of solutions of these nonlinear equations corresponds to ordered systems. Figure 6.10 shows regions of existence of ordered phases $p(2 \times 1)$ for $z = 4$ and $c(2 \times 2)$ for $z = 6$.

Similarly, Eqs. (6.27) enable us to describe the equilibrium state of solid solutions becoming ordered with consideration of the two-body interaction potential for any number of neighbors.

6.5. Influence of a Gas on the State of a Solid

In the contact of a gas with a solid, the molecules of the gas phase can be adsorbed on the surface or dissolve in the bulk. Adsorption of the molecules changes the state of the subsurface region of the solid; outside of the critical conditions, this region is relatively small. Dissolved molecules have a much larger influence (because their number is considerably greater). During short intervals, the absorption process in a number of cases can be considered to proceed in a constant absorbent. The nature of its inhomogeneity depends on the conditions of its preparation [equilibrium in such a situation is described by Eqs. (6.27)]. During prolonged intervals, the absorbed particles change the state in the bulk of the solid. Its short-range and, possibly, long-range orders change.

In the latter case, there is needed a single system of equations describing the interphase equilibrium of the gas-phase molecules and the states of the solid because the atoms of the solid and the absorbed particles are components of the condensed phase with equal rights. The relevant system of equations is a particular case of system of equations (6.27). We shall consider the bulk of the condensed phase to be a set of sublattices each of which can be occupied by particles of a definite species. The number of such sublattices depends on the specific situation. As a minimum, sublattices accommodating atoms of the solid (sites of a substitution lattice) and sites of the interstices (accommodating absorbed particles) must be present. The sites of the substitution lattices, in turn, may also be represented by several sublattices if the alloy is ordered. Similarly, the sites of interstices may also be divided into several sublattices because in addition to the existence of various types of interstices (Sec. 3.7), ordering of the absorbed particles on a specific type of interstices is possible.

If we denote by u the number of sublattices, by s_σ—the number of species of particles in sublattice σ, where $1 \leqslant \sigma \leqslant u$, and by $z_{\sigma\lambda}(r)$ the number of sites in sublattices λ at a distance r from a site in sublattice σ, where $1 \leqslant r \leqslant R$, then system (6.27) will be rewritten as follows [206]:

$$\theta_\alpha^s = \theta_\alpha^i \exp \beta(\nu_\alpha^i - \nu_\alpha^s) \prod_{r=1}^{R} \prod_{\sigma=1}^{u} [S_{\alpha\sigma}^i(r|s_\alpha)]^{z_{\alpha\sigma}(r)}$$

$$S_{\alpha\sigma}^i(r|s_\alpha) = \sum_{j=1}^{s_\sigma} \theta_{\alpha\sigma}^{ij}(r) \exp \beta(\varepsilon_{\alpha\sigma}^{sj}(r) - \varepsilon_{\alpha\sigma}^{ij}(r))/\theta_\alpha^i \qquad (6.35)$$

$$\sum_{j=1}^{s_\sigma} \theta_{\alpha\sigma}^{ij}(r) = \theta_\alpha^i, \quad \sum_{i=1}^{s_\alpha} \theta_{\alpha\sigma}^{ij}(r) = \theta_\sigma^j, \quad \sum_{i=1}^{s_\alpha} \theta_\alpha^i = 1$$

The equations for $\theta_{\alpha\sigma}^{ij}$ retain the form of (6.27), where the type of a site corresponds to the number of the sublattice. It should be noted that all the energies of the bond between the components of the condensed phase "are contained" in the function $S_{\alpha\sigma}^i(r|s_\alpha)$. This is why the parameters ν_α^i do not contain the energy of the bond to the lattice, which was taken into account in them for a nonreorganizing solid. For absorbed particles, if s_α correspond to vacant site α of a sublattice, the quantity $(-\nu_\alpha^i)$ corresponds to the preexponential absorption constant.

Let us consider the process of rearrangement of a binary alloy $A_{0.5}B_{0.5}$ that can be ordered when H_2 molecules are absorbed into octa- (α) and tetrahedral (γ) interstices (here $u = 4$—two types of interstices and two sublattices for the alloy atoms). Each sublattice is occupied by two species of particles of A and B or H and v, therefore the distribution of the particles is given by formulas (6.31). The total fraction of absorbed particles is determined as $\theta = (\theta_\alpha + 2\theta_\gamma)/3$. The number of alloy atoms is constant, and a change in the external pressure changes the occupancies of both species of interstices and alters the ordering of the alloy (see the model calculation in Fig. 6.11). The latter effect is associated with the fact that the absorbed particles repel

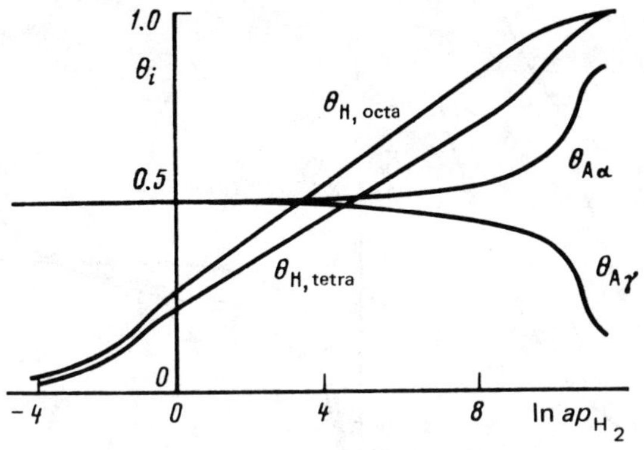

Fig. 6.11. Model calculations of the rearrangement of an alloy $A_{0.5}B_{0.5}$ that can be ordered (sublattices α and γ) in the adsorption of H_2 into the octa- and tetrahedral interstices of an fcc lattice:
$\beta\varepsilon_{AH} = 1.0$; $\beta\varepsilon_{BH} = 1.9$; $a_0 = e^{9.3}\ Pa^{-1.2}$; $\beta\varepsilon_{HH}(1) = -1.0$; $\beta\varepsilon_{HH}(2) = 0.2$; $\beta\omega_{AB} = 0.5$

one another, but at the same time are bonded strongly to the alloy atoms with which they can "form" quasiparticles MeH_n, where $n \leqslant 3$, because in an fcc lattice three interstices fall to the share of each alloy atom. The probability of the formation of such quasiparticles is different for atoms of A and B. Repulsion of the quasiparticles because of the repulsion of particles of H in the model situation being considered causes a sharp change in the distribution of the A and B atoms, which in the absence of H particles have a tendency to stratify.

If the bond between the alloy atoms and the hydrogen particles is relatively weak, an increase in the pressure changes only the short-range order of the alloy components, while the ordering of the H particles depends on the difference of the binding energies in different interstices and on the energy of repulsion between them.

The use of system (6.35) makes it possible to describe the experimental data on the absorption isotherms of the system H_2/Pd_xAg_{1-x}, where $0.6 \leqslant x \leqslant 0.9$, within the temperature interval from 30 to 148 °C using a single set of interaction parameters between the atoms of the alloy, between the alloy and hydrogen atoms, and between the hydrogen atoms [270] (Fig. 6.12). The lattice is fcc, $u = 2$ (sublattice of the alloy atoms and the interstices). Attention must be given to the great dependence of the behavior of the isotherms on

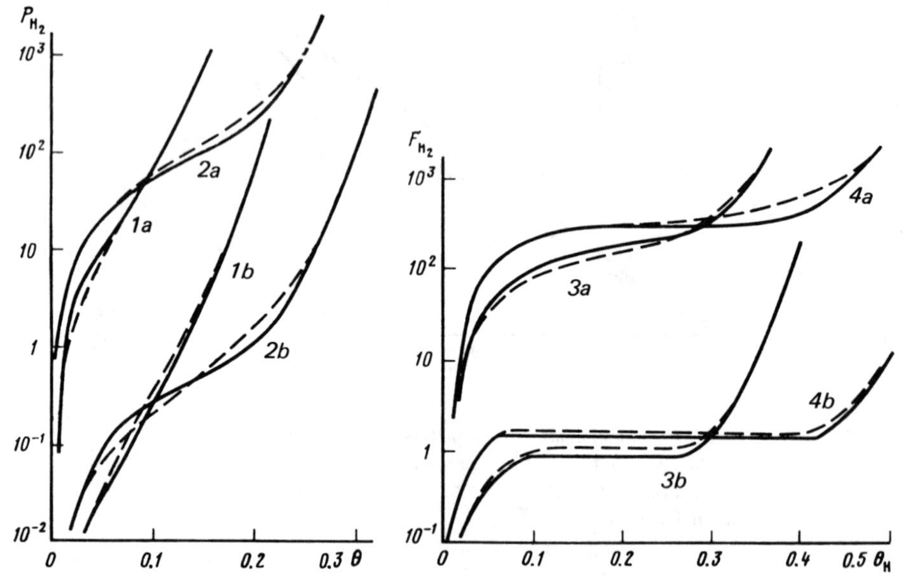

Fig. 6.12. Absorption of hydrogen by the alloy Pd_xAg_{1-x} (fcc lattice):
$\varepsilon_{PdH} = 3.75$; $\varepsilon_{AgH} = 7.5$; $\varepsilon_{HH}(1) = -3.6$; $\varepsilon_{HH}(2) = 1.0$; $\omega_{PdAg} = 0.4$ kJ/mol; $a_0 = e^{11}Pa^{-1.2}$. *1a-4a*—148 °C; *1b-4b*—30 °C: *1*—$x = 0.6$; *2*—0.7; *3*—0.8; *4*—$x = 0.9$

the composition of the alloy, namely, from a sharp growth (condensation) to a very slight growth in the occupancy depending on the external pressure.

System (6.35) can be used in a similar way to describe nonstoichiometric oxide compounds.

Let us see, as a different application of system (6.27), what changes occur in the equations of Secs. 3.8 and 3.9 describing the distribution of the atoms of a solid at the interface with vacuum and a gas phase. In a multilayer model of the subsurface region, a type of site q is brought into correspondence with each layer k, therefore $d(k\{[m]\}R) = 1$, where $m_{kq}(r)$ is the number of neighbors in layer q at a distance r from layer k, where $1 \leqslant r \leqslant R$. The value of q changes from layer k to $k \pm k_R$; the value of k_R differs from the number of the coordination sphere R, $k_R < R$ because the distances between adjacent coordination spheres in which the distances are measured are less than the distance a between adjacent layers. For instance, if $R = 2$, for the three-dimensional lattice $z = 6$ the value of $k_2 = 1$, and it retains this value for both $R = 1$ and 3 (the relevant distances are a, $a\sqrt{2}$, and $a\sqrt{3}$); the value of k_R is 2 only for $R = 4$, when the distance between particles is $2a$. The structure of a lattice sets an unambiguous correspondence between the set of values of $m_{kq}(r)$ and R.

The equations for the concentration profile of the subsurface region of a multicomponent solution with consideration of the direct correlations between R neighbors will be written as follows [195]:

$$\theta_k^s = \theta_k^i \exp \beta(\nu_k^i - \nu_k^s) \prod_{q=k+k_R}^{k-k_R} \prod_r [S_{kq}^i(r|s_k)]^{m_{kq}(r)}$$
$$S_{kq}^i(r|s_k) = \sum_{j=1}^{s_q} t_{kq}^{ij}(r) \exp \beta(\varepsilon_{kq}^{sj}(r) - \varepsilon_{kq}^{ij}(r)) \tag{6.36}$$

The equations for the functions $\theta_{kq}^{ij}(r)$ have the form of (6.27) in which the type of the site corresponds to the layer number. In (6.36), the first product is taken over different layers, and the second over different distances between the particles within the confines of layer q. The sum over the species of the particles in the function $S_{kq}^i(r|s)$ depends on the particles in layer q. Recall that for a flat phase interface, $k = 0$ relates to the adsorbed particles, and the layer $k \geqslant 1$ to the alloy (Secs. 3.8, 3.9). For an alloy-vacuum interface, the sites in the layers $1 \leqslant q \leqslant k_R$ have no neighbors at $q < 1$. Formally, this is easy to take into account if we consider that $m_{kq}(r) = 0$ at $q < 1$. A similar remark relates to the adsorption layer on a flat interface: $m_{kq}(r) = 0$ at $q < 0$.

Table 6.1 presents calculations of the concentration profile of the binary alloy $A_{0.7}B_{0.3}$ in vacuum with consideration of the interaction of the first and

Table 6.1. Influence of the Contribution of the Second Neighbors in a Quasichemical Approximation on the Concentration Profile of the Alloy $A_{0.7}B_{0.3}$

$\beta\omega$	α	Layer number				
		1	2	3	4	5
0.5	0	0.5389	0.6848	0.6986	0.6999	0.7000
0.5	0.125	0.4354	0.6620	0.6950	0.6994	0.6999
−0.5	0	0.5970	0.7048	0.6998	0.7000	0.7000
−0.5	0.125	0.5429	0.7122	0.6991	0.7001	0.7000
−1.0	0.125	0.5320	0.7267	0.6963	0.7005	0.6999
−1.0	−0.125	0.6436	0.7001	0.7000	0.7000	0.7000

second neighbors for a number of values of the exchange parameters $\omega(2) = \varepsilon_{AA}(2) + \varepsilon_{BB}(2) - 2\varepsilon_{AB}(2)$, and $\alpha = \omega(2)/\omega(1)$. It was assumed that the interaction parameters do not depend on the layer number; $\beta\varepsilon_{AA}(1) = 5$ and $\beta\varepsilon_{BB}(1) = 4$. As noted previously, positive values of the exchange parameters correspond to a tendency of stratification, and negative ones—of ordering. It follows from Table 6.1 that the contribution of the second neighbors substantially alters the concentration profile and especially the surface layer of the alloy, therefore when calculating the concentration profile one must not always consider only the interaction of the nearest neighbors.

6.6. Multiparticle Interactions

The concentration dependences of the thermodynamic characteristics of a gas-solid system depend greatly on the interaction of the particles. At present, two-body interaction potentials are chiefly employed. The question of their correctness generally remains open for discussion. This is due to the major difficulties in determining the real potentials. The general state of this problem has been given attention by a number of authors [271-275]. Intensive studying of the nonadditivity of interaction potentials was begun in the 1950's (see, for example, [276]). The inert gases are a classical example of the need to consider three-particle interactions [106]. Efforts aimed at describing them more precisely have not been discontinued to date. Investigations of the influence of three-particle interactions are being conducted for gases [277-279], liquids [280, 281], solids [282-284], adsorbed gases [285-287], etc. It should also be noted that for three-particle interactions accurate solutions are possible for homogeneous lattices [94]. Studies of interparticle potentials with more than three particles have been undertaken in recent years [104,90]. Their influence on the thermodynamic characteristics is only beginning to be determined. In particular, it was found [289] that a system cannot experience phase transitions of the second kind with an increase in the "dimension" of the potential.

The cluster approach enables one to construct a system of equations for the distribution of particles with any models of the interaction potential in a standard way. This procedure, formulated when considering indirect correlations in Sec. 2.2, remains unchanged. We shall discuss the general situation using the example of the arbitrary multiparticle interaction potential between nearest neighbors. We shall set the interaction potentials in the form of the relevant parameters without specifying their nature. For simplicity of the expressions, we shall limit ourselves to binary solutions.

The energy of a system with account taken of the multiparticle potential of interaction between nearest neighbors will be written as

$$H = \sum_{f,\, i} \nu_i \gamma_f^i - \frac{1}{(z+1)!} \sum_{f,\, i} \sum_{g_1 j_1} \cdots \sum_{g_z j_z} \varepsilon_{ij_1 \ldots j_z} \gamma_f^i \gamma_{g_1}^{j_1} \ldots \gamma_{g_z}^{j_z} \tag{6.37}$$

The quantity $\varepsilon_{ij_1 \ldots j_z}$ is the parameter of the $(z + 1)$—particle interaction of the central particle i with the neighboring particles j at sites g_m, where $1 \leqslant m \leqslant z$, of the first coordination sphere of a site f that characterizes the total energy of interaction of particle i with all its neighbors.

The probability of realizing a configuration of particles consisting of particle i at the center and of j_m particles at the neighboring sites g_m on a cluster with one central site is a correlation of the first type, which at a given Hamiltonian (6.37) has the form [291]:

$$\theta^i_{[n\sigma]} = \frac{\text{const }(n\sigma)}{Q} \exp \beta[-\nu_i + E_i(n\sigma)], \; E_i(n\sigma) = \varepsilon_{ij_1 \ldots j_z} \tag{6.38}$$

The quantity $E_i(n\sigma)$ depends on the specific configuration of the particles $j_1 \ldots j_z$. It is characterized, as in Sec. 2.2, by the values of n (the number of particles of A) and $\sigma(n)$ and is not related to the number of the sites at which these particles are. The correlators of the first and second types are related by expressions (2.20) and (2.21). By considering the ratios between the correlators (6.38) for i = A and B with an identical state of occupancy of the coordination sphere sites and excluding the correlators of the first type with the aid of formulas (2.21), we obtain the following system of equations:

$$\begin{aligned} \theta^{\mathrm{B}}_{n\sigma} &= \sum_{k=0}^{z-n} \sum_{\sigma(n+k)} b^{\sigma(n)}_{\sigma(n+k)} M_{n+k,\sigma} \\ &\times \sum_{r=0}^{z-n-k} \sum_{\sigma(n+k+r)} b^{\sigma(n+k)}_{\sigma(n+k+r)} (-1)^r \theta^{\mathrm{A}}_{n+k+r,\sigma} \end{aligned} \tag{6.39}$$

$$M_{n\sigma} = \exp \beta[\nu_{\mathrm{A}} - \nu_{\mathrm{B}} + E_{\mathrm{B}}(n\sigma) - E_{\mathrm{A}}(n\sigma)]$$

We shall limit ourselves to the simplest ways of closing system of equations (6.39) in which we shall ignore the influence of σ on the nature of decoupling of $\theta^i_{n\sigma}$ in terms of θ_{A} and θ_{AA}. Having in view the explicit form of $M_{n\sigma}$, we obtain:

quasichemical approximation (2.13):

$$1 - \theta = \theta \exp \beta(\nu_A - \nu_B) \sum_{k=0}^{z} t^k (1-t)^{z-k} \times \sum_{\sigma(k)} b_{\sigma(k)}^{\sigma(0)} \exp \beta[E_B(k\sigma) - E_A(k\sigma)]$$

$$1 - t = t \exp \beta(\nu_A - \nu_B) \sum_{k=0}^{z-1} t^k (1-t)^{z-1-k} \times \sum_{\sigma(k+1)} b_{\sigma(k+1)}^{\sigma(1)} \exp \beta[E_B(k+1, \sigma) - E_A(k+1, \sigma)] \qquad (6.40)$$

molecular-field approximation (2.17)

$$1 - \theta = \theta \exp \beta \left\{ \nu_A - \nu_B + \sum_{k=0}^{z} \theta^k (1-\theta)^{z-k} \times \sum_{\sigma(k)} b_{\sigma(k)}^{\sigma(0)} [E_B(k\sigma) - E_A(k\sigma)] \right\} \qquad (6.41)$$

The obtained equations can be employed for calculating the thermodynamic characteristics with an arbitrary interaction potential. The latter, in particular, can be set by a list of values of the energies (6.38) of the cluster being considered, as in quantum-chemical calculations. If $E_i(n\sigma) = \varepsilon_{iA} n + \varepsilon_{iB}(z - n)$, formulas (6.39)-(6.41) transform into formulas of Sec. 2.2. A simple approximation of the multiparticle potential is that in terms of the effective two-body potentials $E_i(n\sigma) = n\varepsilon_{iA}(n) + (z - n)\varepsilon_{iB}(n)$, where $\varepsilon_{ij}(n)$ is an arbitrary dependence of the energy contribution of the interaction of the central particle i with the adjacent particle j on the number of particles of A in the coordination sphere of a cluster. The linear approximation of the effective two-body potentials $\varepsilon_{ij}(n) = \varepsilon_{ij} + (n-1)\Delta\varepsilon_{ij}$ transforms expressions (6.41) to expression (2.17) containing the following renormalized parameters:

$$\tilde{\nu} = \nu - z\delta_{AB}, \quad \tilde{\omega} = \omega + z\delta_{AB} + (z-1)\delta\theta$$
$$\delta = \Delta\varepsilon_{AA} - \Delta\varepsilon_{AB} - \delta_{AB}, \quad \delta_{AB} = \Delta\varepsilon_{BA} - \Delta\varepsilon_{BB} \qquad (6.42)$$

Expressions (6.42) make it possible to substantiate the notions on the linear concentration dependence of the interaction parameters in the theory of subregular solutions [292-294]. But these notions are not substantiated for the quasichemical approximation [295].

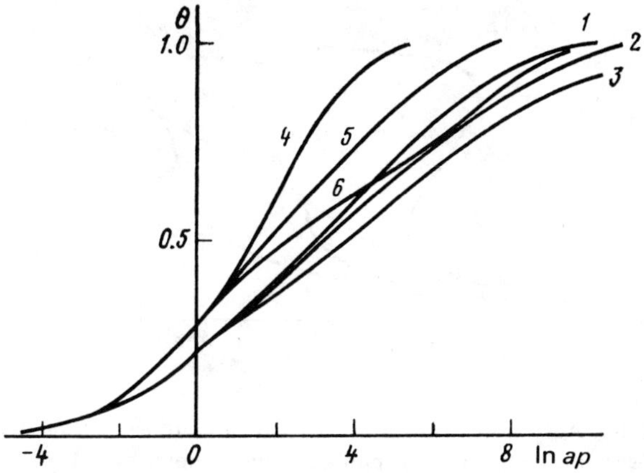

Fig. 6.13. Adsorption isotherms in the linear approximation of effective two-body potentials.

In the molecular-field approximation: *1*—$\beta\varepsilon_{AA} = -1.5$; $\beta\Delta\varepsilon_{AA} = 0$, $z = 4$; *2*—$\beta\varepsilon_{AA} = -1.5$, $\beta\Delta\varepsilon_{AA} = -0.1$, $z = 4$; *3*—$\beta\varepsilon_{AA} = -1.0$, $\beta\Delta\varepsilon_{AA} = -0.1$, $z = 6$; in the quasichemical approximation, $z = 4$: $\beta\varepsilon_{AA} = -1.0$; *4*—$\beta\Delta\varepsilon_{AA} = 0.2$; *5*—0; *6*—$\beta\Delta\varepsilon_{AA} = -0.2$

In sorption problems, $\nu_B = E_B(n\sigma) = 0$, and formulas (6.40) and (6.41) determine the isotherms in the relevant approximations. In particular, in the molecular-field approximation, the isotherm has the form of (2.24), where $\tilde{\varepsilon} = \varepsilon_{AA} + (z - 1)\Delta\varepsilon_{AA}\theta$. The sorption isotherms in both approximations are shown in Fig. 6.13. If the signs of ε_{AA} and $\Delta\varepsilon_{AA}$ coincide, the multiparticle interactions amplify the effects due to two-body interaction, while if their signs are opposite, they weaken the effect. The influence of multiparticle interactions begins to tell at $\theta \geqslant 1/3$, and it reaches a maximum at large densities. It should be noted that in the region of moderate coverages the isotherm can be approximated by relation (2.28). Figure 6.14 presents isotherms of particles that attract one another. For curves similar to curves *2-4*, when $\Delta\varepsilon_{AA} = 0$, the particles do not condense. At a given linear increase in the interaction energy with an increase in the number of neighbors, condensation begins for $\beta\varepsilon_{AA} \approx 0.95$. With the same law of a decrease in the interaction energy, there is no condensation at $\beta\varepsilon_{AA} = 1.435$ (curve *5*) (compare with Table 2.1). A change in the critical temperature depending on the contribution of multiparticle interactions is attended by a change in the critical density, namely, with an increase in $\Delta\varepsilon_{AA} > 0$ its value is displaced into the region $\theta > 1/2$, and

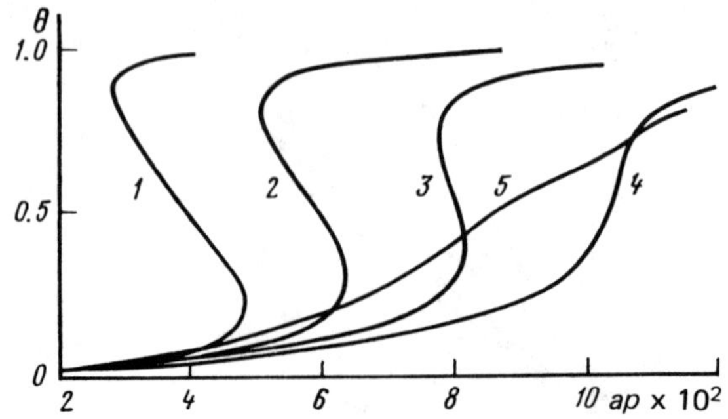

Fig. 6.14. Adsorption isotherms of particles attracting one another in the quasichemical approximation assuming the effective two-body potentials to be linear:

$z = 4$, $\beta\varepsilon_{AA} = 1.43$ (*1*, *5*), 1.25 (*2*), 1.11 (*3*), 1.0 (*4*); $\Delta\varepsilon_{AA}/\varepsilon_{AA} = 0.1$ (*1*-*4*), -0.1 (*5*)

at negative values of $\Delta\varepsilon_{AA}$, into the opposite region. These laws can easily be observed in the molecular-field approximation, for which analytical expressions for the critical quantities are obtained from (6.41):

$$\theta_c = [\{(1 - 2\delta)^2 + 6\delta\}^{1/2} + 1 - 2\delta]^{-1} \quad \delta = (z - 1)\Delta\varepsilon_{AA}/\varepsilon_{AA}$$
$$\beta_c\varepsilon_{AA}z = [\theta_c(1 - \theta_c)(1 + 2\delta\theta_c)]^{-1} \tag{6.43}$$

In Fig. 6.15, relations (6.43) are shown for a lattice $z = 4$. An increase in the contribution of multiparticle interactions increases the values of the critical density and temperature. The limiting value of $\beta_c\varepsilon_{AA}(\delta \to \infty) = 0$, i.e. condensation can be realized only at the expense of multiparticle interactions, and here $\theta_c = 2/3$. If $\Delta\varepsilon_{AA} < 0$, then θ_c decreases with a growth in the repulsion in accordance with $4\theta_c|\delta| = 1$ with a simultaneous increase in $\beta_c\varepsilon_{AA}$ (the particles are attracted as a result of two-body interaction).

The heat of adsorption in the molecular-field approximation changes with the coverage by the formula $Q(\theta) = \varepsilon_A + z\theta\bar{\varepsilon}m$. In the quasichemical approximation, the expression for it is obtained from system (6.40). Typical plots of Q vs. θ are shown in Fig. 6.16. At low densities, the main role is played by two-body interactions. The greatest contribution of multiparticle interactions manifests itself at large coverages. The multiparticle interactions disturb the

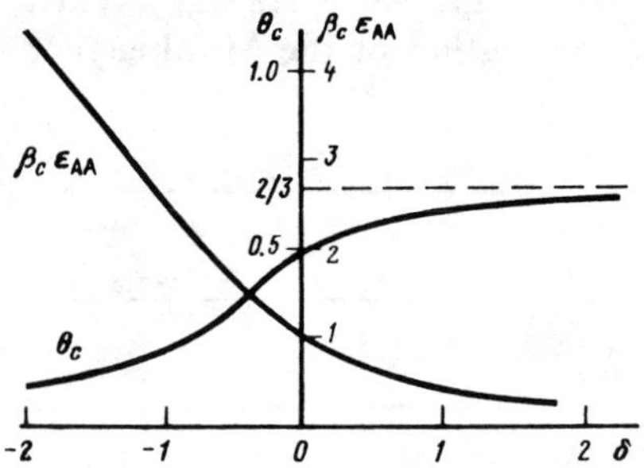

Fig. 6.15. Critical values of the surface coverage θ_c and the temperature $\beta_c\varepsilon_{AA}$ in the molecular-field approximation:

$\delta = (z - 1)\Delta\varepsilon_{AA}/\varepsilon_{AA}$

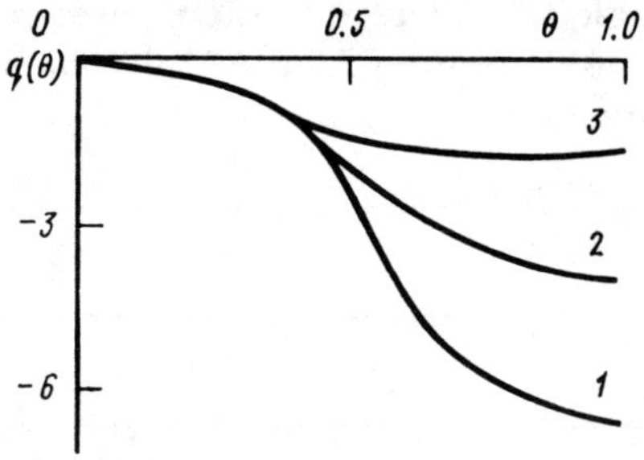

Fig. 6.16. Heat of adsorption in the quasichemical approximation assuming the effective two-body potentials to be linear:

$z = 4$, $\beta\varepsilon_{AA} = -1.0$; $\beta\Delta\varepsilon_{AA} = -0.2$ (*1*), 0 (*2*), and 0.2 (*3*)

antisymmetry of the plot of Q vs. θ relative to the point [$\theta = 0.5$; $Q(0.5)/\Delta Q = 0.5$]. It should be noted that the monotonic nature of the change in Q vs. θ, which is typical when considering the interaction of the nearest neighbors, may be disturbed. A slightly expressed minimum is realized for curve *3* (as a result of the selected parameters). In this circumstance, the quasichemical approximation differs in principle from the molecular-field one.

The expression $E_A(n\sigma) = n\varepsilon_{AA} + m_{n\sigma}\varepsilon_{AAA}$ corresponds to two- and three-body interactions. Here ε_{AAA} is the parameter of three-body interaction between the central particle and the two particles nearest to each other in its coordination sphere; $m_{n\sigma}$ is the number of triple contributions for the configuration $n\sigma$, where $0 \leqslant m_{n\sigma} \leqslant z_3$, z_3 is the maximum number of triple contributions to a completely occupied cluster, $z_3 = z$ for flat lattices and $z = 8$ in the bulk. For three-dimensional lattices, $z_3 = 2z$, if $z = 6$, 12, and $z_3 = 6$ if $z = 4$. In the molecular-field approximation, expressions are obtained similar to those considered above in which $(z - 1)\Delta\varepsilon_{AA}$ has been replaced by ε_{AAA}, which makes it possible to disclose the meaning of the linear approximation of the multiparticle interaction potential.

The cluster approach enables one to describe the distribution of particles with multiparticle interaction in inhomogeneous systems. The author has obtained equations for the concentration profile of the subsurface region of a binary alloy with its aid [195]. As previously, renormalization of the effective two-body interaction parameter depending on the local concentrations of the

Table 6.2. Change in the Concentration Profile of the Alloy $A_{0.7}B_{0.3}$ When Considering the Linear Molecular-Field Approximation of the Multiparticle Potential

α	Layer number				
	1	2	3	4	5
0	0.6647	0.7041	0.6995	0.7001	0.7000
0.25	0.7804	0.6856	0.7026	0.6995	0.7001

particles in various neighboring layers are obtained in the molecular-field approximation with linear approximation of the multiparticle potential. Table 6.2 presents an example of the influence of multiparticle interactions on the shape of the concentration profile. The calculations were performed for $\beta\varepsilon_{AB} = 5$, while the remaining parameters are the same as in Table 6.1 ($\alpha = \Delta\varepsilon_{ij}/\varepsilon_{ij}$). The multiparticle potential can qualitatively change the shape of the concentration profile obtained when only two-body potentials are considered: at the given parameters of the model, the less volatile component A accumulates on the surface.

The potential considering two-body interactions between the first two neighbors and three-body interaction between the nearest neighbors are considered in [195] for the concentration profile of the subsurface region of an alloy. The equations obtained describe the distribution of the particles in the bulk and in the transition region. The structure of the equations differs from that of Eqs. (6.39) in the presence of second sums over the particles in the second coordination sphere.

7 Evolution of a Gas-Solid System

The kind of interaction potential determines the behavior of the concentration dependences for various characteristics of a system. This matter was treated in the preceding chapter for the equilibrium state of a system. Here we shall study the influence of the kind of interaction potential on the kinetic characteristics of a gas-solid system. The structure of the fundamental kinetic equation for the total distribution function (4.2) and the kinetic equations for the correlators (4.3) does not depend on the kind of interaction potential. Its form manifests itself when constructing the probabilities of the occurrence of elementary processes. As in Chap. 6, we shall deal with the two-body potential with a finite radius of interaction and multiparticle interactions.

7.1. Kinetic Equations. Two-Body Interaction Potential

Let us construct a system of kinetic equations for correlators (4.3) when considering the interaction between R nearest neighbors, where R is any preset number of coordination spheres. We shall first consider a homogeneous lattice and see how the expressions change for the probabilities $W_\alpha(\{I\} \to \{II\})$ of one- and two-site processes in comparison with expressions (4.6) and (4.27) and what are the consequences of this change in the expressions of system (4.3). At the end of this section, we shall consider inhomogeneous lattices.

One-Site Reactions A → C [249, 206]. In the absence of an activated complex, the total energy of a system is described by formula (6.1). Assume that a one-site elementary process A → C occurs, where A and C are the components of a lattice solution containing s species of particles. The total energy of a system when one activated complex forms will be written as follows:

$$H_1 = \sum_{f,\ i}{}' \nu_i \gamma_f^i - \frac{1}{2}\sum_{r=1}^{R} \sum_{f,\ g_r}{}' \sum_{i,\ j} \varepsilon_{ij}(r)\gamma_f^i\gamma_{g_r}^j$$

$$+ \overset{*}{\nu}_A \gamma_k^A - \sum_{r=1}^{R} \sum_{j,\ g_r} \overset{*}{\varepsilon}_{Aj}(r)\gamma_k^A\gamma_{g_r}^j \tag{7.1}$$

The primes in the first two addends signify that the contribution of the activated complex at site k has been excluded in them. Its contribution is described by the last two addends. The quantity $\overset{*}{\nu}_A$ has been defined in (1.35), while $\overset{*}{\varepsilon}_{ij}(r)$ is the parameter of interaction of the activated complex formed from particle i with the neighboring particle j at the distance r. The transition of a particle of A to the activated state changes the energy of the system by the amount $\delta h_k^A = H_1 - H$ equal to

$$\delta h_k^A = (\overset{*}{\nu}_A - \nu_A - \sum_{r=1}^{R} \sum_{j,\ g_r} \delta\varepsilon_{Aj}(r)\gamma_{g_r}^j)\gamma_k^A,$$

$$\delta\varepsilon_{ij}(r) = \overset{*}{\varepsilon}_{ij}(r) - \varepsilon_{ij}(r) \tag{7.2}$$

Since this change in the system's energy occurs only when a particle of A is at site k, the probability of the formation of an activated complex at the given site is the probability of a set of events. It is determined as

$$W([\gamma_k^A] \rightarrow [\gamma_k^C]) = \gamma_k^A U_A(k), \ \ U_A(k) = \exp(-\beta\delta h_k^A) \tag{7.3}$$

Formula (7.3) coincides with formula (4.6) in which the quantity δh_k^A has the form of (7.2). As when considering the interaction of the nearest neighbors, it is easy to show that expressions (7.3) satisfy the principle of detailed equilibrium (4.1). It also follows from formula (7.3) that the kinetic equation for the local concentration of particles of A at site f has the form of (4.7) in which the addends are expressed as follows (we assign the number α to the step being considered):

$$U_f^i(\alpha) = K_i(\alpha)\left\langle \gamma_f^i \sum_{r=1}^{R} \prod_{g_r} \sum_{l=1}^{s} \gamma_{g_r}^l \exp \beta\delta\varepsilon_{il}(r \mid \alpha)\right\rangle_T \tag{7.4}$$

The two-body correlators $\theta_{fg}^{ij}(r)$ (2.12) are the next ones in dimension. Three kinds of the right-hand sides of Eqs. (5.3) are possible for them depending on the species of the particles i and j. As an example, we shall write an equation for one of the correlators:

$$\dot{\theta}_{fg}^{AC}(r) = U_{fg}^{\bar{C}C}(r|\alpha) - U_{fg}^{\bar{A}C}(r|\alpha) - U_{fg}^{A\bar{C}}(r|\alpha) + U_{fg}^{A\bar{A}}(r|\alpha) \tag{7.5}$$

where the capping bar signifies particles participating in a reaction. The other particle at a distance of r from the first one affects the magnitude of the activation barrier of step α. This influence varies for different r's. As a result, the system of kinetic equations for the two-body correlators has a larger dimension than when considering the interaction of the nearest neighbors because equations (7.5) are needed for all values of $r \leqslant R$. The addends of Eqs. (7.5) have the form:

$$\begin{aligned} U_{fg}^{\bar{i}j}(r|\alpha) &= K_i(\alpha)\left\langle \gamma_f^i \gamma_g^j \prod_{\varrho=1}^{R} \prod_{h_\varrho} \sum_{l=1}^{s} \gamma_{h_\varrho}^l \exp \beta\delta\varepsilon_{il}(\varrho|\alpha) \right\rangle_\tau \\ &= K_i(\alpha) \exp \beta\delta\varepsilon_{ij}(r|\alpha) \left\langle \gamma_f^i \prod_{\varrho=1}^{R} \prod_{h_\varrho}{}' \sum_{l=1}^{s} \gamma_{h_\varrho}^l \exp \beta\delta\varepsilon_{il}(\varrho|\alpha) \right\rangle_\tau \end{aligned} \tag{7.6}$$

The second equality takes into account that the particle j at a distance of r from the reacting particle i is fixed, while the prime signifies the absence of the multiplier $\gamma_{h_\varrho}^i$ when $h = g$ and $\varrho = r$. If $r > R$, the neighboring particle j does not directly affect the rate of the reaction at site f. When $r < R$, the dimension of correlators (7.6) coincides with that of correlators (7.4). In both cases, it exceeds the dimension of the correlators on the left-hand sides of Eqs. (4.7) and (7.5), consequently for closing the system of kinetic equations correlators (7.4) and (7.6) must be expressed in terms of unary and two-body correlators.

Two-Site Reaction A + B → C + D [296]. The activated complex of the reaction occupies two adjacent lattice sites (we shall designate them by n and k) and interact with the neighboring particles in the r-th coordination sphere, where $1 \leqslant r \leqslant R$, both relative to site n with a particle of A and relative to site k with a particle of B. Let us introduce the concept of a r-th coordination sphere of the dimer molecule AB as a set of sites at the same distance r from the closer of the sites A or B. The coordination spheres of the isolated sites n and k overlap, hence the single r-th coordination sphere contains part of the sites relating only to the r-th coordination sphere of one of the sites n or k, and part of the sites relating to the $(r + \lambda)$ coordination sphere of the adjacent site k or n. The value of λ depends on the dimension of the lattice and the number of nearest neighbors.

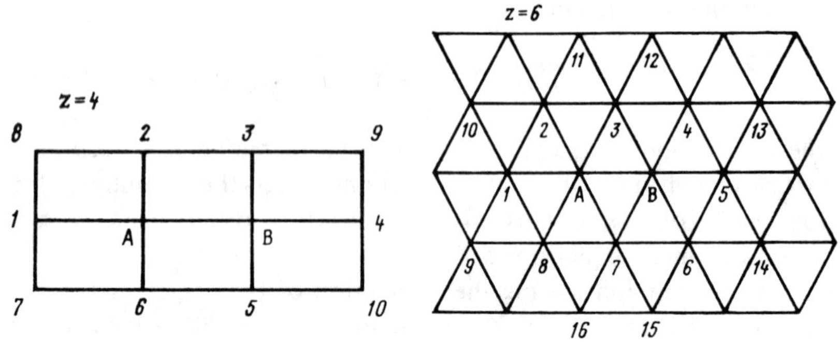

Fig. 7.1. Distribution of the sites of flat lattices for $z = 4$ and 6 among the equivalent groups of the first two coordination spheres.

For $z = 4$, the first coordination sphere includes sites 1-6 ($\pi_1 = 4$), the second—sites 7-10 ($\pi_2 = 2$); site 1 corresponds to $\omega_1 = 1$; sites 2 and 6 to $\omega_1 = 2$; sites 3 and 5 to $\omega_1 = 3$; site 4 to $\omega_1 = 4$; sites 7 and 8 to $\omega_2 = 1$; and sites 9 and 10 to $\omega_2 = 2$. For $z = 6$, the first coordination sphere includes sites 1-8 ($\pi_1 = 5$), the second—sites 9-16 ($\pi_2 = 4$); site 1 corresponds to $\omega_1 = 1$, sites 2 and 8 to $\omega_1 = 2$, sites 3 and 7 to $\omega_1 = 3$; sites 4 and 6 to $\omega_1 = 4$; site 5 to $\omega_1 = 5$; sites 9 and 10 to $\omega_2 = 1$; sites 11 and 16 to $\omega_2 = 2$; sites 12 and 15 to $\omega_2 = 3$, and sites 13 and 14 to $\omega_2 = 4$

We divide the multitude of sites of the r-th coordination sphere into equivalent groups of sites that are set by their orientation. The orientation of the sites is determined by the angle formed by the straight line connecting the central sites and that between the site being considered in the r-th coordination sphere of the dimer molecule and the site at the middle of the distance between the central sites. Let us denote the number of different orientations in the single r-th coordination sphere by π_r and the number of sites with the given orientation ω_r by $\varkappa_{\omega r}$ ($1 \leqslant \omega_r \leqslant \pi_r$). Figure 7.1 shows the distribution of the sites of flat lattices for $R = 2$.

We shall designate the interaction potential of the activated complex with the particle j at a site of the r-th single coordination sphere of a dimer molecule AB with the orientation ω_r by $\overset{*}{\varepsilon}_{ABj}(\omega_r)$. The quantities $\overset{*}{\varepsilon}_{ABj}(\omega_r)$ are expressed in terms of the two-body potentials $\overset{*}{\varepsilon}_{ij}(r)$ as follows: $\overset{*}{\varepsilon}_{ABj}(\omega_r) = \overset{*}{\varepsilon}_{Aj}(r_{nm}) + \overset{*}{\varepsilon}_{Bj}(r_{km})$, where m is the number of the site accommodating the particle j. For instance, for a flat lattice $z = 4$ and $R = 2$, we have:

$$\overset{*}{\varepsilon}_{ABj}(\omega_1 = 1) = \overset{*}{\varepsilon}_{Aj}(1),\ \overset{*}{\varepsilon}_{ABj}(\omega_1 = 2) = \overset{*}{\varepsilon}_{Aj}(1) + \overset{*}{\varepsilon}_{Bj}(2)$$

$$\overset{*}{\varepsilon}_{ABj}(\omega_1 = 3) = \overset{*}{\varepsilon}_{Aj}(2) + \overset{*}{\varepsilon}_{Bj}(1),\ \overset{*}{\varepsilon}_{ABj}(\omega_1 = 4) = \overset{*}{\varepsilon}_{Bj}(1)$$

$$\overset{*}{\varepsilon}_{ABj}(\omega_2 = 1) = \overset{*}{\varepsilon}_{Aj}(2),\ \overset{*}{\varepsilon}_{ABj}(\omega_2 = 2) = \overset{*}{\varepsilon}_{Bj}(2)$$

The quantities $\overset{*}{\varepsilon}_{ABj}(\omega_r)$ for other values of z and R are constructed similarly. If A = B, the symmetric positions of the sites of the single r-th coordination sphere relative to a plane passing through the middle of the straight line connecting the central sites and at right angles to this line result in identical values of $\overset{*}{\varepsilon}_{AAj}(\omega_r)$, where $1 \leqslant \omega_r \leqslant \pi_r/2$, which double values of $\varkappa_{\omega r}$ correspond to.

The change in the total energy of system (6.1) as a result of the formation of an activated complex of a two-site elementary process will be written as follows [296]:

$$\delta h_{nk}^{AB} = [\overset{*}{\nu}_{AB} + \varepsilon_{AB}(1)]\gamma_n^A\gamma_k^B - \nu_A\gamma_n^A - \nu_B\gamma_k^B$$
$$- \sum_{r=1}^{R}\sum_{\omega_r=1}^{\pi_r}\sum_{j,\, g_{\omega r}} \delta\varepsilon_{ABj}(\omega_r)\gamma_n^A\gamma_k^B\gamma_{g_{\omega r}}^j \tag{7.7}$$
$$\delta\varepsilon_{ABj}(\omega_r) = \overset{*}{\varepsilon}_{ABj}(\omega_r) - \varepsilon_{ABj}(\omega_r)$$

The first four addends are identical to those of expression (4.26), while the last one considers the difference between the interactions of the activated complex with the neighboring particle j and of the reactants A and B with the same particle. The parameters $\varepsilon_{ABj}(\omega_r)$ are constructed from the parameters of the two-body interaction potential $\varepsilon_{ij}(r)$ in exactly the same way as the parameters $\overset{*}{\varepsilon}_{ABj}(\omega_r)$ (with omission of the asterisk).

The formation of the activated complex changes the energy of the system by the amount δh_{nk}^{AB} only with the simultaneous location of particle A at site n and particle B at site k. Hence, the probability of the formation of an activated complex of the elementary process A + B is also determined by formula (4.27) in which δh_{nk}^{AB} is expressed by formula (7.7). (That the principle of detailed equilibrium is observed here is proved similarly to how it was done in Sec. 4.3.) It follows that the kinetic equation for the local concentration of particles of A at site f when the two-site reaction being considered proceeds coincides in form with equation (4.29) in which the components are expressed as follows:

$$U_{fg}^{ij}(\alpha) = K_{ij}(\alpha)e^{-\beta\varepsilon_{ij}(1)}\left\langle \gamma_f^i\gamma_g^j \prod_{r=1}^{R}\prod_{\omega_r=1}^{\pi_r}\prod_{h_{\omega r}}\sum_{l=1}^{s}\gamma_{h_{\omega r}}^l e^{\beta\delta\varepsilon_{ijl}(\omega_r|\alpha)}\right\rangle_\tau \tag{7.8}$$

where the subscript h numbers the sites in the equivalent group of them $\varkappa_{\omega r}$; as previously, the step number α has been introduced additionally. The correlator in expression (7.8) has been constructed on a cluster with two central sites f and g containing R coordination spheres of the dimer molecule ij.

The equations for the function $\theta_{fg}^{ij}(r)$ are the following equations of system (4.3). For them, in addition to the differences in the right-hand sides associated with the species of particles i and j [see (5.5)], differences associated with the quantity r appear because the particles react with one another only if they are at neighboring sites of a lattice. Therefore, for example, for the correlator $\theta_{fg}^{AB}(r)$, the kinetic equations will be written as follows:

$$\dot{\theta}_{fg}^{AB}(r) = \Delta_{1r}[U_{fg}^{CD}(\alpha) - U_{fg}^{AB}(\alpha)] - \sum_{h} [U_{hfg}^{\overline{B}AB}(\omega_r|\alpha) - U_{hfg}^{\overline{D}CB}(\omega_r|\alpha)] - \sum_{h} [U_{fgh}^{AB\overline{A}}(\omega_r|\alpha) - U_{fgh}^{AD\overline{C}}(\omega_r|\alpha)] \quad (7.9)$$

where the first addend describes the reaction at the central sites when $r = 1$, while the remaining addends describe the reactions proceeding at one of the central sites and its nearest neighbors; the subscript h in the first sum is taken over all the neighbors of site f except for g, and in the second sum over all the neighbors of site g except for f. The value of ω_r relating to particles not participating in the reaction (a particle of B at site g in the first sum and one of A at site f in the second sum) is unambiguously associated with the number of site h accommodating a particle entering into a reaction with another particle at site f or g; both are capped with a bar.

The addends of formula (7.9) have the form:

$$U_{hfg}^{\overline{i}jl}(\omega_r|\alpha) = K_{ij}(\alpha)e^{-\beta\varepsilon_{ij}(1)}\left\langle \gamma_h^i\gamma_f^j\gamma_g^l \prod_{r=1}^{R} \prod_{\omega_r=1}^{\pi_r} \prod_{\xi\omega r} \sum_{m=1}^{s} \gamma_{\xi_{\omega r}}^m \exp \beta\delta\varepsilon_{ijm}(\omega_r|\alpha)\right\rangle_T \quad (7.10)$$

The dimension of correlators (7.10) coincides with that of correlators (7.8). As for one-site reactions, they must be expressed in terms of unary and two-body correlators to close kinetic equations (4.29) and (7.8)-(7.10).

Approximations [250]. Let us use the approximations considered in Sec. 6.1 when closing the systems of equations for equilibrium correlators. The general relations between the correlators of the first and second types also hold in the absence of equilibrium, as do the normalizing relations.

One-Site Reactions. In the quasichemical and molecular-field approximations, the rates of one-site reactions are expressed as follows:

$$U_f^i(\alpha) = K_i(\alpha)\theta_f^i \begin{cases} \prod_{r=1}^{R} \prod_{g_r} S_{fg}^i(r|\alpha) & \text{QCA} \\ \exp \beta \sum_{r=1}^{R} \sum_{g_r} \sum_{l=1}^{s} \theta_{g_r}^l \delta\varepsilon_{il}(r|\alpha) & \text{MFA} \end{cases} \quad (7.11)$$

$$U_{fg}^{\bar{i}\bar{j}}(r|\alpha) = U_f^i(\alpha)\psi_{fg}^{ij}(r|\alpha), \quad \psi_{fg}^{ij}(r|\alpha) = t_{fg}^{ij}(r)\exp\beta\delta\varepsilon_{ij}(r|\alpha)/S_{fg}^i(r|\alpha)$$

$$S_{fg}^i(r|\alpha) = \sum_{j=1}^{s} t_{fg}^{ij}(r)\exp\beta\delta\varepsilon_{\bar{i}\bar{j}}(r|\alpha) \tag{7.12}$$

Formula (7.12) yields an expression for the addend of equations of type (7.5) describing the change in the two-body correlators. In the molecular-field approximation, the equations for the local concentrations close at the level of the functions θ_f^i, while equations of type (7.5) are absent. It is also possible to use "intermediate" approximations that retain the effects of correlation at distances of $R_1 < R$ and "ignore" them at distance exceeding R_1. This results in a combination of the multipliers from expressions (7.11), namely, the product over r from unity to R_1 and the exponential term containing the sum over r from $R_1 + 1$ to R. In the last case, one must have equations of the functions $\theta_{fg}^{ij}(r)$, where $1 \leqslant r \leqslant R_1$.

In the general case when considering indirect correlations, the dimension of the correlators in terms of which the higher correlators are uncoupled grows, and the dimension of the system of kinetic equations increases. If we employ a two-body approximation when considering indirect correlations, the dimension of the system of kinetic equations remains the same as in the quasichemical approximation, but the form of the right-hand sides becomes more involved [296] (see Appendix 1).

Two-Site Reactions. In the molecular-field approximation, the rate of a reaction has the following form:

$$U_{fg}^{ij}(\alpha) = K_{ij}(\alpha)\theta_f^i\theta_g^j\exp\beta\left[\sum_{r=1}^{R}\sum_{h_r}{}'\sum_{l=1}^{s}\{\theta_{h_r}^l\delta\varepsilon_{il}(r_{fh}) + \theta_{h_r}^l\delta\varepsilon_{jl}(r_{gh})\} - \varepsilon_{ij}(1)\right] \tag{7.13}$$

where when $r = 1$ the subscript h does not equal the subscripts g and f for the first and second addends in braces, respectively.

Formula (7.13) closes the equations for the local concentrations, but the interaction parameter ε_{ij} present in the exponent leads to artificial renormalization of the activation energy of the reaction, as was noted previously in Chap. 4.

The following situation arises when employing the quasichemical approximation. An activated complex AB forms from two neighboring particles, and its interaction with a neighboring particle j makes it necessary to calculate the probability of formation of the closed three-particle configuration ABj. For the strict calculation of such a probability, one must have an equation for three-particle correlators. In the quasichemical approximation, the probabilities of any multiparticle configurations is expressed in terms of the probability of pairs of particles, which does not reflect the closed nature of their configurations. This is why the simplicity of the employed approximation has to be "paid for" by the presence of other shortcomings. The quasichemical approximation can be used in two ways. In the first one, the probability of each pair of particles of a three-particle configuration is considered independently of the probabilities of the two other pairs (a "purely" quasichemical interpretation). In this case, the coordination sphere of each central site is considered independently, and the concept of a single coordination sphere of a dimer molecule formed of central particles does not exist. Hence

$$U_{fg}^{ij}(\alpha) = K_{ij}(\alpha)\theta_{fg}^{ij}(1)\exp(-\beta\varepsilon_{ij}(1))\prod_{r=1}^{R}\prod_{h_r}{}' S_{fh}^{i}(r_{fh}|\alpha)S_{gh}^{j}(r_{gh}|\alpha)$$

$$h_1 \neq g,\ f \qquad (7.14)$$

$$U_{hfg}^{\overline{ijn}}(\omega_r|\alpha) = U_{hf}^{ij}(\alpha)\psi_{fg}^{jn}(r_{fg}|\alpha)\psi_{hg}^{in}(r_{hg}|\alpha).$$

A shortcoming of this way is the fact that the neighboring particle j interacts not with a single activated complex, but separately with various fragments thereof.

The second way consists in consideration of the closed nature of the three-particle configuration ABj by analogy with the two-body approximation when considering the indirect correlations [296]:

$$U_{fg}^{ij}(\alpha) = K_{ij}(\alpha)\theta_{fg}^{ij}(1)\exp(-\beta\varepsilon_{ij}(1))\prod_{r=1}^{R}\prod_{\omega_r=1}^{\pi_r}\prod_{h_{\omega r}} S_{fgh}^{ij}(\omega_r|\alpha) \qquad (7.15)$$

$$S_{fgh}^{ij}(\omega_r|\alpha) = \sum_{l=1}^{s} t_{fgh}^{ijl}(\omega_r)\exp\beta\delta\varepsilon_{ijl}(\omega_r|\alpha)$$

$$U_{hfg}^{\overline{ijn}}(\omega_r|\alpha) = U_{hf}^{ij}(\alpha)\psi_{hfg}^{ijn}(\omega_r|\alpha) \qquad (7.16)$$

$$\psi_{hfg}^{ijn}(\omega_r|\alpha) = t_{hfg}^{ijn}(\omega_r)\exp\beta\delta\varepsilon_{ijn}(\omega_r|\alpha)/S_{hfg}^{ij}(\omega_r|\alpha)$$

In formula (7.15), site h relates to the sites of the equivalent group $\varkappa_{\omega r}$. In formula (7.16), the reaction is considered to proceed at sites h and f, while site g is at distance r with the orientation ω_r relative to the dimer molecule

ij. The function $t_{hfg}^{ijn}(\omega_r)$ is

$$t_{hfg}^{ijn}(\omega_r) = \frac{\theta_{hg}^{in}\theta_{fg}^{jn}}{\theta_h^i\theta_f^j\theta_g^n} = t_{hg}^{in}(r)\xi_{fg}^{jn}(r+\lambda) = t_{fg}^{jn}(r+\lambda)\xi_{hg}^{in} \tag{7.17}$$

where $\xi_{fg}^{ij}(r) = t_{fg}^{ij}(r)/\theta_g^j$ is the contribution of the indirect correlation of molecules i and j at a distance r. The quantity λ indicates the shift in the count to site g from the central site h or f adopted as the reference point. If $r_{hg} > R$, then $\xi_{hg}^{ij} = 1$. For a surface with $z = 6$ in the position $\omega_1 = 3$ (see Fig. 7.1), the particle n is simultaneously the nearest neighbor of the two central particles i and j; this situation is not taken into account by formula (4.33).

The second way reflects the energetics of the process correctly. Approximation (7.17) is exact for noninteracting particles. With increasing interaction, its accuracy diminishes; the latter also depends on the particle concentration. This follows from the consideration of only the normalizing relations between the correlators of different dimension. No account is taken of the different change in the correlators depending on the magnitude of the particle interaction because an equation for the three-body correlators is absent. A shortcoming of this procedure consists in the disturbance of the condition of normalization for three-body correlators $\sum_{n=1}^{s} t_{hfg}^{ijn}(\omega_r) = 1$ with strong interactions.

Formulas (7.14) and (7.17) coincide in the region of large or small concentrations of any of the solution components, but differ in the region of their commensurable concentrations. Notwithstanding the indicated shortcomings of both ways of using the quasichemical approximation, in both cases the self-consistent nature of the description of the kinetics and equilibrium of nonideal reaction systems is retained. As previously, it is possible to use "intermediate" approximations considering the effects of correlation at distances of $R_1 < R$, and also approximations considering the indirect correlations [250].

Equilibrium as the Equality of the Rates of the Forward and Reverse Directions of an Elementary Process [296]. The self-consistent nature of the description of the kinetics of chemical reactions and equilibrium with consideration of the interaction of the nearest neighbors was shown in Secs. 4.2 and 4.3 directly in the quasichemical approximation. By equating the rates of opposite directions of an elementary process, we can exclude the parameters of an activated complex and obtain equilibrium relations between the concentrations of the components of a system. Here this matter is considered in a general form.

Let us rewrite formula (7.4) in the following form: $U_i(\alpha) = K_i(\alpha)V_i(\alpha)$, where

$V_i(\alpha)$ is the concentration component of the rate of the reaction of component i in step α (for a state of equilibrium on a homogeneous lattice, the numbers of the sites may be omitted). For a one-site process A $\rightleftarrows$ C, the constant of chemical equilibrium K_{eq} is obtained from the condition $V_A(\alpha) = V_C(\alpha)$:

$$K_{eq} = K_A(\alpha)/K_C(\alpha) = V_C(\alpha)/V_A(\alpha) \tag{7.18}$$

Let us consider the relation of the correlators of the first type (6.3). We multiply both sides of it by the quantity $\exp \beta \sum_{r} \sum_{l=1}^{s} \varepsilon^*_{Al}(r|\alpha) n_r^l$ and take into account that $\varepsilon^*_{Al}(r|\alpha) = \varepsilon^*_{Cl}(r|\alpha)$ is the parameter of interaction of an activated complex with a neighboring particle l and does not depend on the direction of the reaction, i.e.

$$\theta^{A}_{\{[n]\}R} \exp \beta \sum_{r=1}^{R} \sum_{l=1}^{s} [\varepsilon^*_{Al}(r|\alpha) - \varepsilon_{Al}(r)] n_r^l$$

$$= \theta^{C}_{\{[n]\}R} \exp \beta\left(\nu_C - \nu_A + \sum_{r=1}^{R} \sum_{l=1}^{s} [\varepsilon^*_{Cl}(r|\alpha) - \varepsilon_{Cl}(r)] n_r^l\right) \tag{7.19}$$

Further multiplying the left-hand and right-hand sides of expression (7.19) by $\prod_{r=1}^{R} C_{z_r}^{\{k_r\}}$, summing over all possible numbers of particles of each species for each coordination sphere and over all the coordination spheres, we obtain

$$V_A(\alpha) = \exp \beta(\nu_C - \nu_A) V_C(\alpha) \tag{7.20}$$

The exponent in (7.20) can be excluded by using system (6.8), which we shall rewrite in the following form $(\{n\}R = \{0\}R)$:

$$\theta_A = \theta_C \exp \beta(\nu_C - \nu_A) W_{CA} \tag{7.21}$$

where W_{CA} is a function generally interpreted as the ratio of the activity coefficients of particles of A and C. It depends on the concentration of the components of the system and their interaction in the ground state. As a result, we find from formulas (7.18), (7.20), and (7.21) that the constant of chemical equilibrium is expressed in terms of the equilibrium concentrations and the potentials of particle interactions

$$K_{eq} = \theta_C W_{CA}/\theta_A \tag{7.22}$$

The self-consistent nature of the obtained kinetic expressions for two-site processes A + B ⇄ C + D is also proved simply. The cluster correlators of the first type with two central sites and with their R united coordination spheres are determined as follows:

$$\theta^{il}_{\{[m]\}R} = \frac{\text{const}\,(\{[m]\}R)}{Q} \exp\beta\left[\varepsilon_{il}(1) - \nu_i - \nu_l + \sum_{r=1}^{R}\sum_{\omega_r=1}^{\pi_r}\sum_{j=1}^{s}\varepsilon_{ilj}(\omega_r)m^j_{\omega r}\right] \tag{7.23}$$

By considering their ratio for the central pairs AC and BD, multiplying both sides by $\prod_{r=1}^{R}\prod_{\omega_r} C^{\{m_{\omega r}\}}_{\mathcal{K}\omega r} \exp\beta \sum_r \sum_{\omega_r}\sum_j \overset{*}{\varepsilon}_{ACj}(\omega_r|\alpha)m^j_{\omega r}$, and summing over all the possible numbers of particles of each species for each united coordination sphere with account taken of the nonequivalence of the groups of their sites and over all the coordination spheres, we obtain

$$V_{AB}(\alpha) = V_{CD}(\alpha)\exp\beta[\nu_C + \nu_D - \nu_A - \nu_B + \varepsilon_{AB}(1) - \varepsilon_{CD}(1)] \tag{7.24}$$

where $V_{ij}(\alpha)$ is the concentration component of the rate of two-site elementary process (7.8) determined as $U_{ij}(\alpha) = K_{ij}(\alpha)V_{ij}(\alpha)\exp[-\beta\varepsilon_{ij}(1)]$. Equating the rates of the opposite directions of step α, we express the constant of chemical equilibrium as follows:

$$K_{eq} = K_{AB}/K_{CD} = \exp\beta(\varepsilon_{CD}(1) - \varepsilon_{AB}(1))V_{CD}/V_{AB} \tag{7.25}$$

As a result, using expressions (7.21) and (7.24), we find that

$$K_{eq} = \theta_C\theta_D W_{CA} W_{DB}/(\theta_A\theta_B) \tag{7.26}$$

i.e. the constant of chemical equilibrium determines the relation between the equilibrium concentrations of the components of a system at the given potentials of particle interaction and does not depend on how the state of equilibrium was achieved.

The above proof does not depend on the way of approximate calculation of the multiparticle correlators. It is based on the condition of the self-sustained calculation of the probabilities of all the configurations. But this does not mean that any approximation ensures a self-consistent description of the kinetics and equilibrium. For example, the condition of complete ignoring of the correlation effects (the chaotic and molecular-field approximations) is not compatible with the systems of equations for the correlators (see Sec. 2.2 and [141]), and, consequently, it leads to artificial renormalization of the activation energy of the two-site reactions and to a mismatch in the description of their kinetics and the equilibrium of the system. To satisfy a self-consistent description, it is sufficient to consider the correlation effects between the nearest neighbors.

Inhomogeneous Lattices. The procedure of constructing systems of kinetic equations for inhomogeneous lattices repeats word for word everything said above for their homogeneous counterparts. The differences in the equations are associated with the difference in the molecular parameters contained in the rate constants of the reactions and in the parameters of particle interaction for particles at different sites of a lattice. We shall therefore write out the final expressions for the systems of kinetic equations that are generalizations of equations in Secs. 4.5 and 5.1 when the interaction of the second and following neighbors is considered and of the equations of the present section for inhomogeneous lattices.

Assume that a particle of A participates in the one-site reactions $A \rightleftarrows C_\alpha$ (α is the number of the step, there being Q_q^A steps, where q is the type of a site, its correspondence to the number of the site f is set by the quantity η_f^q that is considered to be known and constant in time), and further assume that a particle of A participates in the two-site reactions $A + B_\alpha \rightleftarrows C_\alpha + D_\alpha$ (the number of steps will be denoted by Q_{qp}^{AB}, where q and p are the types of neighboring sites). Hence the system of kinetic equations for the local concentrations of the particles of A and their pairs AB_1 will be written as follows (the component B_1 participates in one two-site step):

$$\dot{\theta}_f^A(q) = \sum_{\alpha=1}^{Q_q^A} [U_f^C(\alpha) - U_f^A(\alpha)] + \sum_g \sum_{\alpha=1}^{Q_{qp}^{AB}} [U_{fg}^{CD}(\alpha) - U_{fg}^{AB}(\alpha)] \tag{7.27}$$

$$\begin{aligned}\dot{\theta}_{fg}^{AB_1}(qp|r) = {} & \sum_{\alpha=1}^{Q_q^A} [U_{fg}^{\bar{C}B_1}(r|\alpha) - U_{fg}^{\bar{A}B_1}(r|\alpha)] + \Delta_{1r}[U_{fg}^{CD}(1) \\ & - U_{fg}^{AB_1}(1)] + \sum_{h.} \sum_{\alpha=1}^{Q_{q\xi}^{AB}} [U_{hfg}^{\overline{DC}B_1}(\omega_r|\alpha) - U_{hfg}^{\bar{B}AB_1}(\omega_r|\alpha)] \\ & + \sum_h [U_{fgh}^{A\overline{D_1C_1}}(\omega_r|1) - U_{fgh}^{A\overline{B_1A}}(\omega_r|1)].\end{aligned} \tag{7.28}$$

In Eq. (7.27), the first sum corresponds to the contributions of the one-site reactions, and the second one to the contributions of the two-site reactions. The subscript g relates to the nearest neighbors of site f. In Eq. (7.28), the first addend corresponds to one-site reactions at site f, and the second to a two-site reaction at sites f and g when they are the nearest to each other; the last two addends correspond to two-site reactions of particles of A and B_1 with their neighbors at sites h. The number of the adjacent site h is unambiguously related to the orientation ω_r of a particle of B_1 (A) at site g (f) relative to the dimer molecule at hf (gh).

The addends of the system of equations have the following form in the quasichemical approximation:

one-site processes

$$U_f^i(\alpha) = K_q^i(\alpha)\theta_f^i(q) \prod_{r=1}^{R} \prod_{g_r} S_{fg}^i(qp|r|\alpha)$$

$$U_{fh}^{\overline{ij}}(r|\alpha) = U_f^i(\alpha)\psi_{fh}^{ij}(qp|r|\alpha)$$

$$S_{fg}^i(qp|r|\alpha) = \sum_{j=1}^{s_p} t_{fg}^{ij}(qp|r)\exp\beta\delta\varepsilon_{qp}^{ij}(r|\alpha)$$

$$\psi_{fh}^{ij}(qp|r|\alpha) = t_{fh}^{ij}(qp|r)\exp\beta\delta\varepsilon_{qp}^{ij}(r|\alpha)/S_{fh}^i(qp|r|\alpha) \tag{7.29a}$$

$$\delta\varepsilon_{qp}^{ij}(r|\alpha) = \varepsilon_{qp}^{ij*}(r|\alpha) - \varepsilon_{qp}^{ij},\ \theta_{gh}^{ij}(qp|r) = \theta_g^i(q)t_{gh}^{ij}(qp|r)$$

two-site processes

$$U_{fg}^{ij}(\alpha) = K_{qp}^{ij}(\alpha)\theta_{fg}^{ij}(qp|1)e^{-\beta\varepsilon_{qp}^{ij}(1)} \prod_{r=1}^{R} \prod_{\omega_r=1}^{\pi_r} \prod_{h_{\omega r}} S_{fgh}^{ij}(qp\xi|\omega_r|\alpha)$$

$$S_{fgh}^{ij}(qp\xi|\omega_r|\alpha) = \sum_{l=1}^{s_\xi} t_{fgh}^{ijl}(qp\xi|\omega_r)\exp\beta\delta\varepsilon_{qp\xi}^{ijl}(\omega_r|\alpha)$$

$$\delta\varepsilon_{qp\xi}^{ijl}(\omega_r|\alpha) = \varepsilon_{qp\xi}^{ijl*}(\omega_r|\alpha) - \varepsilon_{qp\xi}^{ijl}(\omega_r) \tag{7.29b}$$

$$t_{hfg}^{ijl}(qp\xi|\omega_r) = \theta_{hg}^{il}(q\xi|r_{hg})\theta_{fg}^{jl}(p\xi|r_{fg})/(\theta_h^i(q)\theta_f^j(p)\theta_g^l(\xi))$$

$$U_{hfg}^{\overline{ijn}}(\omega_r|\alpha) = U_{hf}^{ij}(\alpha)\psi_{hfg}^{ijn}(qp\xi|\omega_r|\alpha)$$

$$\psi_{hfg}^{ijn}(qp\xi|\omega_r|\alpha) = t_{hfg}^{ijn}(qp\xi|\omega_r)\exp\beta\delta\varepsilon_{qp\xi}^{ijn}(\omega_r|\alpha)/S_{hfg}^{ij}(qp\xi|\omega_r|\alpha)$$

where $K_q^i(\alpha)$ and $K_{qp}^{ij}(\alpha)$ are the rate constants of the one- and two-site reactions defined in Eqs. (1.57) and (1.60), $\theta_f^i(q)$ is the probability of finding a particle i at a site of type q with the number f, $\theta_{fg}^{ij}(qp|r)$ is the probability of finding particles i and j at sites of type q and p having the numbers f and g at a distance of r, $\varepsilon_{qp}^{ij}(r)$ is the interaction parameter of particles i and j at sites of type q and p at a distance of r, and $\varepsilon_{qp}^{ij*}(r|\alpha)$ is the interaction parameter of the activated complex of step α formed from a particle i at a site of type q with an adjacent particle j at a site of type p at a distance of r. These parameters are used to construct the parameters of interaction of the activated complex of a two-site reaction with its neighbors in exactly the same way as was indicated for a homogeneous lattice, $\varepsilon_{qp\xi}^{ijn*}(\omega_r) = \varepsilon_{q\xi}^{in*}(r_{in}) + \varepsilon_{p\xi}^{jn*}(r_{jn})$. For distances r larger than the radius of interaction, we have $\theta_{fg}^{ij}(qp|r > R) = \theta_f^i(q)\theta_g^j(p)$. The conditions of normalizations for unary and two-body correlators will be written as follows:

$$\sum_{i=1}^{s_q} \theta_f^i(q) = 1, \quad \sum_{i=1}^{s_q} \theta_{fg}^{ij}(qp|r) = \theta_g^j(p), \quad \sum_{j=1}^{s_p} \theta_{fg}^{ij}(qp|r) = \theta_f^i(q) \quad (7.30)$$

In formulas (7.30) and the functions $S_{fg}^i(r|\alpha)$ and $S_{fgh}^{ij}(\omega_r|\alpha)$, summing is performed over the species of particles that can be at the sites of a given type.

On the basis of formulas (7.29), by analogy with formulas (7.11) and (7.13), it is easy to write the expressions in approximations in which the correlations for distances larger than R_1, where $1 \leqslant R_1 \leqslant R$, are disregarded, and also write expressions for the first way of using the quasichemical approximation, replacing $S_{fgh}^{ij}(\omega_r|\alpha)$ with the product

$$S_{fh}^i(r_{fh}|\alpha)S_{gh}^j(r_{gh}|\alpha),$$

$$\text{but } U_{hfg}^{\bar{i}jn}(\omega_r|\alpha) = U_{hf}^{ij}(\alpha)\psi_{hg}^{in}(r_{hg}|\alpha)\psi_{fg}^{jn}(r_{fg}|\alpha)$$

In the following sections, the equations obtained are used for a high surface mobility of the particles (Sec. 7.2) and for surface processes in an alloy that can become rearranged when adsorbed particles dissolve in it (Sec. 7.3). The hierarchy of models of a surface reaction of the type of $CO + 0.5O_2 \rightarrow CO_2$ is considered in Appendix 3.

7.2. Surface Reactions

For a high surface mobility of adsorbed particles, the systems of equations in Sec. 7.1 are simplified. Instead of the kinetic equations for the functions $\theta_{fg}^{ij}(r)$, one can use the algebraic equations of Secs. 6.1 and 6.3 that describe the equilibrium distribution of particles on a surface. The kinetic equations for the functions θ_i contain addends on their right-hand sides that are the rates of the change in the surface concentrations of the adsorbed particles in various surface reactions. The general form of these addends is indicated at the end of the present section. We shall consider first of all the joint influence of the interaction of the first and second neighbors on the kinetics of a very simple surface process, namely, the adsorption of gas phase molecules. We shall use the quasichemical approximation for this purpose. Moreover, we shall briefly discuss the influence of consideration of the indirect correlations, the contributions of farther neighbors, the absence of correlation effects for the second and following neighbors, and also the role of ordering effects.

The distribution of adsorbed particles on a homogeneous surface is described by formulas (6.17). Using the same "simplified" notation, the rates of adsorption and desorption of undissociating molecules, by formula (7.11) are described as follows [207]:

$$U_v = K_v(1 - \theta)(1 + t^1_{BA}y_1)^{z_1}(1 + t^2_{BA}y_2)^{z_2}$$

$$y_r = \exp \beta\varepsilon^*_r - 1 \tag{7.31}$$

$$U_A = K_A\theta(1 + t_1x_1)^{z_1}(1 + t_2x_2)^{z_2}, \quad x_r = \exp \beta\delta\varepsilon_r - 1$$

where $\delta\varepsilon_r = \varepsilon^*_r - \varepsilon_r$, ε_r and ε^*_r are the parameters of interaction between the particles and between the activated complex and a particle at the distance r.

In the region of low coverages, the rates of the processes have the form

$$U_v = K_v(1 - \varkappa\theta), \quad U_A = K_A\theta(1 + \xi\theta) \tag{7.32}$$

where the coefficients $\varkappa$ and ξ consider the influence of interaction. The structure of formulas (7.32) shows that the interactions of the particles mainly affect the rate of adsorption (their influence on desorption is proportional to θ^2, which at low coverages yields a small contribution). The quantity $\varkappa$ determines the concentration dependence of the rate of adsorption ($\varkappa = 1 - z_1y_1 - z_2y_2$). When $\varkappa > 1$, the rate of adsorption diminishes with a growth in θ more rapidly than for noninteracting particles (ε^*_r, $\varepsilon_r = 0$), and when $\varkappa < 1$, more slowly. When $\varkappa > 0$, the rate of adsorption grows with an increasing coverage. When the interaction of the nearest neighbors is taken into account, such a situation can be realized only in a system with attraction (physical adsorption); when the second neighbors are considered, a similar situation also takes place for chemisorption ($\varepsilon_1 < 0$). For this end, observance of the condition $z_2y_2 > 1 - z_1y_1$ is sufficient. If we assume that an activated complex is farther from the surface than an adsorbed particle in the ground state and exchange interaction between the particles rapidly attenuates with an increasing distance, then $\varepsilon_1 < \varepsilon^*_1 < 0$ and for the two limiting cases $|\beta\varepsilon^*_1| > 3$ and

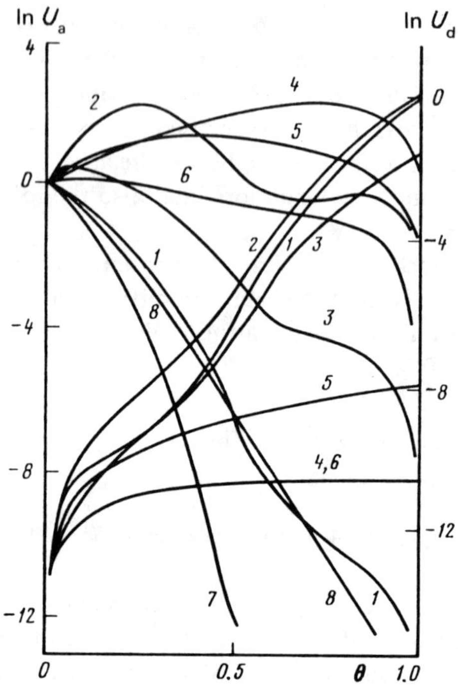

Fig. 7.2. Rates of adsorption and desorption (z = 4):

$\beta\varepsilon_2 = -1.5$ (*1*), 1.5 (*2*), 0.6 (*3*), 0.5 (*4*), -0.5 (*5*, *6*); $\beta\delta\varepsilon_2 = 0.75$ (*1*, *2*), 0.3 (*3*), 0.25 (*4*), -0.25 (*5*), -0.1 (*6*)

$\varepsilon_1^* \approx 0$ we obtain respectively that $\varkappa < 0$ when

$$\beta\varepsilon_2^* \geqslant \ln[(1 + z_1 + z_2)/z_2], \quad \beta\varepsilon_2^* > \ln[(1 + z_2)/z_2] \tag{7.33}$$

The concentration dependences of the rates of adsorption and desorption within the entire range of coverages are shown in Fig. 7.2 for chemisorption systems (curves *1-3*, $\beta\varepsilon_1 = -3.0$, $\beta\varepsilon_1^* = -1.5$) and for physical adsorption (curves *4-6*, $\beta\varepsilon_1 = 1.0$, $\beta\varepsilon_1^* = 0.5$ (*4*, *6*) and $\beta\varepsilon_1^* = 1.5$ (*5*)). The general conclusion consists in that the second neighbors greatly affect the concentration dependences of the rates of adsorption processes; they affect the kinetics of adsorption to a greater extent than the kinetics of desorption. For chemisorption systems, the features noted previously are mainly retained: the presence in the region of moderate coverages of linear dependences of the logarithms of the rates on the magnitude of the coverage and the possibilities of singling out several such linear sections over the entire range of coverages.

For dissociating molecules, all the noted laws of the influence of second neighbors on the concentration components of the rates of adsorption and

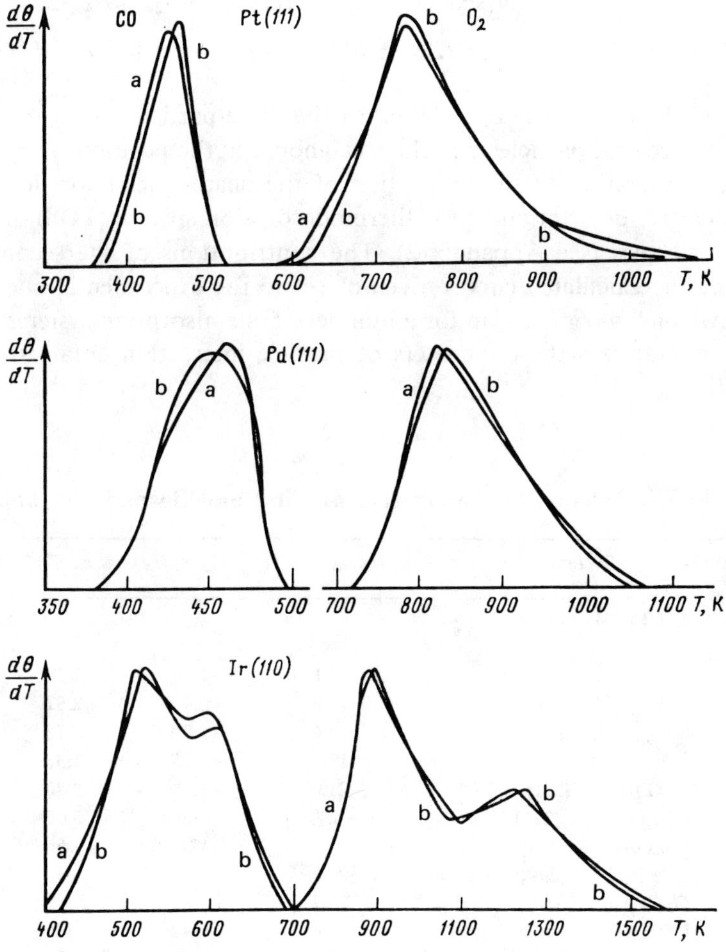

Fig. 7.3. Experimental (*a*) and predicted (*b*) TDS curves for CO and O_2 on platinum group metals.

The interaction parameters are indicated in Table 7.1

desorption (nondissociating molecules) mainly coincide. In Fig. 7.2, curve *7* has been calculated for the same parameters as curve *1*, and curve *8* for parameters that were halved. The numerical distinctions are associated with the increase (virtually two times) of the number of neighbors of the activated complex for the dissociating molecules. For example, for the rate of desorption on a square lattice, by formula (7.15), we have [219, 297]:

$$\begin{aligned} U_A &= K_A\theta t_1 \exp(-\beta\varepsilon_1)(1 + t_1x_1)^2(1 + t_2x_2)^4 S_{AA}^4(2) \\ S_{AA}(r) &= (1 - t_1)(1 - t_r)/(1 - \theta) + t_1t_2 \exp\beta(\delta\varepsilon_1 + \delta\varepsilon_r)/\theta \end{aligned} \tag{7.34}$$

where the multipliers S_{AA} (2) reflect the three-particle configurations formed by two central particles and their neighbors in the position $\omega_1 = 2$ in Fig. 7.1.

Consideration of the interaction of the nearest neighbors leads only to a qualitative description of the thermodesorption spectra (TDS) on homogeneous surfaces (see Appendix 2). The contributions of the second neighbors bring the calculated curves much closer to the experimental ones. Figure 7.3 shows such a comparison for a number of chemisorption systems, while Table 7.1 summarizes the parameters of particle interaction obtained in ref. [219, 297].

Table 7.1. Interaction Parameters of First and Second Neighbors

System	$\varepsilon_{AA}(1)$	$\varepsilon^*_{AA}(1)$	$\varepsilon_{AA}(2)$	$\varepsilon^*_{AA}(2)$
CO/Pt(111) [220]				
Variant 1	−10.1	−1.68	1.38	2.06
2	−12.6	−0.42	2.52	4.70
3	−12.6	−5.90	3.78	7.15
4	−12.6	−4.25	1.31	2.62
O_2/Pt(111) [221]	−21.0	−12.6	8.40	12.6
CO/Pd(111) [227]	−6.7	−5.9	5.05	7.55
O_2/Pd(111) [298]	−21.0	−12.6	7.56	11.6
CO/Ir(110) [299]	−6.7	1.7	0	0
O_2/Ir(110) [300]	−25.2	−16.8	8.4	10.5
CO/Ir(111) [301]	−8.4	−0.42	4.2	6.7
O_2/Ir(111) [301]	−13.8	−6.7	0	0
CO/Pt(110) [302]	−12.6	−5.9	0	0

Note. Variants 2-4 correspond to Fig. 7.4.

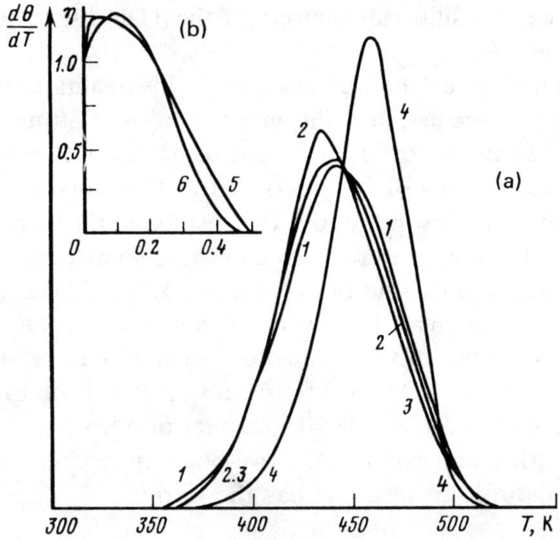

Fig. 7.4. Experimental (*1*) and predicted (*2-4*) TDS curves for CO/Pt (111) corresponding to the parameters of variants 2-4 of Table 7.1 (*a*) and η vs. θ (*b*):
5—experiment; *6*—predicted

In the general case, the selection of the interaction parameters from the experimental data is ambiguous. For the system CO/Pt (111), different sets of parameters ensuring a root-mean-square deviation of the experimental curves from the calculated ones of the order of magnitude of 5-8% are indicated. It should be noted that the parameters of variant 4 have been found from the concentration dependence of the sticking probability defined as $\eta = U_v(\theta)/U_v$ $(\theta = 0)$ at $T = 300$ K [220] (Fig. 7.4), and they have been employed to reproduce the thermodesorption spectrum (curve *4*).

A general parametric analysis of the TDS curves for homogeneous surfaces with consideration of the interaction of the first two neighbors has been performed in ref. [219]. The influence of the molecular parameters of the model (the preexponential factors of the rate constants of desorption, the activation energy, the interaction parameters in the ground and transition states) on the nature of splitting of the TDS was investigated. It was found that the width of the TDS at $\theta(0) \to 1$ is proportional to the quantity $z_1\delta\varepsilon_1 + z_2\delta\varepsilon_2$, i.e. it is determined by the difference in the interaction of the particles in the ground and transition states. The contributions of the second neighbors do not cause additional splitting of the TDS due to the repulsion of the nearest neighbors.

neighbors do not cause additional splitting of the TDS due to the repulsion of the nearest neighbors.

For dissociating molecules, the TDS curves coincide qualitatively with those for nondissociating molecules, but the magnitude of splitting is relatively smaller. In both cases, the contributions of the second neighbors cause noticeable changes in the positions of the peaks of the TDS curves. For instance, for strong repulsion of the first neighbors, the position of the high-temperature peak shifts by the amount $z_2\delta\varepsilon_2$ to the region of lower temperatures in comparison with the absence of contributions of the second neighbors. (The role of each parameter has been treated in greater detail in ref. [219].)

To establish the influence of the approximation employed on the nature of the TDS curve, calculations were conducted with consideration of the indirect correlations [303]. As indicated in [304], the rates of adsorption and desorption of nondissociating molecules in the two-body approximation with consideration of the indirect correlations has the form

$$U_V = K_V(1 - \theta) \sum_{n_1=0}^{z_1} \sum_{n_2=0}^{z_2} \left\{ \prod_{r=1}^{2} C_{z_r}^{n_r} y_r t_{BA}^r \right\} \Phi_{n_1 n_2}$$

$$U_A = K_A\theta \sum_{n_1=0}^{z_1} \sum_{n_2=0}^{z_2} \left\{ \prod_{r=1}^{2} C_{z_r}^{n_r} x_r t_r \right\} \Phi_{n_1 n_2} \qquad (7.35)$$

The functions $\Phi_{n_1 n_2}$ are functions of indirect correlations, and their form like that of the expressions relating the fractional coverage to the functions t_r, where $r = 1, 2$, is given in Appendix 1 (Sec. 3).

A comparison of the TDS curves calculated for a flat square lattice in the quasichemical calculation [see formulas [7.34)] and with consideration of the indirect correlations is shown in Fig. 7.5 (the rate of heating a specimen $b = 50$ K/s). Consideration of the indirect correlations causes the low-temperature peak to grow (because the main contribution of the indirect correlations manifests itself at large coverages) and substantially shifts the high-temperature peak to the region of lower temperatures (because the major part of the particles is desorbed at lower temperatures). The width of the TDS at large initial coverages does not depend on the approximation used. But the nature of splitting of the TDS curve for the given lattice does not change. Hence, consideration of the contributions of the second neighbors does not cause the appearance of an additional fine structure in the TDS when it splits because of interaction of the first neighbors.

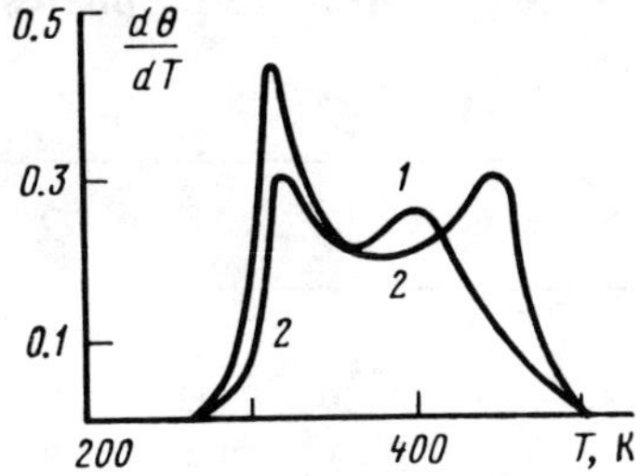

Fig. 7.5. TDS curves with consideration of indirect correlations (*1*) and in the quasichemical approximation (*2*) at $\theta_A(0) = 0.9$, $z = 4$, $\tilde{K}_A = 10^{12}\ s^{-1}$, $b = 50$ K/s, $E_A = 126$, $\varepsilon_{AA}(1) = -16.8$, $\varepsilon^*_{AA}(1) = -8.4$, $\varepsilon_{AA}(2) = 4.2$, $\varepsilon^*_{AA}(2) = 5.25$ kJ/mol

Calculations of the rates of elementary processes, namely, adsorption, desorption, and migration into an adjacent vacant site on a homogeneous surface with one species of adsorbed particles were also performed with consideration of the interactions of up to ten coordination spheres. The Lennard-Jones, exponential, dipole potentials [60] and an arbitrary set of parameters with values decreasing in magnitude and with changing of the sign were employed as the potentials of particle interaction. Table 7.2 shows the division of sites into equivalent groups for a flat square lattice relative to the dimer molecule Av. Counting is performed in numbers of the coordination spheres, the quantities r_1 and r_2 are the distances from the central particles of A and v to the given group of equivalent sites whose number is $\varkappa_{\omega r}$. Table 7.2 also indicates the number of various orientations π_r and the geometric distance ϱ corresponding to the number of the coordination sphere r; the quantity ϱ is measured by the length of the lattice constant. The zero in columns 8-10 signifies that $r_i > 10$, where $i = 1, 2$.

Calculations reveal that consideration of the third and following neighbors does not produce qualitatively novel effects in the concentration dependences of the rates of adsorption, desorption, and migration in comparison with those appearing when the first two neighbors are considered, especially when the parameters of interactions of the first and second neighbors have different signs. Account of the farther neighbors introduces quantitative changes. This conclusion relates only to disordered systems. The contributions of the third

Table 7.2. Equivalent Groups of Sites in Ten Coordination Spheres for a Flat Lattice (z = 4)

Parameter	1	2	3	4	5	6	7	8	9	10
r	1	2	3	4	5	6	7	8	9	10
ϱ^2	1	2	4	5	8	9	10	13	16	17
π_r	4	2	4	4	2	4	4	2	4	2
$[r_1, r_2]$	1, 3	2, 4	3, 6	4, 7	5, 8	6, 9	7, 10	8, 0	9, 0	10, 0
	1, 2	0	3, 4	4, 5	0	6, 7	7, 8	0	9, 10	0
	2, 1	4, 2	4, 3	5, 4	8, 5	7, 6	8, 7	0, 8	0, 9	0, 10
	3, 1	0	6, 3	7, 4	0	9, 6	10, 7	0	10, 9	0
$\varkappa_{\omega r}$	1	2	1	2	2	1	2	4	1	4
	2	0	2	2	0	2	2	0	2	0
	2	2	2	2	2	2	2	4	2	4
	1	0	1	2	0	1	2	0	1	0

and following neighbors affect the conditions of realization of the phase transitions of ordering and stratification in adsorption systems, and this, in turn, affects the kinetics of the surface processes.

If we ignore the correlations of particles at distances of the second and following neighbors, the rates of the adsorption processes will be written as (i, j = A, v):

$$U_i = K_i\theta_i(S_i)^z \exp \beta z\delta\tilde{\varepsilon}_{iA}\theta, \quad \delta\tilde{\varepsilon}_{ij} = \sum_{r=2}^{R} z_r\delta\varepsilon_{ij}(r)/z$$
$$U_{ij} = K_{ij}\theta_{ij}(1)(S_iS_j)^{z-1} \exp \beta[z\theta(\delta\tilde{\varepsilon}_{iA} + \delta\tilde{\varepsilon}_{jA}) - \varepsilon_{ij}(1)] \tag{7.36}$$

The concentration dependences (7.36) describing the contributions of the interaction of particles are the product of similar contributions when considering the interaction of nearest neighbors in the quasichemical and molecular-field approximations (4.16). The quantity $\delta\tilde{\varepsilon}_{ij}$—the effective parameter of close-range action—depends on the structure of a lattice and on the form of the interaction potential. Similar relations of the rates of adsorption are obtained when jointly considering the interaction of the nearest neighbors and the changes in the collective contributions of the electron subsystem in the model of a two-dimensional electron gas [213, 216, 305]. Different models lead to a different physical interpretation of the quantities $\delta\tilde{\varepsilon}_{iA}$, leaving the functional dependences of the rate of the process on the fractional coverage unchanged.

It should be noted that models of the type of (7.36) ensure a self-consistent description of the kinetics of chemical reactions and an equilibrium state of the reaction system. This fact was verified experimentally for the system K/W [158].

In refs. [161, 306] it has shown that the magnitude of the collective contributions of an electron subsystem can be determined from the linear dependence of the decrease in the heat of adsorption on the increase in the coverage at low values of the latter.

Strong interactions change the states of aggregation of the adsorbed particles. The kinetics of the surface processes in this case must be described with consideration of the inhomogeneous distribution of the particles on a homogeneous surface [see formulas (7.29) in which the number of a site has been replaced with subscripts indicating the type of the site]. Let us see how ordering of the particles affects the rates of adsorption-desorption of nondissociating molecules. Assume that a superstructure of the type of $c(2 \times 2)$ is realized on the surface of a face (100) (see Sec. 3.5). The expressions for the rates of adsorption and desorption relating to the entire surface as a whole, as when the interaction of the nearest neighbors is considered, have the form of (4.65), where U^i_α and U^i_γ are the rates of the elementary adsorption processes (i = v) and desorption ones (i = A) in the quasichemical approximation with account taken of the interaction of the first and second neighbors. These rates are written as follows:

$$\begin{aligned} U^{\mathrm{v}}_\sigma &= K_{\mathrm{v}}(1 - \theta_\sigma)(1 + t^{\sigma\lambda}_{\mathrm{BA}}\, y_1)^4(1 + t^{\sigma\sigma}_{\mathrm{BA}}\, y_2)^4, \quad \sigma = \alpha,\ \gamma \\ U^{\mathrm{A}}_\sigma &= K_{\mathrm{A}}\theta_\sigma(1 + t^{\sigma\lambda}_1 x_1)^4(1 + t^{\sigma\sigma}_2 x_2)^4, \quad \lambda \neq \sigma, \end{aligned} \tag{7.37}$$

where x_r and y_r have been defined in (7.31), while the function $t^{\sigma\lambda}_r$ is expressed in terms of the concentration of particles of A on sublattices σ by formulas (6.31); the values of r and the species of the pairs $\sigma\lambda$ are unambiguously related to one another, $t^{\sigma\lambda}_{\mathrm{BA}} = \theta_\lambda(1 - t^{\lambda\sigma}_r)/(1 - \theta_\sigma)$. The total coverage of a surface is determined by formula (3.46).

Figure 7.6 shows the rates of desorption calculated by formulas (7.31) and (7.37) and by the relevant coverages of individual sublattices, which a multitude of solutions of Eq. (6.33) corresponds to. The temperature of ordering and the concentrations of the particles at which their ordering occurs corresponding to it depend on the coordination numbers z_r and on the interaction parameters ε_r. As noted in Sec. 6.4, attraction of the second neighbors in-

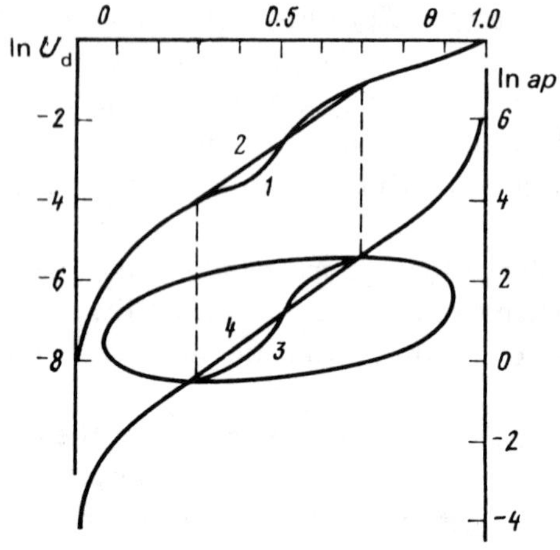

Fig. 7.6. Rate of desorption with consideration (*1*) and without consideration (*2*) of ordering and adsorption isotherms with consideration (*3*) and without consideration (*4*) of particle ordering: $\beta\varepsilon^*_{AA}(r) = \beta\varepsilon_{AA}(r)/2$, $r = 1$ and 2, $\beta\varepsilon_{AA}(1) = -1.0$; $\beta\varepsilon_{AA}(2) = 0.5$

creases the region of existence of ordering of the type $c(2 \times 2)$. In the absence of their contribution, a disordered region corresponds to the parameter ε_1 in Fig. 7.6. As the coverage grows after the point of the disorder-order phase transition, ordering diminishes the rate of desorption because the fraction of the nearest neighboring particles decreases, as does the contribution of their repulsion. When $\theta > 0.5$, the rate of desorption in the ordered state exceeds that of desorption in the disordered state. With an increase in the repulsion of the first neighbors and attraction of the second ones, these distinctions in the rates of the adsorption processes for the ordered and disordered states grow, although the specific shape of the curves of U_v and U_A vs. θ also depends on the parameters ε^*_r.

For dissociating molecules, the rates of adsorption and desorption in the approximation being considered, by formulas (7.29), have the form

$$U_{ii} = K_{ii}\theta^{ii}_{\alpha\gamma}(1)e^{-\beta\varepsilon_{ii}(1)}S^i_{\alpha\gamma}(1)S^i_{\gamma\alpha}(1)[S^i_{\alpha\alpha}(2)S^i_{\gamma\gamma}(2)S^{ii}_{\alpha\gamma\gamma}\, S^{ii}_{\gamma\alpha\alpha}]^2 \quad (7.38)$$

because two nearest particles (i = A, v) are always at sites of different sublattices. One way of arrangement of the particles corresponds to triple configurations, therefore the subscript ω_r has been omitted. To calculate the rate by formulas (7.48) at a given value of the surface coverage, it is necessary to determine the values of θ_σ^A from the solution of Eq. (6.33) and to use formulas (7.29) for $S^i_{\lambda\sigma}$ and $S^{ii}_{\lambda\sigma\sigma}$.

Let us also write out expressions for the rates of adsorption and desorption of nondissociating molecules for other superstructures for which the equilibrium coverages of the sublattices are described by Eqs. (6.34). The overall rate of adsorption and desorption relating to the entire surface is also determined for these superstructures by formulas (4.65), while their components are:

for a lattice $z = 4$ and a superstructure $p(2 \times 1)$

$$U^i_\sigma = K_i\theta^i_\sigma[S^i_{\sigma\lambda}(1)S^i_{\sigma\sigma}(1)[S^i_{\sigma\lambda}(2)]^2]^2, \ \sigma = \alpha, \ \gamma, \ \lambda \neq \sigma \tag{7.39a}$$

For a lattice $z = 6$ and a superstructure $c(2 \times 2)$

$$U^i_\sigma = K_i\theta^i_\sigma[S^i_{\sigma\sigma}(1)[S^i_{\sigma\lambda}(1)S^i_{\sigma\lambda}(2)]^2S^i_{\sigma\sigma}(2)]^2 \tag{7.39b}$$

A distinction of these expressions from (7.37) consists in that there are sites of different sublattices at an identical distance from a type σ site. This feature must be taken into account when calculating the functions $t_{\sigma\lambda}(r)$ by formula (6.31). Such an arrangement of the sublattices causes the appearance of different addends for the rates of the adsorption processes of dissociating molecules [unlike the one term of (7.38) for the superstructure $c(2 \times 2)$]:

$$U_{ii} = \frac{1}{z}[U^{ii}_{\alpha\alpha} + (z - 2)U^{ii}_{\alpha\gamma} + U^{ii}_{\gamma\gamma}] \tag{7.40}$$

where the form of the addends depends on the lattice and on the types of the adjacent sites. For instance, for a hexagonal lattice, the addend $U^{ii}_{\alpha\gamma}$ is written as follows (Fig. 7.7):

$$\begin{aligned} U^{ii}_{\alpha\gamma} &= K_{ii}\theta^{ii}_{\alpha\gamma}(1)e^{-\beta\varepsilon_{ii}(1)}S^i_{\alpha\gamma}(1)S^i_{\gamma\alpha}(1)[S^i_{\gamma\alpha}(2)]^3S^i_{\gamma\gamma}(2)S^i_{\alpha\alpha}(2) \\ &\times [S^i_{\alpha\gamma}(2)]^3S^{ii}_{\gamma\alpha\alpha}(2)S^{ii}_{\gamma\alpha\gamma}(2)S^{ii}_{\alpha\gamma\gamma}(2)S^{ii}_{\alpha\gamma\alpha}(2)S^{ii}_{\alpha\gamma\alpha}(1)S^{ii}_{\gamma\alpha\gamma}(1) \end{aligned} \tag{7.41}$$

The remaining addends have a simpler form because of the symmetry in the arrangement of the neighbors relative to the central pair of sites. The expressions for the rates of adsorption-desorption for the other superstructures are constructed similarly.

General expressions for the rates of surface reactions on an inhomogeneous

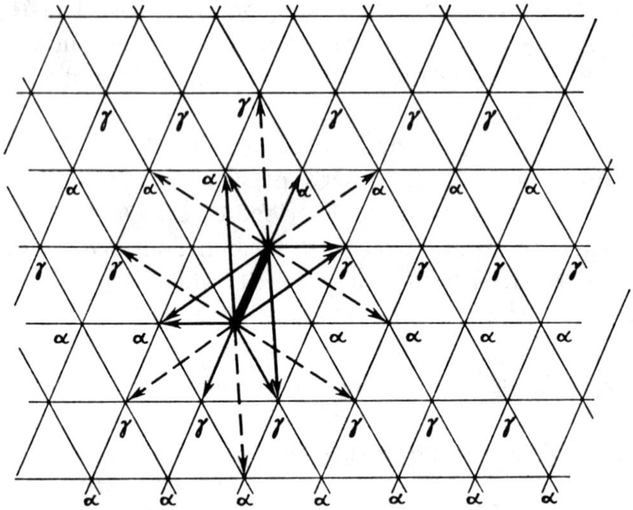

Fig. 7.7. Sites of the first and second (dashed lines) coordination spheres for ordering of the phase $c(2 \times 2)$ for $z = 6$ around a central dimer molecule

surface with consideration of the two-body potential with a radius of interaction of R are obtained by averaging the expressions for $U_f^i(\alpha)$ and $U_{fg}^{ij}(\alpha)$ (7.29) over all the sites of a lattice. The distribution functions $d(q\{[m]\}R)$ needed to describe one-site reactions have been determined in Sec. 6.3. Similar distribution functions are required to calculate two-site reactions. Let us denote them by $d(qp\{[\lambda\sigma]\}R)$. Hence, the rates of a reaction relating to an entire surface can be written as

$$
\begin{aligned}
U_i(\alpha) &= \sum_{q=1}^{t} f_q K_q^i(\alpha) V_q^i(\alpha) \\
V_q^i(\alpha) &= \theta_q^i \sum_{\sigma_{2q}} d(q\{[m]\}R) \prod_{r=1}^{R} \prod_{p=1}^{t} [S_{qp}^i(r|\alpha)]^{m_{qp}(r)} \\
U_{ij}(\alpha) &= \sum_{q,\ p=1}^{t} f_{qp} K_{qp}^{ij}(\alpha) \exp(-\beta\varepsilon_{qp}^{ij}(1)) V_{qp}^{ij}(\alpha) \\
V_{qp}^{ij}(\alpha) &= \theta_{qp}^{ij}(1) \sum_{\sigma_{2qp}} d(qp\{[\lambda\sigma]\}R) \prod_{r=1}^{R} \prod_{\omega_r=1}^{\pi_r} \prod_{\xi=1}^{t} [S_{qp\xi}^{ij}(\omega_r|\alpha)]^{\chi_{qp\xi}(\omega_r)}
\end{aligned}
\qquad (7.42)
$$

where the functions $S^i_{qp}(r|\alpha)$ have the form of the expressions in (7.29) in which the local means $t^{ij}_{fg}\ (qp|r)$ have been replaced by the macroscopic means $t^{ij}_{qp}\ (r)$. The latter are determined from the solution of the systems of equations (6.27). The same also relates to the functions $S^{ij}_{qp\xi}\ (\omega_r|\alpha)$. If the distribution functions of the clusters obey an equilibrium distribution, relations similar to expressions (3.9), (4.61), and (6.29) hold for them and simplify formulas (7.42). The latter take on a form similar to formulas (4.63).

For a methodological purpose, let us compare the results of calculations employing approximations (7.14) and (7.15) for the rates of two-site processes. We shall limit ourselves to a process of dissociative adsorption proceeding on a homogeneous surface. Since both approximations yield the same expression for an adsorption isotherm, it is sufficient to consider, for instance, the ratio of the desorption rates. We shall consider the interactions of the first and second neighbors. For a square and hexagonal lattices, by (7.34), the relation of the rates has the form

$$\xi_4 = [\varphi(2)]^4,\ \xi_6 = \xi_4\xi_6(1),\ \xi_6(1) = [\varphi(1)]^2, \qquad (7.43)$$
$$\varphi(r) = S_{AA}(r)/(S_A(1)S_A(r))$$

where all the quantities have been determined above in (6.17) and (7.31). In the region of low coverages, the quantity $\varphi(r)$ is represented as $\varphi(r) = 1 + \varkappa_r\theta$, $\varkappa_r = (1 - H_1)(1 - H_r)$, where $H_r = \exp\beta\varepsilon_r^*$, i.e. it is determined only by the parameters of interaction of an activated complex with its neighbors. The use of formula (7.14) lowers the rate of a process ($\varkappa_r > 1$) at identical signs of ε_r^* and increases it ($\varkappa_r < 1$) at opposite ones. Concentration dependences of ξ_4 and $\xi_6(1)$ are shown in Figs. 7.8 and 7.9. The curves with $\gamma = 1/2$, where $\gamma = \varepsilon_r^*/\varepsilon_r$, $r = 1$ and 2, are symmetrical relative to the line $\theta = 1/2$ (in Fig. 7.9 they are shown on the section $0 \leqslant \theta \leqslant 0.5$). The curves with $\gamma = 0.75$ and 0.25 coincide with each other when moving along θ from zero and unity (see Fig. 7.8, curves *5* and *6*). For any interactions, $\xi_6(1) > 1$, while the value of ξ_4 can be either larger or smaller than unity. An analysis of relations (7.43) for $0 \leqslant \gamma \leqslant 1$ reveals that the differences in the approximations do not exceed an order of magnitude of 10% only when $|\beta\varepsilon_r| \leqslant 1$ and $\gamma \approx 1/2$. In the remaining cases, these deviations are larger. The case corresponding to the chemisorption of the dissociating molecules deserves special attention. Using the example of the chemisorption of oxygen molecules on platinum metals (see Table 7.1), we can readily appraise the quantities ξ_z, where $z = 4$ and 6, in the region $0.2 < \theta < 0.35$. The values of $t_1 \approx 0$ and $t_2 \approx 1$ correspond to the parameters indicated in the table, which leads to $\xi_z \approx H_1^4 \lll 1$. In other

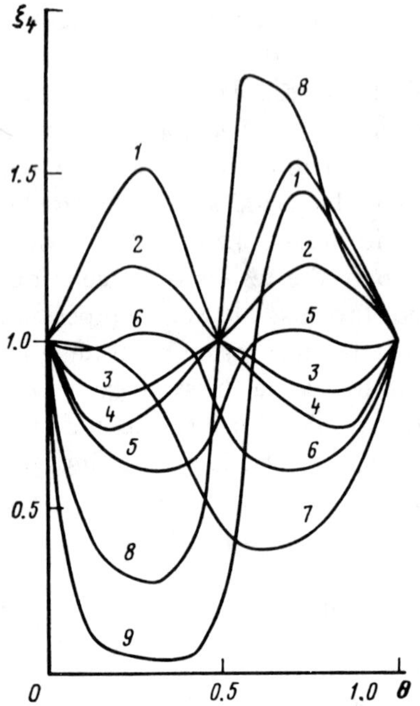

Fig. 7.8. Concentration dependences of ξ_4 at $\beta\varepsilon_1 = -4.0$ for $\gamma = 1/2$ ($\beta\varepsilon_2 = -2$ (*1*), -1 (2), 1 (*3*), 2 (*4*)) and at $\beta\varepsilon_2 = 1$ for $\gamma = 0.75$ (*5*), 0.25 (*6*), and 0 (*7*):

$\beta\varepsilon_1^* = -2.0$, $\beta\varepsilon_2 = 1.0$, and $\beta\varepsilon_2^* = 1.5$ correspond to curve *8*, and $\beta\varepsilon_1^* = -3.0$, $\beta\varepsilon_2 = 1.0$, and $\beta\varepsilon_2^* = 2.25$ to curve *9*

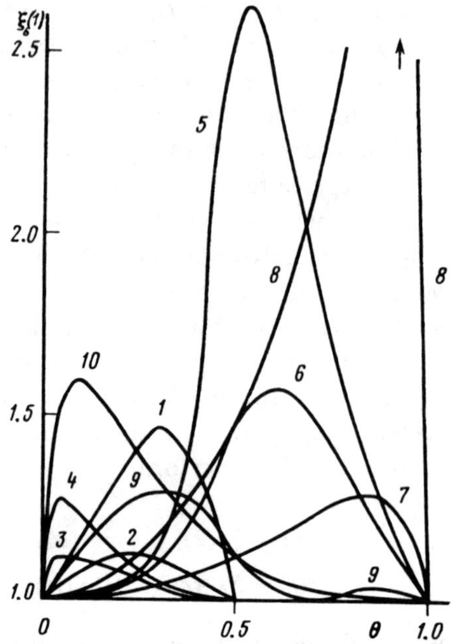

Fig. 7.9. Concentration dependences of ξ_6 (1) for $\gamma = 1/2$ [$\beta\varepsilon_1 = -4$ (*1*), -2 (*2*), 2 (*3*), 3 (*4*)], $\gamma = 0$ [$\beta\varepsilon_1 = -4$ (*5*), -2 (*6*), 1 (*7*), 2 (*8*)], and $\gamma = 0.75$ [$\beta\varepsilon_1 = -2$ (*9*), 2 (*10*)].

The arrow indicates the position of the peak equal to 3.12

words, formula (7.14) gives a substantial (up to one or two orders of magnitude) overstating of the rate of the process (see Fig. 7.8, curves *8* and *9*).

Similar conclusions also relate to more complicated situations. For example, for the reaction A + B with arbitrary values of s on an inhomogeneous lattice for small coverages of a surface, we have

$$\varphi_{fgh}^{AB}(\omega_r) = 1 + \sum_{m=1}^{s-1} \theta_h^m (1 - H_{fh}^{Am}(r_{fh}))(1 - H_{gh}^{Bm}(r_{gh}))$$

$$H_{fh}^{im}(r) = \exp \beta\varepsilon_{fh}^{im*}(r) \tag{7.44}$$

and, if $\varepsilon_{fh}^{Am*}(r_{fh})$ and $\varepsilon_{gh}^{Bm*}(r_{gh})$ are of the same sign, the addend with the superscript m is positive, while if they are of opposite signs, it is negative. (Quantitative appraisals require numerical analysis.)

Consequently, the approximation determined by formula (7.14) is a rough one, and it is not good to use it for describing the kinetics of two-site processes. When $R > 2$, the number of neighbors that simultaneously react with both reactants grows, and the differences in the calculations performed by formulas (7.14) and (7.15) grow. In the limiting case for long-range potentials, formulas (7.14) are also not applicable at low interactions because the interaction potential is noticeably distorted. For instance, in a system with ion-dipole interaction, their use actually corresponds to a transition to ion-ion interaction.

7.3. Processes with Slow Particle Mobility

Let us consider the process of adsorption on a homogeneous surface and the process of a change in the state of a solid alloy when gas phase molecules dissolve in it as examples of using the kinetic equations of Sec. 7.1 for constructing closed models of surface processes.

Process of Adsorption on a Homogeneous Surface. The kinetic scheme of the process for nondissociating molecules consists of two steps, namely, (1) $A + vZ_f \rightleftarrows AZ_f$ and (2) $AZ_f + vZ_g \rightleftarrows vZ_f + AZ_g$.

The equations for the local concentration of particles of A at the sites of a two-dimensional lattice will be written in the form:

$$\dot{\theta}_f^{A} = U_f^{v}(1) - U_f^{A}(1) - \sum_g [U_{fg}^{Av}(2) - U_{fg}^{vA}(2)],$$
$$\theta_f^{A} + \theta_f^{v} = 1 \tag{7.45}$$

The addends of the equations have been determined in (7.11) and (7.15). To close system of equations (7.45), it must be supplemented with kinetic equations for two-body correlators at different distances:

$$\dot{\theta}_{fg}^{AA}(r) = U_{fg}^{\bar{v}A}(r|1) + U_{fg}^{A\bar{v}}(r|1) - U_{fg}^{\bar{A}A}(r|1) - U_{fg}^{A\bar{A}}(r|1)$$
$$- \sum_h [U_{hfg}^{\overline{vA}A}(\omega_r|2) - U_{hfg}^{\overline{Av}A}(\omega_r|2)]$$
$$- \sum_h [U_{fgh}^{A\overline{Av}}(\omega_r|2) - U_{fgh}^{A\overline{vA}}(\omega_r|2)] \tag{7.46}$$

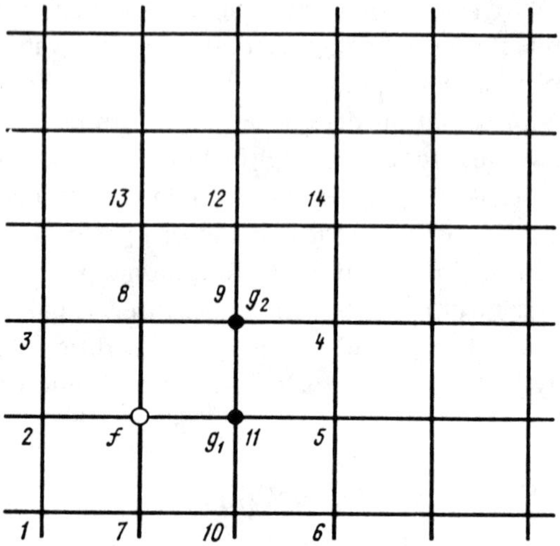

Fig. 7.10. A pair of sites fg (11) at the distance of the first neighbors (the first neighbors h = 2, 7, 8, 9, 10, 5; the second neighbors h = 1, 3, 4, 6) and at the distance of the second neighbors fg (9) (the first neighbors h = 2, 7, 8, 4, 11, 12, the second neighbors h = 1, 3, 10, 5, 13, 14)

where the subscript h in the first sum relates to the nearest neighbors of site f, and in the second one to the nearest neighbors of site g; when $r = 1$, the subscript h in the indicated sums does not equal g and f, respectively.

The addends of Eqs. (7.46) have been determined in (7.12) and (7.16). Let us see what multipliers are contained in the addends of Eqs. (7.45) and (7.46) using the example of a square lattice at $R = 2$. If the central particle is at point f (Fig. 7.10), the functions $S_{fg}^{i}(r|1)$ with the numbers g = 2, 8, 11, and 7 for $r = 1$ and 1, 3, 9, 10 for $r = 2$ are present in formula (7.11). In formula (7.12), a neighboring particle not participating in an elementary processes with $\alpha = 1$ may be at a distance of the first (e.g. g_1) and second (e.g. g_2) neighbors. The multiplier $\psi_{fg}^{iA}(r|2)$, where $g = g_1$ when $r = 1$ and $g = g_2$ when $r = 2$, corresponds to this.

In formulas (7.15), if the migrating particles are at sites f and g_1, the multipliers $S_{fh}^{i}(r|2)$ and $S_{g_1h}^{i}(r|2)$, where h = 2 and 5 at $r = 1$ and h = 1, 3 and 4, 6 at $r = 2$, respectively, are present, and also the multipliers S_{fgh}^{ij}, where h = 7-10; at these sites the neighboring particles interact with both central particles. In formulas (7.16), it is assumed that a central particle can react with a neighboring particle at h, where $h = 2$, 8, and 7 if $g = g_1$, or $h = 2$, 8, 11, and 7 if $g = g_2$.

Matters are similar for the central particle g. In the first case, $h = 9$, 5, 10, and in the second $h = 8$, 12, 4, 11. Depending on the number of h, the quantity ω_r characterizing the mutual arrangement of the particles f, g, and h varies. In the example being considered, two ways of arranging the particles are possible for each distance between sites f and g: for $r = 1$, sites $h = 2$, f, and g_1 are on one straight line, while sites $h = 7$ or 8, f, and g_1 make a right angle; for $r = 2$, sites $h = 2$ or 7, f, and g_2 make an obtuse angle, and sites $h = 8$ or 11, f, and g_2—an acute angle. In the first configurations, the particles at sites h and g do not interact with one another, therefore in formulas (7.16) the functions $\psi_{hfg}^{ijn}(\omega_r|\alpha)$ transform into the functions $\psi_{fg}^{ij}(r|\alpha)$ determined in (7.12), while in the second configurations the particles at sites h and g interact with one another in addition to the interaction of the particles at sites f and g. This leads to the appearance of the multipliers in (7.16).

As an example, let us write out some addends of Eq. (7.46) in which the numbers of h and g correspond to Fig. 7.2 (the distance between sites h and g is indicated instead of the index ω_r):

$$\begin{aligned} &U_{2f11}^{\bar{v}AA}(3/2) = U_{2f}^{vA}(2)\psi_{f11}^{AA}(1|2),\ U_{7f11}^{\bar{v}AA}(2|2) = U_{7f}^{vA}(2)\psi_{7f11}^{vAA}(2|2) \\ &U_{2f9}^{\bar{v}AA}(4/2) = U_{2f}^{vA}(2)\psi_{f9}^{AA}(2|2),\ U_{8f9}^{\bar{v}AA}(1|2) = U_{8f}^{vA}(2)\psi_{8f9}^{vAA}(1|2) \end{aligned} \tag{7.47}$$

The described structure of the addends in the kinetic equations of models with a limited mobility of the particles also remains the same for other kinetic schemes of processes and for any number of components of the reaction system. This fact, and also the hierarchy of the mathematical models for the surface reactions when considering the interactions of the first and second neighbors are illustrated by the material in Appendix 3. Here we shall limit ourselves to a transition from the distributed model (7.45), (7.46) to a point model considering the ordered state of the adsorbed particles $c(2 \times 2)$ on a square lattice. The kinetic equation for the concentration of the particles on the surface has the form

$$\dot{\theta}_\sigma^A = U_\sigma^v(1) - U_\sigma^A(1) - z[U_{\sigma\lambda}^{Av}(2) - U_{\sigma\lambda}^{vA}(2)],\ \sigma = \alpha,\ \gamma \quad \lambda \neq \sigma \tag{7.48}$$

where the addends $U_\sigma^i(1)$ have been determined in (7.37), and the migration terms are

$$U_{\sigma\lambda}^{\mathrm{Av}}(2) = K_{\mathrm{Av}}(2)\theta_{\sigma\lambda}^{\mathrm{Av}}(1)S_{\sigma\lambda}^{\mathrm{A}}(1|2)S_{\lambda\sigma}^{\mathrm{v}}(1|2) \times [S_{\sigma\sigma}^{\mathrm{A}}(2|2)S_{\lambda\lambda}^{\mathrm{v}}(2|2)S_{\sigma\lambda\lambda}^{\mathrm{Av}}(2)S_{\lambda\sigma\sigma}^{\mathrm{vA}}(2)]^2 \qquad (7.49)$$

In point models, there are no macroscopic flows of the particles, and their migration consists in local redistribution. In the given case, this relates to the redistribution of the particles between different sublattices, and equations (7.48) reflect the relaxation nature of the process on the entire surface. To close system of equations (7.48), equations are needed for the two-body correlators, namely, one equation for the nearest neighbors and two for the second neighbors:

$$\begin{aligned}
\dot{\theta}_{\alpha\gamma}^{\mathrm{AA}}(1) = {} & U_{\alpha\gamma}^{\bar{\mathrm{v}}\mathrm{A}}(1|1) + U_{\alpha\gamma}^{\mathrm{A}\bar{\mathrm{v}}}(1|1) - U_{\alpha\gamma}^{\bar{\mathrm{A}}\mathrm{A}}(1|1) - U_{\alpha\gamma}^{\mathrm{A}\bar{\mathrm{A}}}(1|1) \\
& - [U_{\gamma\alpha\gamma}^{\overline{\mathrm{vA}}\mathrm{A}}(3|2) - U_{\gamma\alpha\gamma}^{\overline{\mathrm{Av}}\mathrm{A}}(3|2) + 2(U_{\gamma\alpha\gamma}^{\overline{\mathrm{vA}}\mathrm{A}}(2|2) \\
& - U_{\gamma\alpha\gamma}^{\overline{\mathrm{Av}}\mathrm{A}}(2|2)) + U_{\alpha\gamma\alpha}^{\mathrm{A}\overline{\mathrm{Av}}}(3|2) - U_{\alpha\gamma\alpha}^{\mathrm{A}\overline{\mathrm{vA}}}(3|2) \\
& + 2(U_{\alpha\gamma\alpha}^{\mathrm{A}\overline{\mathrm{Av}}}(2|2) - U_{\alpha\gamma\alpha}^{\mathrm{A}\overline{\mathrm{vA}}}(2|2))] \\
\dot{\theta}_{\sigma\sigma}^{\mathrm{AA}}(2) = {} & 2[U_{\sigma\sigma}^{\bar{\mathrm{v}}\mathrm{A}}(2|1) - U_{\sigma\sigma}^{\bar{\mathrm{A}}\mathrm{A}}(2|1) - 2(U_{\lambda\sigma\sigma}^{\overline{\mathrm{vA}}\mathrm{A}}(4|2) \\
& - U_{\lambda\sigma\sigma}^{\overline{\mathrm{Av}}\mathrm{A}}(4|2) + U_{\lambda\sigma\sigma}^{\overline{\mathrm{vA}}\mathrm{A}}(1|2) - U_{\lambda\sigma\sigma}^{\overline{\mathrm{Av}}\mathrm{A}}(1|2))]
\end{aligned} \qquad (7.50)$$

The addends of Eqs. (7.50) are expressed in terms of elementary processes (7.37), (7.49) by formulas (7.12), (7.16), and (7.47), in which the numbers of the sites have been replaced with their types.

Dissolution of Adsorbed Particles and Rearrangement of an Alloy. In Chap. 5, we discussed models of surface processes considering the restricted mobility of the components of a system on a homogeneous surface, in the subsurface region of a solid, and also the simultaneous proceeding of surface reactions and a change in the concentration profile of an alloy. In many systems, the adsorbed particles dissolve and change the state of the solid. Such processes are induced especially frequently by hydrogen and oxygen atoms [236-239] (see also [45-48]). The dissolved atoms occupy the interstices or vacancies in the lattice of the atoms of the solid. The latter atoms may also be in the interstices.

For a systematic description of the evolution of the states of such a system, consideration of the interaction of the nearest neighbors is insufficient because its components are at different distances. The distances between the atoms and dissolved particles are less than those between the atoms and between the dissolved particles. To explain the meaning of a "systematic description", we remind our reader that in the model of absorption of gas-phase molecules by an absorbent that does not reorganize itself (see Sec. 5.6), the rate constant of particle migration includes the activation energy of the process. This energy characterizes the energy binding a particle to the atoms of the solid. In the model of the change in the state of the subsurface region of a solid (see Secs. 5.4, 5.5), the expression for the rate of migration of the solid's atoms contains a preexponential factor of the rate constant, while the energies of the bonds between neighboring atoms of the alloy are contained in the function S^i_{kq}. The values of these functions depend not only on the interaction energies, but also on the concentrations of the neighboring particles.

In other words, the concept of the activation energy of migration as of a constant characteristic of a system is absent. For each local composition, this quantity has a relevant value (the latter does not exclude the introduction of the effective activation energy of migration depending on the concentrations of the alloy's atoms). When describing the processes of absorption and the change in the state of a solid from single standpoints, the activation energies of migration both for the particles being dissolved and for the alloy atoms will also be absent in the model. But in a limit transition to the absence of migration of the alloy atoms, the overall model must describe the absorption of particles in an inhomogeneous solid with the same parameters of interaction between the particles and "decipher" the energy of activation of particle migration in terms of the energy of the bond between a particle and an atom of the solid (it is assumed here that multiparticle interaction is absent in the system).

Let us consider the process of absorption of gas-phase molecules through the flat surface of a multicomponent alloy. We shall denote the dissolving particles by A. The adsorption process is described by equations similar to those set out above, therefore in the following we shall consider only the steps of migration of the dissolved particles and the atoms of the alloy. For simplicity, we shall limit ourselves to the assumptions that the dissolved particles occupy one type of interstices and the alloy is disordered, and also that the dissolved particles do not occupy vacancies in the lattice of the alloy's atoms, and these atoms do not emerge into the interstices (immiscibility of the particles and atoms of the alloy). If we introduce the concepts of a sublattice of the interstices (γ) whose sites are occupied by particles and vacancies and of a substitution sublattice (δ) accommodating s components of the alloy including vacancies, then the problem being considered is one of describing the processes of migration of various species of particles in various sublattices.

Assume that an alloy has an fcc lattice and that the dissolved particles occupy octahedral interstices. In this case, the number of sites in sublattices γ and δ is the same in each layer in the subsurface region and in the bulk of the solid, and the kinetic equations for the concentration of particles i at the sites of sublattice σ in layer k, where $i = \mathrm{A}$ for $\sigma = \gamma$, and $i = 1, \ldots, (s - 1)$ for $\sigma = \delta$, will be written as follows:

$$\dot{\theta}_k^i(\sigma) = \sum_q z_{kq}[U_{kq}^{vi}(\sigma\sigma) - U_{kq}^{iv}(\sigma\sigma)], \quad q = k \pm 1 \tag{7.51}$$

These equations coincide in their structure with Eqs. (5.27). The indices in parentheses indicate the sublattice along whose nearest sites migration of the dissolved particles or alloy atoms occurs. The inhomogeneity of particle distribution in the system is due to the following factors: the difference in the interactions of the particles, which "singles out" different sublattices, and the presence of an interface, which "singles out" the consecutive arrangement of the layers in the subsurface region of the solid and in its bulk, and can also change the molecular parameters of the particles (see Secs. 3.8 and 5.6). All these factors are taken into account by formulas (7.29).

The model being considered is a one-dimensional one; averaging has been performed over the sites of each sublattice inside each layer, therefore the distribution of the particles is described by the functions $\theta_k^i(\sigma)$ and $\theta_{kq}^{ij}(\sigma\lambda|r)$. The last function characterizes the probability of finding particles i and j at sites of type σ and λ in layers k and q at a distance of r. Accordingly, in formulas (7.29) the number of a site must be replaced with the number of the layer containing these sites, while the type of site is determined by the type of sublattice. The addends of (7.51) have the form

$$\begin{gathered} U_{kq}^{iv}(\sigma\sigma) = \hat{K}_{kq}^{iv}\,\theta_{kq}^{iv}(\sigma\sigma) \prod_{r=1}^{R} \prod_{\omega_r=1}^{\pi_r} \prod_{\lambda} [S_{kqh}^{iv}(\sigma\sigma\lambda|\omega_r)]^{\varkappa_{kgh}(\sigma\lambda|\omega_r)} \\ S_{kqh}^{iv}(\sigma\sigma\lambda|\omega_r) = \sum_{j=1}^{s_\lambda} t_{kqh}^{ivj}(\sigma\sigma\lambda|\omega_r) \exp \beta\delta\varepsilon_{kqh}^{ivj}(\omega_r) \end{gathered} \tag{7.52}$$

where the number of layer h and its type λ are unambiguously related by the structure of the lattice to layer numbers k and q and their type σ, and the values of r and ω_r. The maximum deviations of layer h from these layers are $\pm k_R$, where k_R has been determined in Sec. 6.5; $\varkappa_{kqh}(\sigma\lambda|\omega_r)$ is the number of sites of type λ in layer h at a distance of r and with an orientation ω_r from the dimer molecule iv at sites of type σ; the species of the particles at site h depend on its type λ; the interaction parameters and the preexponential factor of the rate constant depend on the numbers of the layers in the subsurface region; these indices may be omitted for the bulk; the function $t^{inj}_{kqh}(\sigma\sigma\lambda|\omega_r)$ is determined in terms of the function θ^i_k and $\theta^{ij}_{kq}(\sigma\lambda|r)$ as indicated in formula (7.29).

Equations (7.52) are supplemented by equations for two-body correlators. In the general case, it is necessary to consider all the pairs appearing between particles on different sublattices (between dissolved particles on the sublattices of interstices, or between atoms of the solid on the sublattices of substitution), because all these pairs are contained in the function $S^{ij}_{kqh}(\sigma\sigma\lambda|\omega_r)$. The overall structure of these equations is as follows:

$$\dot{\theta}^{ij}_{kq}(\sigma\lambda|r) = -\sum_h \varkappa_{kqh}(\sigma\sigma|\omega_r)[U^{\overline{vi}j}_{hkq}(\sigma\sigma\lambda|\omega_r) - U^{\overline{iv}j}_{hkq}(\sigma\sigma\lambda|\omega_r)]$$

$$-\sum_h \varkappa_{kqh}(\lambda\lambda|\omega_r)[U^{i\overline{jv}}_{kqh}(\sigma\lambda\lambda|\omega_r) - U^{i\overline{vj}}_{kqh}(\sigma\lambda\lambda|\omega_r)] \quad (7.53)$$

where in the first sum, the index h relates to the nearest neighbors of type σ counted from site σ in layer k, the first index in the argument of the addend corresponds to them; in the second sum h relates to the nearest neighbors of species λ of site λ in layer q, the last index in the argument of the addend corresponds to them. The particles participating in an elementary process are capped with a bar. The addends of Eqs. (7.53), by (7.29), equal

$$\begin{aligned} U^{\overline{ij}n}_{hkq}(\sigma\sigma\lambda|\omega_r) &= U^{ij}_{hk}(\sigma\sigma)\psi^{ijn}_{hkq}(\sigma\sigma\lambda|\omega_r), \\ \psi^{ijn}_{hkq}(\sigma\sigma\lambda|\omega_r) &= t^{ijn}_{hkq}(\sigma\sigma\lambda|\omega_r)\exp\beta\delta\varepsilon^{ijn}_{hkq}(\omega_r)/S^{ij}_{hkq}(\sigma\sigma\lambda|\omega_r) \end{aligned} \quad (7.54)$$

The normalizing relations have the form

$$\sum_{j=1}^{s_\lambda}\theta^{ij}_{kq}(\sigma\lambda|r) = \theta^i_k(\sigma), \quad \sum_{i=1}^{s_\sigma}\theta^{ij}_{kq}(\sigma\lambda|r) = \theta^j_q(\lambda), \quad \sum_{i=1}^{s_\sigma}\theta^i_k(\sigma) = 1 \quad (7.55)$$

As a result, the system of equations (7.51)-(7.55) is closed and solved with the given initial and boundary conditions.

The constructed system of equations is generalized with consideration of the specific features of lattice structure and the mechanisms of the migration processes in the following cases: (1) when the number of interstices does not coincide with the number of sites of the lattice of alloy atoms (e.g. for tetrahedral interstices in an fcc lattice), (2) when the dissolved particles migrate along different types of interstices, or the alloy is in the ordered state, or both,

(3) when mixing of particles of different sublattices occurs—the alloy atoms emerge into the interstices and migrate along them, while the dissolved particles occupy vacancies in the lattice of the alloy atoms. These factors affect the number of neighboring sites between which the exchange of particles and vacancies is possible, the length of these jumps, the number of sublattices in the system, and the number of species of particles in different sublattices. The structure of kinetic equations (7.51) and (7.53) remains unchanged. The boundary conditions depend on the mechanisms of the surface reactions; the rules for constructing these equations are similar to the examples considered above.

7.4. Multiparticle Interactions in the Kinetics of Adsorption

It has been shown in Sec. 6.6 that multiparticle interactions substantially change the equilibrium characteristics of adsorption. Let us see how they affect the kinetic characteristics of adsorption. We shall limit ourselves to multiparticle interaction between nearest neighbors. (The case of two-body interaction of the first two neighbors and three-body interaction of nearest neighbors has been considered by the author in [132].) Let us assume that the adsorbed particles of A are distributed in equilibrium on a homogeneous surface. Our task is to obtain expressions for the concentration components of the rates of adsorption and desorption of the undissociating molecules.

The overall structure of the kinetic equations does not depend on the form of the interaction potential, therefore the kinetic equation for the concentration of the adsorbed particles will be written as usually:

$$\dot{\theta}_{\mathrm{A}} = U_{\mathrm{v}} - U_{\mathrm{A}}, \quad U_i = \langle \gamma_f^i \exp(-\beta \delta h_f^i) \rangle, \quad i = \mathrm{v}, \mathrm{A} \tag{7.56}$$

where the deviation in the energy of the system when an activated complex forms, by (6.37), is

$$\begin{aligned} \delta h_f^i &= \left(\overset{*}{\nu_i} - \nu_i + \sum_{g_1,\, j_1} \cdots \sum_{g_z,\, j_z} \delta\varepsilon_{ij_1 \ldots j_z} \gamma_{g_1}^{j_1} \cdots \gamma_{g_z}^{j_z} \right) \gamma_f^i \\ \delta\varepsilon_{ij_1 \ldots j_z} &= \overset{*}{\varepsilon}_{ij_1 \ldots j_z} - \varepsilon_{ij_1 \ldots j_z} \end{aligned} \tag{7.57}$$

where $\overset{*}{\varepsilon}_{ij_1 \ldots j_z}$ is a multiparticle parameter of interaction of the activated complex formed from particle i with the neighboring particles $j_1 \ldots j_z$ at sites $g_1 \ldots g_z$. The values of the interaction parameters do not depend on the numbers of the sites, and they can be characterized by the number of particles of A in the coordination sphere and by their configuration $\sigma(n)$, $E_i^*(n\sigma) = \varepsilon^*_{ij_1\ldots j_z}$.

As a result, the addends of Eq. (7.56) will be rewritten in the following form:

$$U_i = K_i \sum_{n=0}^{z} \sum_{\sigma(n)} b_{\sigma(n)}^{\sigma(0)} \theta^i_{[n\sigma]} \exp \beta\delta E_i^*(n\sigma) \tag{7.58}$$

In obtaining the formula, the projection properties of the quantities γ_f^i were used, as set out in detail in Sec. 4.2. The sum over $\sigma(n)$ appears because of the difference in the energy contributions of various configurations of adjacent particles in the coordination sphere of the reacting particle (see Sec. 6.6).

To calculate the rates by (7.58), one must evaluate the functions $\theta^i_{[n\sigma]}$. The formulas for this purpose are indicated in Sec. 6.6. In the quasichemical approximation, expressions (7.58) are:

$$\begin{aligned} U_A &= K_A\theta_A \sum_{n=0}^{z} t^n(1 - t)^{z-n} \sum_{\sigma(n)} b_{\sigma(n)}^{\sigma(0)} \exp \beta\delta E_A(n\sigma) \\ U_v &= K_v(1 - \theta) \sum_{n=0}^{z} (t_{BA})^n(1 - t_{BA})^{z-n} \sum_{\sigma(n)} b_{\sigma(n)}^{\sigma(0)} \exp \beta E_A^*(n\sigma) \end{aligned} \tag{7.59}$$

where it has been taken into account that $E_v(n\sigma) = 0$, while the function t_{BA} is related to the function $t \equiv t_{AA}$ by formula (2.18). The dependence of t on θ is given by Eq. (6.40).

In the molecular-field approximation, expressions (7.58) have the form

$$\begin{aligned} U_v &= K_v(1 - \theta) \exp \beta \sum_{n=0}^{z} \theta^n(1 - \theta)^{z-n} \sum_{\sigma(n)} b_{\sigma(n)}^{\sigma(0)} E_A^*(n\sigma) \\ U_A &= K_A\theta \exp \beta \sum_{n=0}^{z} \theta^n(1 - \theta)^{z-n} \sum_{\sigma(n)} b_{\sigma(n)}^{\sigma(0)} \delta E_A(n\sigma) \end{aligned} \tag{7.60}$$

The obtained expressions can be used with any potential of particle interaction (see Sec. 6.6). In the linear approximation of effective two-body interaction potentials

$$\begin{aligned} E_A^*(n\sigma) &= n\varepsilon^*_{AA} + n(n - 1)\Delta\varepsilon^*_{AA} \\ E_A(n\sigma) &= n\varepsilon_{AA} + n(n - 1)\Delta\varepsilon_{AA} \end{aligned} \tag{7.61}$$

formulas (7.60) will be rewritten as follows:

$$U_v = K_v(1 - \theta) \exp \beta z\theta\tilde{\varepsilon}^*, \quad \tilde{\varepsilon}^* = \varepsilon^*_{AA} + \omega^*\theta, \quad \omega^* = (z - 1)\Delta\varepsilon^*_{AA}$$
$$U_A = K_A\theta \exp \beta z\theta(\tilde{\varepsilon}^* - \tilde{\varepsilon}) \qquad (7.62)$$

where $\tilde{\varepsilon}$ has been determined in Sec. 6.6. Expressions (7.62) are formulas (4.16) with a concentration dependence of the parameters of nearest neighbor interaction, which leads to a quadratic dependence of the logarithms of the rates of adsorption and desorption on the surface coverage in the region of moderate and large coverages.

Similar expressions are obtained in the molecular-field approximation if the two-body and three-body interactions $E^*_A(n\sigma) = n\varepsilon^*_{AA} + m_{n\sigma}\varepsilon^*_{AAA}$ are considered, where ε^*_{AAA} is the parameter of three-body interaction between an activated complex and the two particles nearest to each other in the coordination sphere of the complex; the quantity $m_{n\sigma}$ has been determined in Sec. 6.6. In this case, $\omega^* = \varepsilon^*_{AAA}$.

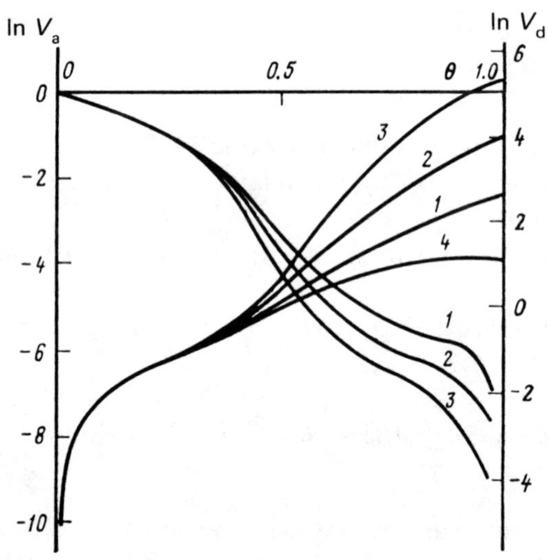

Fig. 7.11. Rates of adsorption and desorption in the quasichemical approximation with linear approximation of the effective two-body potentials for $z = 4$, $\beta\varepsilon_{AA} = -1.5$, $\beta\Delta\varepsilon_{AA} = -0.1$ (*3*), 0 (*2*), 0.1 (*1*), and 0.2 (*4*)

Figure 7.11 shows the concentration dependences of the rates of adsorption and desorption in the quasichemical approximation with the potential (7.61). As for equilibrium characteristics, if ε_{AA} and $\Delta\varepsilon_{AA}$ are of the same sign, the contribution of the multiparticle interactions increases the "effect" of the two-body potential, while if ε_{AA} and $\Delta\varepsilon_{AA}$ are of opposite signs, they diminish it. The maximal contribution of multiparticle interactions manifests itself at large coverages. Formulas (7.59), unlike formulas (7.62), when potential (7.61) is used may lead to nonmonotonic dependences of the rates on the coverage even when interaction between the nearest neighbors is taken into consideration (see, for instance, curve *4* for the rate of desorption).

Similarly, the two-body interaction of the first two neighbors and the three-body interaction between the nearest neighbors are considered in the kinetics of adsorption [132]. Account of the second neighbors increases the number of various configurations of the particles in the coordination spheres of a cluster. In the expressions for the rates of adsorption and desorption, an additional sum over the number of particles in the second coordination sphere appears, as in formulas (7.35).

The results obtained point to the importance of considering the contributions of multiparticle interactions that change the concentration dependences of the rates of elementary steps, and to the need of studying in greater detail the nature of the particle interaction of adsorbed particles.

8 Conclusion

The development of the cluster approach made it possible to obtain a solution of the problem formulated in Sec. 1.2 on the self-consistent description in time and in an equilibrium state of the distributions of interacting particles between the sites of homogeneous and inhomogeneous lattices. and this enables one to solve a large number of matters in the practical description of processes at a gas-solid interface.

The studied examples of very simple processes, namely, adsorption, absorption, the diffusion of particles, and surface reactions, indicated above and in Appendix 2 point to the fundamental role of the cooperative effects due to the interaction between the components of the reaction system in the kinetics of these processes. Nonideal models of surface processes made it possible to explain the following facts.

1. Splitting of the thermodesorption spectra for homogeneous surfaces as a result of repulsion of the nearest chemisorbed particles.

2. The empirical laws of adsorption processes of Becker-Langmuir, Roginsky-Zeldovich, and Benham-Barta-Kwan, but unlike these laws, the constructed models also describe the processes correctly when $\theta \to 0$ and when $\theta \to 1$.

3. The varying order of magnitude of the rate of desorption including its zero value that is associated with the existence of two or more phases of the adsorbed molecules on a surface.

4. An increase in the sticking probability of adsorbed particles with a growth in the coverage at low coverages of a surface.

5. The phenomenon of adsorbed particle ordering and the change in the rates of the surface processes in ordered systems; the relation of the characteristics of these processes to data on low-energy electron diffraction.

6. Jumps, breaks, and peak-like behavior of the concentration dependences of the diffusivity in structural phase transitions.

7. The negative values of the diffusivities in processes of particle aggregation and a change in the rates of processes under conditions of stratification of the adsorption system.

8. The strong mutual influence of adsorbed particles of various species on the rates of elementary processes.

9. The dependence of the kinetic characteristics of processes on the conditions of formation of the initial states of the reaction system.

10. The plurality of steady states of the reaction systems and isothermal self-oscillating modes of proceeding of reactions.

It is quite natural that in more involved processes too the role of the cooperative effects is substantial, and they must be considered correctly. To appraise the possibilities and practical significance of the developed theory, we shall revert again to the oxidation of CO molecules on platinum metals, which has become a "proving ground" both for testing new experimental procedures and for verifying theoretical approaches.

This process is a very simple one. Adsorbed particles of two species participate in it—CO molecules and oxygen atoms. By modern notions [167], the mechanism of a process consists of three steps (see Sec. 4.4). The third step is considered to be irreversible at $T \geqslant 300\text{-}350$ K because of the low energy binding the CO_2 molecules to the surface [307]. The question of whether there is a fourth step $ZO + CO \rightarrow Z + CO_2$ for low pressures ($P \leqslant 10^{-4}$ mmHg), which has been discussed for a long time, has been solved in favor of a three-step mechanism. Experiments with molecular beams have shown the low probability of there being a fourth step. If we assume that this step is not an elementary one [308], but is a process of seating a CO molecule onto a free center and a rapid reaction with an adjacent atom of oxygen, at large coverages by oxygen atoms it is difficult to distinguish the contribution of this channel from the third step.

The introduction of the fourth step was associated with the oxidation reaction at partial oxygen pressures of the order of magnitude of $10^{-5}\text{-}10^{-4}$ mmHg, at which, by the law of mass action, the surface must be completely covered by oxygen, and the adsorption of CO is impossible. But experimental data [167] on various properties of the reaction system $CO + O_2$ on metals of the platinum group studied by low-energy electron diffraction, photoemission spectroscopy, thermodesorption spectra, etc. point to the strong mutual influence of the adsorbed particles. In particular, several kinetic relations are proposed for the rate of CO oxidation for various concentration regions of the reactants. The author has shown elsewhere [309] that when considering the lateral interaction between adsorbed particles at pressures up to 10^{-4} mmHg, the surface coverage does not exceed 0.5, i.e. the surface has vacant centers on which the adsorption of CO molecules is possible, and the need of introducing the fourth step vanishes.

Notwithstanding the simplicity of the kinetic scheme of the process, it is characterized by a complicated dynamic behavior [167, 310-315]: breaks in the rate of formation of CO_2 under steady conditions when the temperature of the system changes, hysteresis of the rate of steady formation of CO_2 at relatively low temperatures, and complicated and chaotic vibrational modes of the reaction of CO oxidation. The latter phenomenon attracted the greatest attention of researchers. Dagonnier *et al.* [316, 317] and Lagos *et al.* [318] considered that the reaction proceeds by an adsorption mechanism on a homogeneous surface, but the process is not isothermal. The temperature of the surface changes because of the dissipation of the heat evolved in the catalytic reaction [319].

Suhl [320] assumed the existence of two types of adsorption centers for the oxygen atoms. The temperature of the surface was considered to be constant. The kinetic equations were also constructed on the basis of the law of mass action. A similar model was considered by Sales *et al.* [321]. In this model, the adsorbed oxygen and the oxygen forming the oxide blocking a part of the surface were singled out. Mukesh *et al.* [322] explained the fluctuations of the rate of CO oxidation and their chaotic nature by a model permitting the existence of reactant islets. Their description obeys the equations of a two-dimensional topochemical process (see also [323]) that are similar to the equations of three-dimensional processes [42-44]. Close notions were advanced by Borker and Wicke [324], while Rickert [325] considers that the topochemical process relates to the oxide phase. The self-oscillating mode was also associated with the presence of adjacent free centers to avoid steric hindrances in the formation of the product from the reactants [326]. A detailed review of early publications has been made by Yablonsky *et al.* [327].

All these models ignored the fundamental real factors of a reaction system: interaction between adsorbed particles, their ordering, and the rearrangement of the platinum atoms, and all of them obtained good agreement with experimental data on the rate of formation of CO_2 molecules.

Only in experiments on single crystals on the basis of simultaneous measurements of low-energy electron diffraction, the work function, and the rate of oxidation of CO did it become possible to ascertain the nature of the self-oscillating mode and relate the fluctuations of the reaction rate to the rearrangement of the surface layer of the metal [40, 72, 224, 328-331]. The free surface of the platinum face (100) was shown to be metastable, and the atoms of the first surface layer (with $z = 4$ nearest neighbors) transform into a phase with a hexagonal packing where $z = 6$. The adsorption of CO and O_2 molecules on this hexagonal phase leads to the reverse reorganization into a phase with $z = 4$. Studying of the process on the stable platinum face (111) revealed the absence of self-oscillations. The two phases differ sharply in their properties, in particular the sticking probability of oxygen on the hexagonal phase is about two orders of magnitude lower than on the phase with $z = 4$. Precision measurements at $T \approx 500$ K have shown that the organization process undulates with excitation of a trigger nature at the boundary of a specimen. Scanning tunnel microscopy has uncovered the nature of the redistribution of the platinum atoms in reorganization of the surface from the phase with $z = 6$ to that with $z = 4$ and vice versa [332]. Since the density of the hexagonal phase is higher than that of the phase with $z = 4$, then when going over from the phase with $z = 6$ to that with $z = 4$, the excess platinum atoms migrate into the upper layer and form islets that grow in the course of the phase transition.

It should be noted that the above process is not the only one causing the appearance of self-oscillations of the rate of the CO oxidation reaction. At high pressures, a close picture is produced by the redistribution of impurity silicon atoms between the bulk and surface [333] (see also [334]).

The treated example indicates the intricacy of the real nature of surface processes due to the interaction of particles and the essential factors that must be considered by the theory of surface processes.

First, the theory must take into account the interaction between adsorbed particles. This interaction determines the nature of the mutual distribution of the particles and, consequently, the probability of their participating in elementary reactions. In particular, it determines the realization of an ordered distribution of the particles and the formation of islets of the ordered phase. Ordering is chiefly due to the repulsion of particles at adjacent adsorption centers, and for the formation of islets of the ordered phase it is at least essential that there be attraction between particles at the distance of second neighbors (see Sec. 6.4). Here we are speaking of the superstructure $c(2 \times 2)$ observed on phases with $z = 4$ and 6 in the adsorption of CO. Their intensities were used to watch the state of the surface during a catalytic reaction.

Second, the theory must reflect the broad spectrum of mobilities of the particles in the reaction system. We have already mentioned the distributed nature of CO oxidation and the experimental data on the adsorption of individual gases CO and O_2 [227-229] pointing to the restricted mobility of the adsorbed particles and to the influence of the nature of the mobility on the kinetics of desorption depending on how the same initial coverage was formed (see Sec. 5.2).

Third, the theory must consider the possibility of a change in the surface and subsurface region of a solid under the influence of adsorption of molecules of the gas phase.

The oxidation of CO is at present also described on the basis of the law of mass action [331, 335]: "partial" relations are constructed for the rates of elementary reactions that hold for separate concentration regions, and "switches" are introduced for changing these relations with a change in the coverages of the reactants. An ordered arrangement of the adsorbed particles is postulated here. Imbihl *et al.* [331] described the process by a system of differential equations of the diffusion type, while Moller *et al.* [335] described it by a system of conventional differential equations for a two-dimensional set of elementary areas. The calculations formally enable one to reproduce almost all the experimentally observed relations and effects, but such a description does not seem to be convincing from a physical viewpoint. This is why it becomes necessary to employ the molecular-kinetic theory of processes in condensed phases so that all the real properties of a system would "appear" in a natural way on the basis of knowledge of the potentials of interaction between particles.

The theory of surface processes constructed on the basis of the cluster approach considers all the indicated factors. It uses a common principle for describing all the components of a system, namely, the adsorbed particles and particles (atoms or ions) of the solid. The distinctions associated with the nature of the various components are reflected only in the energies of their interactions with the other components. This manifests itself in their rate constants of migration and chemical transformation (in terms of the energies of interaction of the elementary processes) that are traditionally considered in ideal models, and in taking account of the contributions resulting from the nonideal nature of the reaction system.

The general circle of kinetic models of an atomic and molecular level covered by the theory is shown in Fig. 8.1. The models reflect the basic factors of experimental systems — interactions between particles and the state of aggregation of the components, surface inhomogeneity, a different mobility of the particles, and reorganization of the subsurface region of a solid. The figure singles out models in which in addition to "simple" consideration of the interaction between the components of a system, additional conditions are employed to consider their state of aggregation. The models of a molecular level operate with the states of occupancy of individual sites, and the phase a site belongs to must be determined additionally from macroscopic conditions. Equations (2.30) for condensing particles are an example of such conditions.

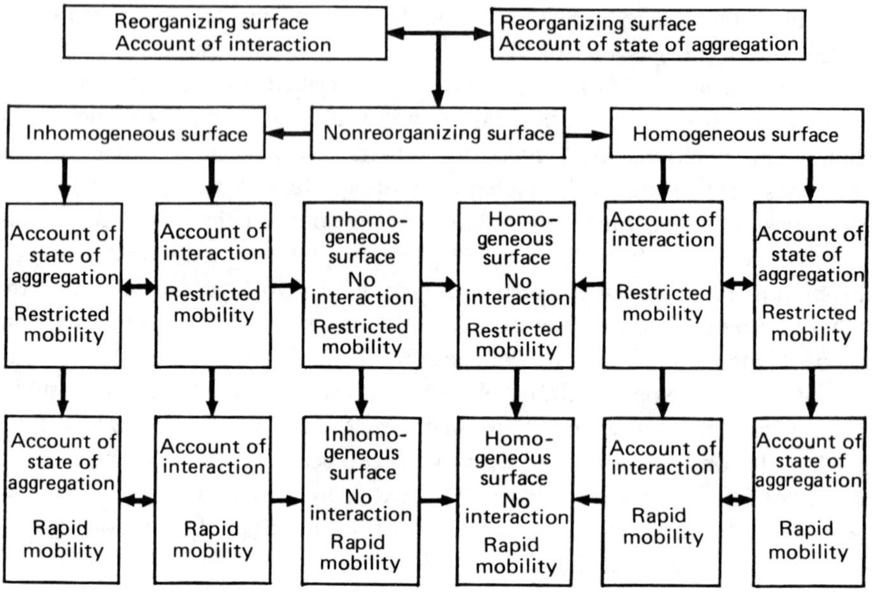

Fig. 8.1. Hierarchy of surface process models

Transitions between different models are performed when the relevant restrictions are superposed in the more general models. For instance, the condition of the absence of migrations of a solid's particles leads to a model with a nonreorganizing surface. This makes it possible within the scope of a common method of description to consider the processes of formation of the surface themselves and relate their characteristics (e.g., catalytic activity and resistance to corrosion) to these processes. The latter can be related not only to the characteristics of the surface itself, but also to the bulk characteristics of a solid (e.g., the defectiveness of a crystal depending on the mode of its growth).

Appendix 3 considers the hierarchy of mathematical models of surface reactions for a homogeneous surface differing in how the description of particle distribution and their mobility has been elaborated. Similar models can be constructed for inhomogeneous nonreorganizing surfaces and for reorganizing ones, including surfaces with a moving boundary, on the basis of system of equations (7.27)-(7.29). It should be noted that the form of the systems of kinetic equations, in addition to the indicated factors, depends on the approximations used to consider interaction and on the interaction potential.

The following must be noted when discussing the applied aspects of the obtained results. When speaking of a solid, we generally used concepts of an alloy and solid solutions. The constructed systems of equations in Secs. 6.5 and 7.3 for lattices composed of a number of sublattices relate equally to a broad class of nonstoichiometric compounds playing an important role in catalysis [29, 336].

The molecular-statistic theory of nonstoichiometric compounds is traditionally based on the existence of several sublattices, at least one of which is completely occupied, while the others have a varying composition [204, 337-339]. This approach is convenient only with small degrees of nonstoichiometry, and a separate theory must be constructed for each type of nonstoichiometry. System (6.35) can serve as the grounds for constructing a more general theory of nonstoichiometric compounds [340]. Every sublattice in it has equal rights and allows any occupancy by each species of particles. This enables us to construct a general theory of nonstoichiometric compounds both for an excess and for a lack of any component. The negation of the existence of at least one completely filled sublattice makes it possible to take account of the thermal expansion of a structure by increasing the number of vacancies and consider the simultaneous existence of defects of various types.

The use of system (6.35) signifies that what is being studied is the distribution of the particles on a multitude of sublattices: a cationic, anionic, and several sublattices of interstices, each of which contains vacancies as one of the components. Such a formulation is standard for the theory of solutions and alloys considered previously, to which it reduces the molecular-statistical theory of nonstoichiometric compounds. Any specific structure is given by a definite set of sublattices, the numbers $z_{\lambda\sigma}(r)$, and the states of occupancy of each sublattice.

System (6.35) operates with strictly definite states of the lattice sites. With its aid, we can avoid the introduction of conditional processes of the formation of large complexes and of the formation from "zero" of sites of cationic and anionic sublattices that were used traditionally [204, 205, 341].

Generally, the use of such symbols is associated with disturbance of the balance with respect to the number of sites in different sublattices. At the same time, it should be noted that the introduction of large complexes is a natural result of the inadequate application of the law of mass action [342]. Large complexes reflect the need to consider the correlated arrangement of the particles on different sublattices. But this correlated nature must arise from a general analysis of particle distribution depending on the temperature and external pressure of the gas (e.g., for oxides). The possibility of such a correlated distribution follows from calculations similar to those shown in Fig. 6.9.

Let us consider as an example the nonstoichiometry of wustite (FeO). The structure of wustite can be represented with the aid of three sublattices: α — contains ions of doubly charged and triply charged iron and vacancies, δ — ions of triply charged iron and vacancies, and γ — ions of oxygen and vacancies [341]. The elementary processes in wustite can be expressed as follows:

$$2\mathrm{Fe}_\alpha^{+2} + \mathrm{v}_\gamma + 1/2\mathrm{O}_2 \rightleftarrows 2\mathrm{Fe}_\alpha^{+3} + \mathrm{O}_\gamma^{-2}, \quad \mathrm{v}_\delta + \mathrm{Fe}_\alpha^{+3} \rightleftarrows \mathrm{Fe}_\delta^{+3} + \mathrm{v}_\alpha$$

where v_δ are vacancies, the subscript indicates the sublattice containing the particle. The first process is implantation of the gas phase oxygen into the wustite lattice, and the second is redistribution of the triply charged iron ion between the sublattices (a conventional migration process following a vacancy mechanism). The total amount of iron in wustite is determined as follows:

$$\theta_{\mathrm{Fe}} = [\theta_\alpha^{\mathrm{Fe}^{+2}} + \theta_\alpha^{\mathrm{Fe}^{+3}} + 2\theta_\delta^{\mathrm{Fe}^{+3}}]/3$$

where it is taken into account that the number of tetrahedral interstices (δ) is double the number of sites of the cationic sublattice (α). Equations (6.35) must be supplemented by the condition of electroneutrality $\sum_{\sigma=\alpha}^{\gamma} \sum_{i=1}^{s_\sigma} \theta_\sigma^i e_\sigma^i = 0$, where e_σ^i is the charge of particle i on sublattice σ, and by the normalization conditions. Other nonstoichiometric compounds can be described in a similar way.

For nonstoichiometric compounds, the kinetic equations of Secs. 7.1 and 7.3 make it possible to describe the kinetics of surface reactions, the processes of ion transfer in a solid, and the processes of formation of the compounds themselves.

In completing our brief discussion of the theory, let us consider the possible trends of its further development and use.

The practical application of the present theory and other models of surface processes that are its particular cases is restricted at present to very simple processes (the material of Chaps. 2-7* and Appendix 2) and does not reflect all the possibilities it contains. Homogeneous nonreorganizing surfaces are

* The author has used calculations performed by A. M. Kolchin (Fig. 5.3), O. A. Kuznetsova (Figs. 4.5, 5.5), V. M. Pak (Fig. 2.12), V. A. Pindyurin (Figs. 6.11, 6.12), S. Yu. Surovtsev (Fig. 6.10), and O. V. Chelnokova (Figs. 3.13-3.15). The questions illustrated by these figures are considered in greater detail in separate communications [490-496].

mainly treated. The interactions of first and second neighbors are beginning to be considered quite actively. To date, the joint influence of particle interaction and surface inhomogeneity has virtually not been investigated. Models of reorganizing surfaces are meanwhile limited to an explanation of their thermodynamic aspects [34, 343-348] (and this only in the molecular-field approximation), while the dynamics of reorganization is studied only in problems involving crystal growth (see, for instance, [197, 198, 208]).

Equations (7.27)-(7.29) make it possible to describe processes of particle redistribution between the layers of the subsurface region and of the change in their state of aggregation in each layer. This reflects most of the possible mechanisms of the reorganization processes, including the change in the structure (symmetry) and the lattice parameters (recall that the equations are constructed on an arbitrary type of inhomogeneous lattices). It is natural that subsequently the models describing processes on nonreorganizing inhomogeneous and on reorganizing surfaces will acquire widespread application. The same can be said with respect to the joint consideration of various real factors and their applications to multistep processes, including practically important catalytic processes, because matters relating to the change in the activity of catalysts (thermal stability, sintering, poisoning, etc.) can be investigated on their basis. Close problems arise when studying the nature of the surface segregation of alloys in vacuum and in the gas phase, in the growth of crystals from the gas phase and their sublimation, in the formation and removal of protective films, in etching of a surface and acting on it with streams of particles varying in energy, and so on.

In addition to a broad range of applied questions, the set out theory of processes at a gas-solid interface contains great "internal" reserves. This relates primarily to the elaboration of the description of particle distribution because of consideration of the contributions of the indirect correlations, especially because of the search for approximations of these contributions (a simple use of the contributions made by configurations of the particles is only slightly effective already for three coordination spheres). It follows from Fig. 2.9 and results described elsewhere [132] that the joint consideration of correlations and changes in the state of aggregation of particles leads to practical coincidence with the exact result for the temperature dependence of the two-body correlation function of the nearest neighbors (the latter does not relate to phase diagrams in the vicinity of critical points where scaling approximations have to be employed [154, 349]).

The more accurate consideration of the correlation effects also plays an important role in the kinetics of elementary processes. We shall note the possibility of the correlated appearance of "activated complexes" of a reaction in the critical and subcritical regions. Tovbin and Fedyunin have shown [350] that the consideration of such correlation gives a different value for the rate of a reaction. This situation is not unexpected. Though traditionally every elementary process is described independently of other elementary processes, the fact of conjugated chemical transformations is well known [55, 56, 351].

This phenomenon is explained for gases by the presence of intermediate compounds formed during the main reaction that stimulate accompanying reactions [351]. In a condensed phase, conjugation effects are possible as a result of the particle interaction when a particle passes from the ground state to a transition one, and their role can be quite appreciable. General expressions have been obtained for the rates of reactions on homogeneous and inhomogeneous surfaces for rapid mobility of the reactant when correlation between the activated complexes has been taken into account [171, 212, 249, 296]. Such effects may cause local heating in highly exothermal reactions.

More accurate consideration of the geometry of the bond between adsorbed particles and atoms of the solid is a different trend in the development of the theory [132]. Here questions appear as to how to consider the orientation of the adsorbed particles (this problem is solved by a method described herein if each orientation of the particles is characterized by its own molecular parameters and is treated as a separate species of particles) and the blocking by large particles of several surface atoms (see Appendix 2).

A close but still more involved problem appears if an adsorbed particle and surface atoms are considered to form several types of adsorption centers (several localization regions at distances of the order of the lattice constant). Quantum-chemical calculations make it possible to specify these types of adsorbent-adsorbate bonds more precisely [352, 353]. For instance, on the faces (100) and (111) of single crystals, three types of bonds are singled out that differ in the arrangement of an adsorbed particle: (1) a particle over an atom of the solid; (2) a particle over the bond between adjacent atoms; and (3) a particle over an interstice of atoms of the solid. The coexistence of adsorbent-adsorbate bonds is considered by a transition to an auxiliary inhomogeneous lattice whose constant is smaller than that of the lattice of the solid's atoms and to distribution thereon of larger particles [132].

It also seems natural to reject the traditional assumption in the kinetics of surface processes that all the parameters of particle interaction in the system adsorbed particles-atoms of the solid do not depend on their internal degrees of freedom. Account of the indicated relation leads to a temperature dependence of the interaction parameters [148, 354], which corresponds to a real situation. Finally, the considered theory can be generalized for charged adsorbed or absorbed particles and the redistribution of the charges in the solid associated with their presence. Consideration of the charges requires the addition of equations relating to the distribution of the charges to the constructed systems of equations for the distribution functions and their simultaneous solution. Such a general system of equations has been constructed for bulk solutions containing a set of sublattices [355]. It is not difficult to generalize it for the subsurface regions. Though the mathematical difficulties involved in solving such a problem grow quite appreciably, the theory can have a great significance for describing processes on the surfaces of many insulators and semiconductors [25-28].

APPENDIX 1

Approximate Systems of Equations

The distribution of interacting particles within broad ranges of variation in the concentrations of the components and the temperature of a system is described on the basis of various approximate approaches. An approximation considers the following questions.

1. The method of correlation functions enables one to obtain new approximate systems of equations in comparison with the method of the statistical sum. One of the simplest of such approximations is the polynomial one [109]. When considering the interaction of the nearest neighbors, it takes into account the correlation effects, though all the correlators are approximated in terms of the unary correlators. This lowers the dimension of the system of equations because equations for the two-body correlators are absent.

2. With an increase in the number of components, systems of equations in the polynomial approximation become cumbersome, while the dimension of the systems in the quasichemical approximation sharply grows. Of importance for practical purposes is the matter of the possibility of reducing the dimension of the systems of equations in the second case. This is done by replacing the variables.

3. Refinement of the description of the correlation effects requires consideration of the indirect correlations; for this end, the ways of closing the systems of equations are discussed.

The Polynomial Approximation

Equations in the polynomial approximation are obtained on the basis of the correlators given in Sec. 2.1. If the states of occupancy of the central sites of a cluster are averaged in the functions (2.3) and (2.6), the obtained correlators of the first and second types characterize the probabilities of the states of occupancy of the sites in the coordination sphere of a cluster:

$$\theta_{\{[n]\}} = \sum_{i=1}^{s} \theta^{i}_{\{[n]\}}, \quad \theta_{\{n\}} = \sum_{i=1}^{s} \theta^{i}_{\{n\}} \qquad \text{(Ap. 1.1)}$$

Binary System (components A and B). Instead of the ratio of the correlators $\theta^{A}_{[n]}$ and $\theta^{B}_{[n]}$ in (2.7), we can consider the relative contribution of a central A particle, for instance $\theta^{A}_{[n]}/(\theta^{A}_{[n]} + \theta^{B}_{[n]})$. This enables us to obtain a relation between the correlators $\theta^{A}_{[n]}$ and $\theta_{[n]}$:

$$\theta^{A}_{[n]} = N^{A}_{n}\theta_{[n]}, \ N^{A}_{n} = \{1 + M_n\}^{-1} \tag{Ap. 1.2}$$

where M_n has been defined in (2.7). There is a similar relation between the functions $\theta^{B}_{[n]}$ and $\theta_{[n]}$ when $N^{B}_{n} = 1 - N^{A}_{n}$. A system similar to system (2.10) exists for the functions $\theta_{[n]}$. Inserting it and a system of the type of (2.10) into (Ap. 1.2), we obtain a system of equations between the correlators of the second type similar to system (2.12):

$$\theta^{A}_{n} = \sum_{k=0}^{z-n} C^{k}_{z-n}N^{A}_{n+k} \sum_{r=0}^{z-n-k} (-1)^{r}C^{r}_{z-n-k}\theta_{n+k+r} \tag{Ap. 1.3}$$

System (Ap. 1.3) with a view to regrouping of the right-hand sides and symmetrization of the correlators ($F^{A}_{n} = \theta^{A}_{n}z!/(z - n)!$) coincides with results obtained by Tyablikov and Fedyanin [108]. This shows the equivalence of the cluster approach and an approach based on the use of equations of motion for the Green functions. The given system and system (2.12) are equivalent. Tyablikov and Fedyanin obtained an exact solution on the basis of this system by the consecutive coupling of the sites of a one-dimensional solution [108]. For lattices $z > 2$, closed equations in a polynomial approximation are obtained with the following assumption: $\theta_n = \theta^{n}_{A}$, which signifies that the particles in a coordination sphere are independent of one another, while the correlation effects are retained between them and the central particle. This yields the following expression for the correlators:

$$\theta^{A}_{n} = \sum_{k=0}^{z-n} C^{k}_{z-n}\theta^{n+k}_{A}(1 - \theta_{A})^{z-k-n}N^{A}_{n+k} \tag{Ap. 1.4}$$

where the unknown quantity θ_A is determined by solving nonlinear equation (Ap. 1.5) following from expression (Ap. 1.4) when $n = 0$:

$$\theta_{A} = \sum_{k=0}^{z} C^{k}_{z}N^{A}_{k}\theta^{k}_{A}(1 - \theta_{A})^{z-k} \tag{Ap. 1.5}$$

Knowing the expressions for the correlators of the second type, it is easy to obtain expressions for those of the first type by formulas (2.10). The first-type correlators are used in formulas (4.14) for calculating the rates of adsorption and desorption [145].

An equation for the adsorption isotherm is obtained on the basis of (Ap. 1.5) if account is taken of the determinations of parameters (1.23), (2.7), and (2.23) corresponding to adsorbed A particles and vacant B sites, i.e.

$$N_k^{\mathrm{A}} = y/(y + \exp(-\beta k\varepsilon_{\mathrm{AA}}), \; y = (a_{\mathrm{A}}P_{\mathrm{A}})^{1/m_{\mathrm{A}}} \qquad \text{(Ap. 1.6)}$$

The average number of adjacent pairs of adsorbed particles is calculated after finding the value of $\theta_{\mathrm{A}}(P_{\mathrm{A}})$ by formula (Ap. 1.4) at $n = 1$.

The isosteric heat of adsorption is expressed by formula (2.32) in which [356]

$$\varphi = \sum_{k=0}^{z} kC_z^k\theta_{\mathrm{A}}^k(1 - \theta_{\mathrm{A}})^{z-k}N_k^{\mathrm{A}}(1 - N_k^{\mathrm{A}}) \Bigg/ \sum_{k=0}^{z} C_z^k\theta_{\mathrm{A}}^k(1 - \theta_{\mathrm{A}})^{z-k}N_k^{\mathrm{A}}(1 - N_k^{\mathrm{A}}) \qquad \text{(Ap. 1.7)}$$

Multicomponent System [141]. The relative contribution of a central particle i determined as $\theta^i_{\{[n]\}}/\sum_i \theta^i_{\{[n]\}}$ leads to the following relation between the correlators:

$$\theta^i_{\{[n]\}} = N^i_{\{n\}}\theta_{\{[n]\}}, \; N^i_{\{n\}} = y^i_{\{n\}} \Bigg/ \sum_{i=1}^{s} y^i_{\{n\}}$$
$$y^i_{\{n\}} = \exp\beta\left[-\nu_i + \sum_{j=1}^{s-1} n_j(\varepsilon_{ij} - \varepsilon_{is}) + z\varepsilon_{is}\right] \qquad \text{(Ap. 1.8)}$$

Relations similar to (2.40) hold for the functions $\theta_{\{[n]\}}$, hence by expressing the correlators of the first type in terms of those of the second type and inserting them into relations (Ap. 1.8), we obtain a linear system relative to the correlators of the second type, and its solution relative to the correlators $\theta^i_{\{n\}}$ yields

$$\theta^i_{\{n\}} = \sum_{k_1=0}^{\tilde{z}} \cdots \sum_{k_{s-1}=0}^{\tilde{z}-l} N^i_{\{n+k\}}C_{\tilde{z}}^{\{k\}} \sum_{r_1=0}^{\tilde{\tilde{z}}} \cdots \sum_{r_{s-1}=0}^{\tilde{\tilde{z}}-\sigma} (-1)^{\tilde{r}}C_{\tilde{\tilde{z}}}^{\{r\}}\theta_{\{n+k+r\}}$$
$$\tilde{\tilde{z}} = \tilde{z} - \tilde{k}, \; \sigma = \sum_{m=1}^{s-2} r_m, \; \tilde{r} = \sum_{m=1}^{s-1} r_m \qquad \text{(Ap. 1.9)}$$

where the abridged notation defined in formula (2.35) is employed.

The polynomial approximation for a multicomponent system is formulated with the aid of the approximation

$$\theta_{\{n\}} = \prod_{m=1}^{s-1} (\theta_m)^{n_m} \qquad \text{(Ap. 1.10)}$$

Insertion of it into the right-hand side of system (Ap. 1.9) enables us to evaluate the correlators $\theta^i_{\{n\}}$:

$$\theta^i_{\{n\}} = \sum_{k_1=0}^{\tilde{z}} \cdots \sum_{k_{s-1}=0}^{\tilde{z}-l} C_{\tilde{z}}^{\{k\}} N^i_{\{n+k\}} \theta_s^{\tilde{z}-l} \prod_{m=1}^{s-1} (\theta_m)^{n_m+k_m} \qquad \text{(Ap. 1.11)}$$

where $\theta_s = 1 - \sum_{m=1}^{s-1} \theta_m$. The unknowns θ_i, $1 \leqslant i \leqslant (s-1)$, are determined by solving the following system of nonlinear equations ($\{n\} = \{0\}$):

$$\theta_i = \sum_{k_1=0}^{z} \cdots \sum_{k_{s-1}=0}^{z-l} C_z^{\{k\}} N^i_{\{k\}} \theta_s^{z-l} \prod_{m=1}^{s-1} (\theta_m)^{k_m} \qquad \text{(Ap. 1.12)}$$

The correlators of the first type are obtained on the basis of expressions (Ap. 1.11) by formula (2.40). The partial adsorption isotherms are obtained from system (Ap. 1.12) at $\nu_v = \varepsilon_{iv} = 0$, where $1 \leqslant i \leqslant s$.

Inhomogeneous Lattices. It is not at all difficult to generalize the polynomial approximation for inhomogeneous lattices. The distribution of particles of two species on a lattice containing an arbitrary number of site types is treated in [172]. Systems of equations for an arbitrary number of components are obtained in a similar way. The general structure of the equations is illustrated for three species of particles (A, B, C) and two types of sites. The closed system of equations has the form:

$$\theta^i_q = \sum_{\sigma_{2q}} d(q[m_{q_1} m_{q_2}]) \sum_{n^A_{q_1}=0}^{m_{q_1}} \sum_{n^B_{q_1}=0}^{m_{q_1}-n^A_{q_1}} \sum_{n^A_{q_2}=0}^{m_{q_2}} \sum_{n^B_{q_2}=0}^{m_{q_2}-n^A_{q_2}} \prod_{p=1}^{2} C^{n^A_{qp} n^B_{qp}}_{m_{qp}}$$

$$\times (\theta^A_p)^{n^A_{qp}} (\theta^B_p)^{n^B_{qp}} (\theta^C_p)^{n^C_{qp}} y^{i\{[n]\}}_{q[m_{q_1} m_{q_2}]} \Big/ \sum_{j=A}^{C} y^{j\{[n]\}}_{q[m_{q_1} m_{q_2}]} \qquad \text{(Ap. 1.13)}$$

$$y^{i\{[n]\}}_{q[m_{q_1} m_{q_2}]} = \exp \beta \left[-\nu^i_q + \sum_{j=A}^{B} n^j_{qp} (\varepsilon^{ij}_{qp} - \varepsilon^{iC}_{qp}) + m_{qp} \varepsilon^{iC}_{qp} \right]$$

This approximation has been used by the author [192] when calculating the adsorption of molecules of two species on the flat surface of a reorganizing binary alloy.

Lowering the Dimension of a System of Equations in the Quasichemical Approximation

A closed system of equations in the quasichemical approximation consists of equations relative to the functions θ_q^i and $\theta_{qp}^{ij}(r)$ (6.27), where $1 \leqslant i$ and $j \leqslant s$. This system generalizes the systems of equations for considering the interaction of nearest neighbors on homogeneous [see formulas (2.43) and (2.45)] and inhomogeneous [see formulas (3.51) and (3.52)] lattices and for considering an infinite number of neighbors on a homogeneous lattice [see formulas (6.11)]. An increase in the number of components of a system raises the dimension of the system of equations. The number of equations for θ_q^i grows linearly with increasing s, while that of the equations for two-body functions even for a homogeneous lattice at $R = 1$ increases as $s(s-1)/2$. A growth in R and in the number of site types in a lattice increases the dimension of a system of equations still more.

It was already noted in early publications [357, 358] that for $R = 1$ the dimension of a system of equations can be lowered sharply if we go over from equations in the variables θ_i and θ_{ij} to new variables X_i with the aid of the following relation:

$$\theta_{ij} = a_{ij}X_iX_j, \quad a_{ij} = \exp \beta\varepsilon_{ij} \tag{Ap. 1.14}$$

In this case, Eqs. (2.45) transform into identities $1 \equiv 1$, while after the insertion of the expressions

$$\theta_i = X_i \sum_{j=1}^{s} a_{ij}X_j \tag{Ap. 1.15}$$

into Eqs. (2.43) instead of the functions θ_i, the functions X_i are the unknown ones; their number is s, i.e. the same as in the molecular-field and polynomial approximations or in the absence of interaction. If the quantities θ_i are considered to be given, the unknown X_i's are found by solving the system of equations (Ap. 1.15). This procedure is traditionally used in the quasichemical approximation for describing thermodynamic solutions [359], diffusion processes [357, 358], and phase interfaces [82, 169]; it can also be used in problems describing the kinetics of surface reactions under steady-state conditions [360, 361]. In the general case, the choice of the coefficients a_{ij} is ambiguous, for example, Elenin *et al.* [360, 361] have used $a_{ij} = \exp(\beta\omega_{ij}/2)$, where ω_{ij} has been determined in (2.45).

The condition of normalizing the functions θ_i enables us to express one of them in terms of the others, i.e. in terms of the functions θ_i the system of equations has the dimension $(s - 1)$. A similar lowering of the dimension of a system is also possible for the system (Ap. 1.15), if from the functions X_i we go over to the functions [217]

$$Y_i = a_{is}X_i \Big/ \sum_{j=1}^{s} a_{js}X_j \tag{Ap. 1.16}$$

or to the functions [362]

$$\tilde{Y}_i = X_i/X_s \tag{Ap. 1.17}$$

The replacement of the variables in (Ap. 1.16) is symmetrical relative to the values of the quantities X_i, but leads to cumbersome expressions. The asymmetry of the replacement in (Ap. 1.17) yields unnormalized values of the functions $\tilde{Y}_i$, but leads to simpler equations. Its use is preferable if it is known that the concentration of the component s (vacant surface sites or a "solvent" in a solution) is not vanishingly small. As a result, the system of equations with the dimension $s(s + 1)/2$ is reduced to the system of equations of the dimension $(s - 1)$.

In the general case, the structure of equations (6.27) also makes it possible to substantially lower the dimension of the system of equations to $(s - 1)Rt$, where R is the interaction radius and t is the number of site types, and obtain its approximate solution. This is achieved by converting the system of equations for the two-body functions into the identities $1 \equiv 1$ with the aid of the substitution

$$\theta_{qp}^{ij}(r) = a_{qp}^{ij}(r)X_q^i(r)X_p^j(r), \quad a_{qp}^{ij}(r) = \exp \beta\varepsilon_{qp}^{ij}(r) \tag{Ap. 1.18}$$

Here the new variables $X_q^i(r)$, where $1 \leqslant i \leqslant s$ and $1 \leqslant q \leqslant t$, are determined by solving the first equations of system (6.27) or by solving the following system of equations

$$\theta_q^i = X_q^i(r) \sum_{j=1}^{s} a_{qp}^{ij}(r)X_p^j(r) \tag{Ap. 1.19}$$

The further lowering of the dimension of a system of equations is associated with the natural generalization of expressions (Ap. 1.16) and (Ap. 1.17):

$$\begin{aligned} Y_q^i(r) &= a_{qp}^{is}(r)X_q^i(r) \Big/ \sum_{j=1}^{s} a_{qp}^{js}(r)X_q^j(r) \\ \tilde{Y}_q^i(r) &= X_q^i(r)/X_q^s(r) \end{aligned} \tag{Ap. 1.20}$$

We shall note in concluding that the indicated lowering of the dimension of the system of equations in the quasichemical approximation can be used not only in problems describing the equilibrium and kinetics of surface processes for the rapid surface mobility of particles in steady-state conditions, but also in nonsteady conditions. In the latter case, the derivatives of the functions $Y_q^i(r)$ or $\tilde{Y}_q^i(r)$ on the left-hand sides of the equations are linearly related to one another, and for integration of the system of equations with respect to time they must be determined preliminarily from the relevant system of equations. Notwithstanding this circumstance, the indicated replacement of the variables noticeably diminishes the calculation difficulties in solving the problem.

Consideration of Indirect Correlations

Interaction of Nearest Neighbors (lattice $z = 6$, two species of particles). Consideration of the indirect correlations leads to the following system of equations in the polynomial approximation:

$$\theta_{n\sigma}^i = \sum_{k=0}^{z-n} N_{n+k}^i \sum_{\sigma(n+k)} b_{\sigma(n+k)}^{\sigma(n)} \sum_{r=0}^{z-n-k} (-1)^r \times \sum_{\sigma(n+k+r)} b_{\sigma(n+k+r)}^{\sigma(n+k)} \theta_{n+k+r,\sigma} \qquad \text{(Ap. 1.21)}$$

If we disregard the indirect correlations, system (Ap. 1.21) transforms into system (Ap. 1.4). As in the quasichemical approximation, the system is closed with the aid of three lower correlators [161, 363]: θ_{A}, θ_{AA}, θ_{33}^{A} (but the third correlators differ in the polynomial and quasichemical approximations). These functions allow the higher correlators to be expressed as follows:

$$\theta_{21}^{\text{A}} = \theta_{\text{A}}^2, \ \theta_{32}^{\text{A}} = \theta_{\text{A}}\theta_{\text{AA}}, \ \theta_{31}^{\text{A}} = \theta_{\text{A}}^3$$

$$\theta_{43}^{\text{A}} = \theta_{33}^{\text{A}}\theta_{\text{AA}}/\theta_{\text{A}}, \ \theta_{41}^{\text{A}} = \theta_{\text{AA}}^2, \ \theta_{42}^{\text{A}} = \theta_{33}^{\text{A}}\theta_{\text{A}}$$

$$\theta_{51}^{\text{A}} = (\theta_{33}^{\text{A}})^2/\theta_{\text{A}}, \ \theta_{61}^{\text{A}} = (\theta_{33}^{\text{A}}/\theta_{\text{A}})^3$$

Interaction of First and Second Neighbors. A closed system of sorption equations with account taken of indirect correlations is obtained from system of equations (6.16). For the sorption of molecules of one species, we have

[250, 256]:

$$y = \frac{\theta}{1-\theta}\Sigma_0,\ \Sigma_r\xi_r(1-\theta)\exp(-\beta\varepsilon_r) = (1-\xi_r\theta)\Sigma_0,\ r = 1,\ 2 \qquad \text{(Ap. 1.22)}$$

where the following notation has been introduced:

$$\Sigma_0 = \sum_{k_1=0}^{z_1}\sum_{k_2=0}^{z_2} P_{k_1k_2}\Phi_{k_1k_2},\ P_{k_1k_2} = \prod_{r=1}^{2} C_{z_r}^{k_r}(x_r\theta\xi_r)^{k_r}$$

$$\Sigma_1 = \sum_{k_1=0}^{z_1-1}\sum_{k_2=0}^{z_2} P_{k_1k_2}\Phi_{k_1+1k_2}\left(1-\frac{k_1}{z_1}\right) \qquad \text{(Ap. 1.23)}$$

$$\Sigma_2 = \sum_{k_1=0}^{z_1}\sum_{k_2=0}^{z_2-1} P_{k_1k_2}\Phi_{k_1k_2+1}\left(1-\frac{k_2}{z_2}\right)$$

Here x_r have been defined in (6.18), and $\Phi_{k_1k_2}$ are functions of indirect correlations. Their specific form depends on the structure of a lattice, the interaction radius, and on the particle configuration. We shall limit ourselves to the use of a two-body approximation of indirect correlations, when the dimension of the systems of equations remains the same as in the quasichemical approximation [250, 304]. The contributions introduced by the indirect correlations are determined in terms of $\xi_r = \theta_{AA}(r)/\theta_A^2$, which make it possible to consider the closed nature of the configurations of the particles in exactly the same way as is done in the superposition approximation [77, 105, 106].

For example, if sites g_1 and g_2 belong to the first coordination sphere of site f and the distance between them on a square lattice is $a\sqrt{2}$, we have

$$\langle\gamma_f^A\gamma_{g_1}^A\gamma_{g_2}^A\rangle = \frac{\langle\gamma_f^A\gamma_{g_1}^A\rangle\langle\gamma_f^A\gamma_{g_2}^A\rangle\langle\gamma_{g_1}^A\gamma_{g_2}^A\rangle}{\theta_A^3} = \theta_A t_{AA}^2(1)\xi_2$$

The functions of indirect correlations are expressed only in terms of ξ_r. They can be represented as

$$\Phi_{n_1n_2} = \sum_{m_1m_2} C_{n_1n_2}^{m_1m_2}\xi_1^{m_1}\xi_2^{m_2}\Big/\sum_{r=1}^{2} C_{z_r}^{n_r} \qquad \text{(Ap. 1.24)}$$

$$\sum_{m_1m_2} C_{n_1n_2}^{m_1m_2} = \prod_{r=1}^{2} C_{z_r}^{n_r}$$

where $C^{m_1 m_2}_{n_1 n_2}$ is the weight of a configuration containing n_1 and n_2 particles in the first and second coordination spheres whose arrangement is characterized by the presence of m_1 first and m_2 second neighbors between them. The total number of configurations of adjacent particles is $2^{z_1+z_2}$. For simple situations, e.g. for $z_1 = z_2 = 4$ the coefficients are easily calculated without a computer (see the relevant table proposed by Surovtsev *et al.* [268]). For a larger number of sites in both coordination spheres, the weights of the configurations were calculated with the aid of a simple algorithm for determining the number of interacting pairs of particles at a distance of r. The values of the weight coefficients when solving system (Ap. 1.23) were stored in the memory of a computer.

The principles of constructing the functions of indirect correlations for problems with a large number of components, for the kinetics of adsorption of nondissociating and dissociating molecules, and also when calculating the concentration profile of a binary alloy have been treated in [174, 250, and 304].

APPENDIX 2

Particle Interactions in the Kinetics of Surface Processes (a Review)*

The simplest kinetic models of heterogeneous processes for homogeneous surfaces and noninteracting adsorbed particles were not able to explain many experimental results [85, 150]. Models were developed along two parallel paths. In some of them, the main role was relegated to the biographical inhomogeneity of a surface [33, 136], while it was considered that the adsorbed particles do not interact. In other models, the main role was relegated to the interaction of particles, while the surface was assumed to be homogeneous. The task of considering particle interaction was found to be much more involved than the consideration of surface inhomogeneity, hence it was just these models that were developed the most [85-88, 116] (the physical fundamentals of these models have been considered by the author elsewhere [140]).

The perfection of experimental equipment [95-97] showed that the consideration of particle interaction must not be ignored in models of the kinetics of surface processes. Illustrative in this respect are investigations of the thermodesorption spectra of H_2 and CO on the W (100) face [364] and the reactions of CO oxidation on platinum catalysts [167]. Under real conditions, the inhomogeneity of a surface and particle interaction simultaneously affect the kinetics of surface processes. This is why information on the contributions of particle interactions and how they affect the rates of processes is required when studying surface processes.

The review briefly treats the development of nonideal models of the kinetics of surface processes and their present state. The formal way of taking into account the nonideal nature of reaction systems [134] that consists in the introduction of activity coefficients is not dealt with herein because for their theoretical calculation it is all the same necessary to involve model notions on particle interactions [3, 365].

The model of an elementary event is important from the standpoint of chemical kinetics. Models of collisions and the transition state [134] are employed most frequently in the kinetics of surface processes. In a collision model, the rate of a reaction is proportional to the probability of collisions

* A detailed review is given in [497].

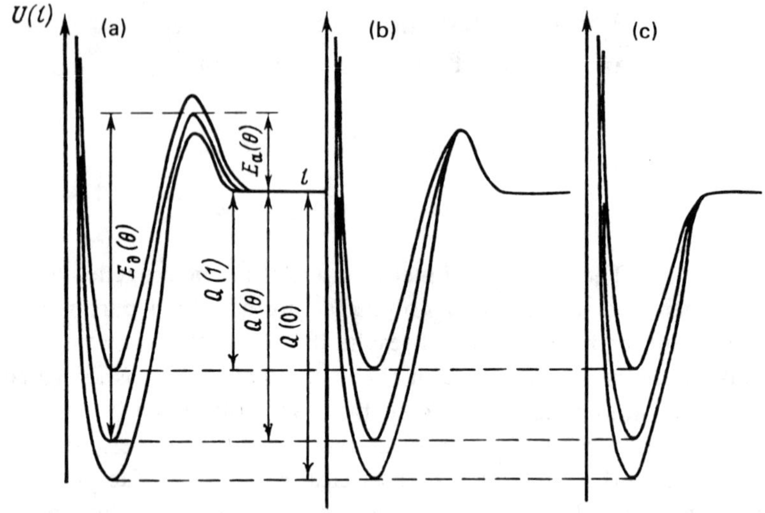

Fig. Ap.1. Potential curves of U vs. l in transition state (a) and collision models for $E_A > 0$ (b) and $E_a = 0$ (c):

E_a and E_d—adsorption and desorption activation energy, Q = heat of adsorption, l = distance from the surface

of solid spheres of particles counted on the basis of the molecular kinetic theory. The model of a transition state in addition to the geometric factor considers the role of energy redistribution over the internal degrees of freedom. In lattice models, the concept of a collision model signifies that the rate of a reaction is determined by the properties of the reactants and the height of the activation barrier, but does not depend on the properties of the reactants in the transition state. In nonideal systems, this signifies the absence of interaction of the particles in the transition state with neighboring particles. For this reason, we shall relate all models in which there is no activated complex-neighboring particle interaction, including the "intermediate" case when the concept of an activated complex is introduced for an ideal system, to collision models.

Figure Ap. 1 shows the difference between the effective potential curves of adsorption in the collision and transition state models for various surface coverages θ. These curves are averaged values of similar potential curves relating to specific local surroundings of particles participating in a chemical reaction. (Figure 2.3 shows local configurations of neighboring particles around a site at which adsorption without dissociation occurs, for lattices with $z = 4$ and 6, where z is the number of nearest sites.) Most authors in averaging used an equilibrium distribution function of the particles over a surface.

First Studies. Particle interaction in the kinetics of adsorption was taken into consideration for the first time by Langmuir [366]. By assuming the existence of dipole-dipole interaction between adatoms, he explained the exponential dependence of the rate constant of desorption of Th, Cs, and O from the surface of W on θ:

$$K_d = \hat{K}_d \exp[-\beta E_d(\theta)], \quad E_d(\theta) = E_d - H\theta \qquad \text{(Ap. 2.1)}$$

where $\hat{K}_d$ is the preexponential factor of the rate constant, E_d is the activation energy at $\theta \to 0$, and H is a constant. Langmuir [366] employed the collision model, $E_d(\theta) = Q(\theta)$.

The existence of a more complicated dependence of E_d on θ was presumed by Konozev and Gelfeld [367]. They considered that interaction can be greater than at the expense of van-der-Waals or polarization forces. Using a two-dimensional analogue of the van-der-Waals equation, they obtained an expression for the rate of desorption (collision model) which they used to describe the desorption of propylene from the surface of copper at various temperatures. Subsequently [368], they used these notions for H = const to describe the desorption of CO, H_2, C_3H_6, and $(CN)_2$ from copper and of NH_3 from iron specimens.

Roberts [369, 370] was the first to employ the model of a lattice gas in the quasichemical approximation (QCA). The interaction of the nearest neighbors was considered. Roberts used the collision method and with the aid of a kinetic approach, equating the rates of adsorption and desorption, obtained an adsorption isotherm. Similarly, Chang [371] obtained adsorption isotherms in the quasichemical and molecular-field approximations for the ordered distribution on particles on a lattice with $z = 4$ [the superstructure $c(2 \times 2)$].

The consideration of interaction in the molecular-field approximation (MFA) was proposed by Okamoto *et al.* [372] when describing the rate of hydrogen desorption. Account was taken of repulsion between the adatoms and between them and the activated complex. A more complete interpretation of the consideration of adsorbed particle interaction in the kinetics of surface reactions on the basis of the transition state model was given by Temkin [136]. Account was taken both of the internal degrees of freedom of an activated complex, and of the distinction of its interaction with its surroundings from the interaction of the reactants. Using the virial theorem for a multicomponent system, he obtained expressions for the rates of surface processes in the molecular-field approximation.

The notions on particle interactions were used to describe the rate of the reaction of *ortho-para* conversion of hydrogen on tungsten catalysts [373-375]. The results were not successful because a simplified interpretation of the model was used [370].

The mutual influence of adsorbed particles was also considered in phenomenological models via the change in the properties of the electron subsystem, in particular via the change in the position of the Fermi level. The model of a two-dimensional electron gas [264] for the adsorption of molecules on the surface of a metal is such a model. For a gas of noninteracting electrons, the Fermi level changes linearly with a change in θ (see the corrections for correlation between electrons [265, 266]). Generalization of this model for a mixture of adsorbed particles [376] allows us to consider that the particles of one of the species are activated complexes. This results in expressions for the rates of the processes similar to those given by Temkin [136]. Such a model was employed for simulating the processes of methane conversion [377].

In another model [25, 378, 379] for the surface of semiconductors, the position of the Fermi level is a function of the fraction of charged particles that coexist with the neutral form of the adsorbed particles. The rates of adsorption and desorption also have the form of formula (Ap. 2.1). In the general case, the relation $E(\theta)$ can be diverse.

Collision Model. The theories of nonideal adsorption systems on the basis of the lattice gas model began to develop rapidly when a report by Toya [380] appeared indicating that when the interaction of the nearest neighbors is taken into account (QCA), the splitting of the thermodesorption spectrum (TDS) of H_2 on a homogeneous W (100) surface can be explained. Goymour and King [381] obtained expressions (QCA) for the rates of reactions AZ + BZ (AZ is a site occupied by an A particle, and Z is a vacant site) and $A_2 + 2Z \rightleftarrows 2AZ$ considering the internal degrees of freedom of an activated complex, but not considering its interactions with neighbors. The model has been constructed for two species of adsorbed particles provided that $\varepsilon_{AA} = \varepsilon_{AB} = \varepsilon_{BB}$, which reduces the problem to a quasibinary system described by formulas (2.14) for "component" D, whose concentration is $\theta_D = \theta_A + \theta_B$. The model was employed for calculating the β peaks of the TDS for the associative desorption of the system CO/W [382], ($z = 4$, $\varepsilon_{DD} = -20$ kJ/mol; everywhere below, this dimension of the energy has been omitted). The theory is consistent qualitatively with experimental data with an arbitrary CO/O ratio: an increase in this ratio shifts the β_2 peak to lower temperatures and the β_1 peak to a smaller extent to higher ones. The order of the desorption kinetics depends on the temperature and concentration.

Notwithstanding the appreciable limitations of the model proposed by Goymour and King [381], it has been used quite widely for describing the TDS [383-390]. For CO molecules on the surface Re (0001), a qualitative description has been obtained at $\varepsilon_{COCO} = -12.3$; for the mixture O_2 + CO, the description is less satisfactory [383]. A qualitative explanation of the TDS for molecules of O_2, CO, NO, and H_2O on the surface Re (0001) has been given by Ducros *et al.* [384]; an appraisal of $\varepsilon_{NN} = 3.36$ has been obtained for nitrogen atoms. Calculations of the TDS for the system O_2/W [385] have resulted in the following appraisals: $\varepsilon_{OO} = -3.36$ ($z = 6$), -4.62 ($z = 4$), and -7.76 ($z = 2$) for various faces. Wilf and Dawson [386] consider that the model is not consistent with the observed kinetics of desorption of the system O_2/Pt (110). For the systems O_2/Ag (110) [387] ($\varepsilon_{OO} = 14$) and Hg/W (100) [388, 389] ($\varepsilon_{HgHg} = 5.85$), on the other hand, consistence is considered to be satisfactory. An analysis of the TDS for attracting particles [388, 389] has shown that the order of the desorption kinetics may change from the first to the zero one when a two-phase adsorption system forms and that with an increase in the initial coverage, the TDS peak shifts to higher temperatures. Zhdanov [390] has considered the influence of the interaction $\varepsilon_{AA} = 0$, ± 4.2, -8.4 on the TDS of the products of the reaction AZ + BZ for various initial coverages θ_i, where i = A, B. An increase in both the repulsion of particles and in $\theta_A(0)$ leads to a growth in the width of the TDS. If $\theta_A(0) = \theta_B(0)$, when $\theta_A(0)$ diminishes, the TDS curves fit into one another. At repulsions of the order of -8.4, the TDS split into two peaks.

Bridge and Lambert [391, 392] used the condition $\varepsilon_{BA} = \varepsilon_{BB} = 0$ and $\varepsilon_{AA} \neq 0$ instead of the condition $\varepsilon_{AA} = \varepsilon_{AB} = \varepsilon_{BB}$. For the face Pt (110) ($z = 2$), a good description of the TDS for HCN and C_2N_2 ($\varepsilon_{CN—CN} = -13.2$) has been obtained, but with a larger value of the preexponential factor for C_2N_2 (10^{23}) [391]. Bridge and Lambert consider that the model should take into account the two-dimensional nature of the surface and other types of adsorption centers. The calculations of the TDS for the mixture N + O on the surfaces Pt (110), Pt (111), Ru (101), and Ni (111) [392] point to the strong repulsion of the oxygen atoms. The parameter ε_{NN} is estimated to be close to zero or slightly positive for the surfaces of Pt and Ni, but strong repulsion should take place for Ru.

A detailed analysis of the TDS for various combinations of ε_{AA}, ε_{AB}, ε_{BB} (attraction, repulsion, and the absence of interaction) indicates that their position, shape, and width may change appreciably, but the TDS split only for the strong attraction of particles of one species (apart from this, $\theta_i(0)$ were varied [393]). A formal comparison of the calculated TDS with the experimental curves of the system N + O for Co (0001), Ni (100), Ru (101), Rh (110), Re (0001), and Ir (110) is apparently hardly justified, especially for $\varepsilon_{OO} > 0$. From all the available data for various surfaces, it follows that $\varepsilon_{OO} < 0$. It is more logical to determine ε_{OO} and ε_{NN} separately from the TDS of oxygen and nitrogen, and ε_{NO} from the TDS for the mixture.

The attraction of particles of various species (the system CO + K) changes the TDS greatly in comparison with the TDS for the individual components — the process is noticeably accelerated ("cooperative" desorption) [394]. The calculations were performed in the molecular-field approximation. The latter was also employed by Alnot *et al.* [385] for describing the molecular desorption of the system O_2/Pt at low temperatures ($z\varepsilon_{O_2O_2} = 3.3$).

Transition State Model. In a condensed phase, the reactants "constantly" interact with the surrounding particles, therefore a change in the local composition is attended by a change in the level of the activation barrier (Fig. Ap. 1). In 1974, a number of authors [358, 395, 396] independently considered this fact for the migration of the components of a binary alloy in the bulk [306] and for adsorption processes. Adams [395] obtained (QCA) a rough description of the TDS for the systems H_2/W (100) ($\varepsilon_{HH} = -7.55$, $\varepsilon^*_{HH} = -4.2$) and CO/W (210) ($\varepsilon_{COCO} = -40$, $\varepsilon^*_{COCO} = 0$) having two peaks; the description is better for the TDS of the system N_2/W (100) ($\varepsilon_{NN} = -8.8$, $\varepsilon^*_{NN} = 0$) having one peak. To fit the curve for the TDS of CO, the phenomenological approximation $\varepsilon_{COCO} = \varepsilon_2 + (\varepsilon_1 - \varepsilon_2)\theta$ was introduced. This made it possible to improve the description of the TDS for $\varepsilon_1 = -40$ and $\varepsilon_2 = -10.5$, though this procedure contradicts the quasichemical approximation. It should be noted that part of the indirect correlations have been included in the expression for associative desorption. This leads to overstating of the low-temperature peak. Calculations (for $\varepsilon^*_{AA} = 0$) have shown that the splitting of the TDS is caused by even slight repulsion of the particles: $|\varepsilon_{AA}| \approx 0.05E_d$. The formulas proposed by Adams [395] were used (at $\varepsilon^*_{AA} = 0$) to determine the average distance between particles in the presence of dipole-dipole interaction in terms of the parameters ε_{AA} [397].

The results of the author's work [396] relating to the kinetics of the adsorption-desorption process were published later [158, 213, 305]. The rates were expressed in terms of multiparticle correlators, which made it possible to compare different approximations for considering the correlation effects (see also [145]). In all the approximations, the logarithms of the rates depend linearly on θ in the region of moderate coverages (the slopes of the curves depend on the type of approximation), i.e. the Becker-Langmuir empirical relations for the rate of desorption and the Zeldovich-Roginsky emprical relations for the rate of adsorption can be explained by considering lateral interactions.

The conditions have also been determined for the applicability of the Benham-Bart-Kwan desorption equation [95], the approximation of the dependence of the logarithms of the rates on θ by several linear sections (which is frequently interpreted as a change in the mechanism of the process at various coverages [215]), and the separation of the chemisorption process into a "rapid" and "slow" ones [150]. In addition to the contributions of the interaction of the nearest neighbors, account has also been taken of the contributions of the electron subsystem in the approximation of a two-dimensional electron gas [264]. Their joint influence has made it possible to describe self-consistently the equilibrium and kinetics of desorption in the polynomial approximation of the system K/W [158].

Studying of the TDS of chemisorption systems provided a number of important results on the role of the correlation effects and the interaction of an activated complex with its neighbors [145]. The splitting of the TDS is due to the correlation effects (QCA, PA) — there is no splitting of the TDS when interaction is taken into account in the chaotic and molecular-field approximations. (The statements of Adams [395] and of Zhdanov and Zamaraev [398] are not correct — splitting is due to the nonlinear dependence of the effective activation energy of desorption on θ; in the chaotic approximation this relation is highly nonlinear, but splitting is absent.) To split the TDS, in addition to the correlation effects, a definite difference between ε^*_{AA} and ε_{AA} is required: there is no splitting when the differences are small. The quantity ε^*_{AA} plays the same important role as ε_{AA}. The TDS of the systems CO/W (210), CO/Mo (100), Cs/W, and CO on a number of metals have been described qualitatively. The important role of the correlation effects manifests itself for the kinetics of dissociative adsorption [216] — the failure to consider them leads to shifting of the desorption activation energy and to disturbance of the agreement between the kinetic and equilibrium characteristics of a system. Tovbin and Fedyanin [158, 213, 216, 305, 399] have additionally considered the contribution of the electron subsystem.

Benziger and Madix [400] and Benziger [397], using the molecular-field approximation, have studied the influence of the interaction of intermediate particles in processes of the decomposition of formic acid on Ni ($\omega = 5.9$, face (100), and 11.3, face (111), $\omega = z\delta\varepsilon_{AA}$) and of the methoxy group in the oxidation of methanol on Ag (110) ($\omega = -10$). In the former system, islets of intermediate particles form. The found values of ω were used to appraise the distance between neighboring particles with the assumption that dipole-dipole interaction exists; good agreement was obtained for Ag (110).

A theory of the absolute rates of reactions on homogeneous surfaces for multicomponent systems has been constructed by the author in a different publication [212]. General expressions have been obtained for the rates of processes proceeding at one and two neighboring sites, as well as their expressions in various approximations. This made it possible to "reproduce" all the previously obtained models and show how they interact and in what they differ. It has been proved that the expressions obtained in [212] for the rates in the forward and reverse directions of each step are self-consistent with the conditions of thermodynamic equilibrium of the reaction system.

For the particular case of two species of particles ($s = 3$), the formulas in the quasichemical approximation for the rates of adsorption and desorption, similar to formula (4.16), have been obtained by Zhdanov [401, 402]. Formulas for the rates of adsorption-desorption of dissociating molecules and the reactions AZ + BZ ($s = 3$) [402, 403] do not formally contain the exponential multiplier of formula (4.31). Its presence is revealed in the expressions for Φ_{Ad} (23) and Φ_{LH} (36) [403]. The same way of writing has been used by Zhdanov [404, 405], Stiles and Metiu [406], and Silverberg and Ben-Shul [407]. The given multiplier must also be present in the collision model expressions considered above. Its absence leads to the artificial shifting of the activation energy, namely, instead of E_{ij} what is obtained is $E_{ij} - \varepsilon_{ij}$, which causes a mismatch with the equilibrium conditions. For surface migration, we have $\varepsilon_{iv} = 0$, and its absence does not play any role.

Surface Reactions and Diffusion. The value of the rate of a reaction in nonideal adsorption systems differs from that in ideal systems because of the differences in the interactions in the ground and transition states [the functions S_i (4.16)] and because of the change in the probability of the reactants encountering one another ($\theta_{ij} \neq \theta_i\theta_j$). The nonideal nature of a system greatly affects the nature and length of the transition modes of multistep processes [408] (plots of θ_i vs. τ, where τ is the time, see [231, 409, 410]; it also affects the number of steady states of the reaction system [225]. For instance, the very simple process $A + Z \rightleftarrows AZ$, $AZ + B \rightarrow Z + C$ in the absence of interaction has one steady state, while when interaction is considered (quasichemical approximation, transition state model), the system can have up to three steady states at ε^*_{AA}, $\varepsilon_{AA} \leqslant (3\text{-}5)kT$. The same situation holds in the molecular-field approximation, collision model for this process (at parameters of the model differing from those given by Tovbin and Cherkasov [225] and for the process $A_2 + 2Z \rightleftarrows 2AZ$, $AZ + B \rightarrow Z + C$ [411]. Moreover, Kaza and Sunderesan

[411] discuss the conditions of appearance of self-oscillating modes of the isothermal process A + Z ⇄ AZ, B + Z ⇄ BZ, AZ + BZ → 2Z + C.

Nonideal models were used in describing the three-step mechanism of CO oxidation on platinum metals [167, 309]: CO + Z ⇄ ZCO, O_2 + 2Z ⇄ 2ZO, and ZCO + ZO → 2Z + CO_2 with independent determination of the parameters of the model relating to the first two steps. Zhdanov [404, 405] performed the calculations in the collision model with known values of the coverages of the preliminarily adsorbed molecules of CO and O_2 on the Ir(110) and (111) faces. Initially, the TDS of the systems CO/Ir(110) ($\varepsilon_{COCO} = -9.25$), O_2/Ir(110) ($\varepsilon_{OO} = -15.1$), CO and O_2/Ir(111) ($\varepsilon_{COCO} = -2.94$, $\varepsilon_{OO} = -3.36$) were described, while ε_{OCO} were determined from the rate of the reaction of CO_2 formation [-7.14 for the face (110) and -2.3 for the face (111)]. The TDS of the CO and O_2 molecules and the oxidation reactions are reproduced roughly. With additional assumptions on the rate of adsorption of CO and O_2 [414], the model was employed for describing the rate of the reaction of steady-state oxidation of CO/Ir(111).

A complete model of a steady-state oxidation process in the system CO/Pt (quasi-chemical approximation, transition state model) has been considered by Elenin *et al.* [217]. The parameters $\varepsilon_{COCO} = -9.24$, $\varepsilon^*_{COCO} = -0.84$, $\varepsilon_{OO} = -21$, and $\varepsilon^*_{OO} = -12.6$ were determined [219] independently from the TDS of CO and O_2 on a Pt(111) surface. The experiment has been described satisfactorily [218] for $\varepsilon_{OCO} = -14.7$ and $\varepsilon^*_{ij}(3) = \varepsilon_{ij}/2$. It was shown that the influence of $\varepsilon^*_{ij}(3)$ on the reaction rate is commensurable with that of ε_{ij}. The model describes qualitatively the rate of the reaction of steady-state CO oxidation on polycrystalline platinum [222], while in the region of pressures of about 1 mmHg it predicts two peaks in the temperature dependence of the oxidation rate (instead of one at low pressures, as was discovered experimentally by Hori and Schmidt [415].

An attempt to describe the oxidation process in the system H_2/Pt(111): H_2 + 2Z ⇄ 2HZ, O_2 + 2Z ⇄ 2ZO, ZO + ZH → ZOH + Z, ZOH + ZH → 2Z + H_2O has been made by Zhdanov [416]. But owing to the use of the collision model for the rates of H_2 and O_2 adsorption, phenomenological relations with empirical coefficients were introduced. Success was achieved in qualitatively reflecting the peak of the rate of steady-state oxidation of H_2 with increasing temperature. The value of $\varepsilon_{HH} = -2.94$ was determined from the TDS of the system H_2/Pt(111) [417], while the remaining parameters are not indicated. This model was employed for pressures of 5-100 Pa [418]. For a stoichiometric mixture, the rate of the reaction was found to vary regularly with the temperature (the magnitudes are overstated by an order).

The decomposition of formic acid on the nickel (111) face was treated by Benziger and Schoof [419] (quasichemical approximation, collision model), while Kustov and Pekarchuk [420] employed the transition state model for the kinetics of the reaction of chemical sputtering of the surface of a solid in active gas phases. The model is contradictory, however: in the molecular-field approximation, the average number of interacting particles is appraised via the polynomial approximation.

Surface diffusion is determined by the rate of the transition of a particle to an adjacent vacant site. Interactions between particles affect the rate of an elementary process. They also greatly affect the macroscopic characteristics of transfer processes that are described by diffusivities. The latter can be used to study the blurring of the concentration profiles [421] or the local fluctuations of the densities of the adsorbed particles [422]. The calculations were performed using the collision model (quasichemical approximation), and their results are consistent with calculations by Monte Carlo methods. Bowker and King [421] discuss the applicability of the model to diffusion in the system O/W (110) [246]. In the region of moderate coverages, the relations $D(\theta)$ have a gently sloping maximum ($\varepsilon < 0$) or minimum ($\varepsilon > 0$).

Diffusion of the particles in the transition state model has been treated by Tovbin and Fedyanin [242, 350] and by Gokhman *et al.* [245]. In the region of moderate coverages for $\varepsilon > 0$, the diffusivities may be negative — two-dimensional condensation is realized. Data have been described [246] for the region $0.6 < \theta < 1.0$ at $\varepsilon_{OO} = -1.64$ and $\varepsilon^*_{OO} = -1.47$. These values agree very well with the results obtained by Nartikoev [247].

The relation between $D(\theta)$ (quasichemical approximation, collision model) and the thermodynamics of adsorbed particles under conditions of local equilibrium were discussed by Reed and Ehrlich [423]. The need was stressed of knowing the thermodynamic properties of a system throughout the entire interval of coverages to obtain correct appraisals of the rates of migration of the particles from an experiment.

Tarasenko and Chumak [424] (quasichemical approximation, collision model) found that in addition to jumps of individual particles, account must be taken of the collective motion of pairs, while collective jumps of a larger number of particles may be disregarded.

Internal Degrees of Freedom. The experimental data for most systems [95-98] are not described by ideal models. To retain the customary notions of the rate constants of reactions, the effective activation energies $E(\theta)$ and effective preexponential factors $\hat{K}(\theta)$ are introduced. With the development of nonideal models, they began to be used for calculating $\hat{K}(\theta)$ and $E(\theta)$. For instance, Johansson [425] and Leuthausser [426] obtained satisfactory agreement of the values of $\hat{K}(\theta)$ calculated in the collision model for the desorption of the system CO/Ru(001) ($\varepsilon_{COCO} = -19$) with $\hat{K}(\theta)$ found experimentally [428] (the same results were obtained by Zhdanov [428]). The same "terminology" has been employed in other studies [398, 429], hence it is necessary to recall that all the models indicated above were constructed proceeding from the assumption that the internal degrees of freedom of the reactants are independent of their concentrations, i.e. the preexponential factor must be constant, while the quantity $E(\theta)$ is determined by equating the expression for the reaction rates when interaction is considered and for the phenomenological form of writing. For example, for desorption and a reaction, we have:

$$U_d = \hat{K}_d \exp[-\beta E_d(\theta)]\theta_A = K_d \theta_A (S_A)^z$$

$$U_{AB} = \hat{K}_{AB}\theta_A\theta_B \exp[-\beta E_{AB}(\theta_A\theta_B)]$$

$$= K_{AB}\theta_{AB} \exp(-\beta\varepsilon_{AB})(S_A S_B)^{z-1}$$

Figure Ap. 2 shows the effective activation energies of the reaction of oxidizing CO molecules on platinum [217].

Hellsing and Mallo [354] have proposed a model taking account of the change in the vibrational degrees of freedom of the particles depending on the composition of the surroundings (quasichemical approximation, collision model). In this model we can speak about $\hat{K}(\theta)$. The evaluated TDS are qualitatively consistent with experimental data for the system H/Ni(100) [430] ($\varepsilon_{HH} = -6.3$). Consideration of the internal degrees of freedom for surface diffusion (molecular-field approximation, collision model) has shown the importance of taking these contributions into account for $D(\theta)$ — in the region of particle ordering they lower the values of the diffusivity [431].

In the model considered by Asada [432], the particles oscillate along a normal to the surface and along the surface in the field of the varying relief of the support. The interactions between particles are described by the Lennard-Jones potential. Here too we can speak about $\hat{K}(\theta)$. Calculations using the quasichemical approximation for various temperatures correspond qualitatively to data on the diffusion of methane on the basalt plane of graphite [433].

For the migration of atoms of H, D, and T on the tungsten (110) surface, an important role is played by tunneling effects [434]. Calculations employing the Monte Carlo method, in which the influence of particle interactions on the height of the activation barrier (collision model) has been taken into account [31], are consistent qualitatively with experimental data [434] and confirm the anisotropy of diffusion in different directions. (Zhdanov's appraisals

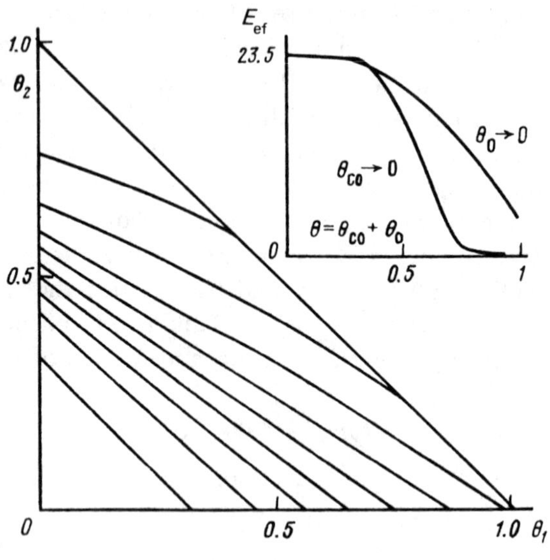

Fig. Ap.2. Effective activation energies $E_{ef}(\theta_A, \theta_B)$ of the reaction of oxidation of CO molecules on Pt:

the distance between the level lines is 9.9 kJ/mol, $E_{ef}(0, 0) =$ 98.7 kJ/mol, ($\sim$23.5 kcal/mol), $E_{ef}(0, 1) = 0$ kJ/mol. The limiting cases $\theta_A \to 0$ and $\theta_B \to 0$ are shown in the right-hand top corner

of the tunneling contributions [435] contradict the results of Tringides and Gomer [31].)

The model of Stechel and Knotek [436] considers the delocalization of particles when θ grows because of the particle interaction, which lowers the yield of desorption stimulated by the external field. And if the energy evolved in associative desorption does not have time to be completely absorbed by the solid, the particle interactions can change the energy carried off by a particle by an amount of the order of $z\delta\varepsilon_{AA}$ [437].

Presorption State. Particle interaction and the presorption state have been considered jointly by King and Wells [438] and by Singh-Bparai *et al.* [439] for the systems N_2/W(110), (310), (320), and (411) (quasichemical approximation, collision model). A satisfactory description of the TDS and the sticking probability was obtained for $\varepsilon_{NN} = -6$ (instead of -9 [395]). The great temperature dependence of ε_{NN} was noted, which resulted in a constant value of $\beta\varepsilon_{NN}$.

The schemes of the transitions between the ground and presorption states for adsorption processes have been considered in detail by Cassuto and King [440]. They presume that their models explain the temperature independence of the sticking probability for the system $H_2/W(001)$. Zhdanov [403] described the rates of adsorption and desorption of the system CO/Ni(111) (quasichemical approximation collision model) at $\varepsilon_{COCO} = -2.5$.

Nagai [412, 441] and Nagai and Hirushima [442] have proposed a two-layer model of the presorption state. The model (molecular-field approximation, transition state model), on the one hand, does not distinguish cases when a particle in the presorption state (second layer) is over a vacant or occupied site of the first layer, and is not correct when θ tends to unity (which produced a remark by Zhdanov [413]), and on the other, correctly poses the problem of considering the intrinsic size of particles in various states. For $\varepsilon_{XeXe} = -0.37$, good agreement of the predicted values for the system Xe/CO/W (110) with the experimental values [443, 444] was obtained.

The rate of desorption when the adsorbate is in the two-phase state is of the zero order. Nagai [412] used this circumstance to construct a curve of the coexistence of the gas-liquid two-dimensional system according to TDS data. Qualitative agreement has been obtained with the experimental thermodesorption spectra in the systems Hg/W(100) ($\omega = -31.6$), Cu/Ru(0001) ($\omega = -40.2$), Ag/W(110) ($\omega = -33.4$), Cu/W(100) ($\omega = -11.6$), H/Zn(0001) ($\omega = -11.5$), and H/Ni(111) ($\omega = -5$), where $\omega = z\varepsilon_{AA} + b\varepsilon^*_{AA}$, b is the number of neighbors in the second layer over a site in the first layer. We shall also note multilayer models (molecular-field approximation, collision model) of the desorption kinetics explaining its zero order [445, 446] — desorption occurs from the rarefied phase of the upper layer.

Two-Particle and Multiparticle Interactions. Consideration of the interaction of the nearest neighbors in the best case enables one to obtain qualitative agreement of theory with experimental results. In some situations, this discrepancy is associated with the inhomogeneity of the surface, nonequilibrium distribution of the particles, etc., in others with the approximate nature of the interaction potential used. Considered below is the influence of the interaction potential on the kinetics of surface processes: multicenter particles, two-particle interaction of the first and following neighbors, and multiparticle interactions.

Multicenter particles are defined as particles whose solid sphere blocks more than one surface site — the limiting case of strong repulsion of the nearest neighbors. The joint consideration of the size of a solid sphere and the lateral interaction of neighboring particles introduces substantial complications into the problem of calculating various configurations of the particles. Snagovsky [447] calculated the configurations of multicenter particles on a one-dimensional lattice. The kinetics of the adsorption processes of such a system was solved analytically by the author in [448]. With an increase in the size of the particles n in the region of low values of θ, the region of coverages in which the rates of processes behave as in an ideal model grows. A sharp change in the rate occurs at $\theta^* \approx n/(n+1)$. At $\theta < \theta^*$, it is difficult to distinguish between these two models, especially with strong repulsion of the particles, while at $\theta^* < \theta$ they differ sharply. The author in [448] has summarized the equations of the kinetics of desorption (collision model) for $n = 2$ [449] used to described the TDS of the system C_2N_2/Pt (110) ($\varepsilon_{CNCN} = -5$ instead of -13.2 when $n = 1$). Splitting of the TDS occurs at $\theta(0) > 2/3$ instead of $\theta(0) > 1/2$ when $n = 1$. Though the ratio of β_1/β_2 of the peaks could not be reproduced, preference is given to the model with $n = 2$.

For the diffusion of particles on a lattice with $z = 6$, having in view that a particle cannot migrate to an adjacent vacant site if the passage is blocked by other neighboring particles, the peak of $D(\theta)$ ($\varepsilon < 0$) shifts to $\theta < 0.5$ [450] (quasichemical approximation, collision model). The model is qualitatively consistent with the data on the diffusion in the system Li/W (110) at large coverages. The distinction at $\theta < 0.5$ are associated with phase transitions [451, 452].

Two-Body Interaction of First and Second Neighbors. It has been found for diffusion on a one-dimensional lattice that $D(\theta)$ depends greatly on the contributions of the second neighbors, especially with their attraction (transition state model [453], collision model [454]). It should be noted, as indicated by Zwerger [454], that Asada and Masuda [453] defined the diffusivity incorrectly. The rates of adsorption processes of nondissociating molecules with the consideration of the interaction of R neighbors have been obtained by the author in [304] (transition state model), while a theory of the surface processes for multicomponent adsorption systems (transition state model) has been constructed by the author [248, 249, 296]. Elementary adsorption-desorption processes with and without dissociation of the molecules, reactions of isomerization, exchange, and by a collision and adsorption mechanisms were considered. The rates of the processes are expressed in terms of multiparticle configurations, and various ways of approximation have been constructed for them, including consideration of indirect correlations.

The obtained expressions for the reaction rates in the forward and reverse directions for one- and two-site processes have been proved to be self-consistent with the conditions of equilibrium of the reaction system [296].

The contributions of the second neighbors play an appreciable role in the kinetics of adsorption processes, especially when attraction prevails [219, 297]. The growth in the sticking probability of CO and O_2 molecules on a Pt (111) surface with an increase of θ at low coverages can be explained with their aid. The contributions of the second neighbors do not change the nature of TDS splitting, but quantitatively greatly alter the position, width, and shape of the peaks. The distinctions in the TDS for desorption with association of the particles and without it have a quantitative nature.

Formulas (7.36) in the collision model are used for calculating the TDS [455] and $D(\theta)$ [410, 456]. Zhdanov [455] found that the contribution of the second neighbors only slightly affects the shape of the TDS. This contradicts the results obtained by Surovtsev and Tovbin [219]: the curves were analyzed for a constant width of the TDS when the contributions of the first and second neighbors were changed simultaneously.

The contributions of three neighbors were considered by Tringides and Gomer [434] when calculating the root-mean-square fluctuations of the density of oxygen atoms with account taken of the anisotropy of their diffusion of the W(110) surface.

Indirect Correlations and Ordering Effects. An improvement in the accuracy of describing the distribution of interacting particles is associated with the consideration of the correlations between particles in the coordination spheres of the reactants (for the consideration of the differences in θ_i (2, 1) and θ_i (2, 2) for $z = 6$, for example, see Fig. 2.3). Consideration of the indirect correlations [425] (collision model) leads to the appearance of two jumps of $\hat{K}(\theta)$ at $\theta = 1/3$ and $2/3$ on a lattice $z = 6$ instead of one at $\theta = 1/2$ for the quasichemical approximation. The improvement in the accuracy of describing the indirect correlations is associated with the increase in the size of the cluster on which the configurations of the particles are being considered. The transition to a cluster consisting of three sites from one consisting of two sites improved the agreement of the desorption kinetics [426] (collision model) with experimental results. A further transition to clusters consisting of six and twelve sites introduced only small distinctions. The transition from two to four sites in a cluster also resulted in quantitative distinctions of $D(\theta)$ on a lattice with $z = 4$ [454].

General expressions for the rates of surface processes with account taken of indirect correlations were obtained by the author [248, 249, 296] (transition state model). The consideration of the indirect correlations for the TDS of nondissociating molecules ($R = 2$, $z = 4$) increases the intensity of the low-temperature peak and alters the shape of the curve, but no additional splitting of the TDS occurs in comparison with the quasichemical approximation [303].

Strong repulsion of the adsorbed particles results in their ordered arrangement on a surface (observed in low-energy electron diffraction), and this substantially affects the kinetics of the surface processes. For flat lattices with $z = 3$ and 4 [the superstructure $c(2 \times 2)$] at points of phase transitions of ordering, jumps appear on the relation of D versus θ that also reduce the values of D in the ordered state at $\theta \approx 1/2$ (quasichemical approximation, collision model, $R = 1$) [451]. When the interaction of three neighbors was considered [452] (molecular-field approximation, collision model), the superstructure $p(2 \times 1)$ was taken into account in addition to $c(2 \times 2)$. The additional structural transformations lead to additional peaks and a jump in the values of D on the curve of D vs. θ, where $\theta < 1/2$ (the curve of D vs. θ is symmetrical relative to $\theta = 1/2$).

The kinetics of surface processes has been generalized for ordered systems for any values of R using the transition state model [206]. Ordering of particles of species $c(2 \times 2)$ on a lattice with $z = 4$ lowers the rate of desorption of nondissociating molecules at $\theta < 1/2$ and increases it at $\theta > 1/2$ ($R = 2$) [297].

The rates of desorption in the molecular-field approximation (collision model, $R = 2$) have been constructed by Zhdanov [457]; for dissociating molecules, the number of nearest neighbors has been taken into account incorrectly. A comparison of the calculations of $D(\theta)$ in the molecular-field approximation and by the Monte Carlo method has revealed that they coincide in the region of high temperatures (collision model, $R = 2$) [458].

Multiparticle Interactions. The consideration of three-body interactions is a natural generalization of the two-body interaction potential. They have been taken into account when calculating $D(\theta)$ by the Monte Carlo method (collision model) [31, 459, 460], and for adsorption processes (transition state model) [132]. The latter publication also considers the case when multiparticle interactions between nearest neighbors are taken into account (see Sec. 7.4).

Consideration of three-body interaction changes the relation between the TDS peaks evaluated in the collision model for monatomic desorption on a one-dimensional lattice [461].

The form of the interaction potential appreciably affects the concentration dependences of the reaction rates, therefore a detailed investigation of the nature of particle interactions is important.

Inhomogeneous Surfaces. Most surfaces contain inhomogeneities, hence to describe real surface processes, one must consider jointly the inhomogeneity of surfaces and the lateral interaction of particles.

The theory of surface processes when considering the interaction between the nearest neighbors of a multicomponent adsorption system for the equilibrium distribution of particles over inhomogeneous sites has been developed by the author in a number of publications [171, 177, 226, 462]. Expressions have been obtained for the rates of reactions (transition state model) proceeding at one and two neighboring sites on a surface in the same approximations that were used for a homogeneous surface (consideration of the inhomogeneity of a surface did not lead to a loss of accuracy dealing with particle interaction). The theory allows one to "reflect" a broad range of compositions and structures of surfaces. The mathematical models of the kinetics of surface reactions are simplified if the molecular field approximation is employed and if it is assumed that the structure of a surface is chaotic and that the parameters of particle interaction depend on the types of sites accommodating particles [193].

The thermodesorption spectra for a surface with two types of sites (their distribution is patchwise, chaotic, and regular) for a one-dimensional lattice and the collision model have been calculated in the chaotic approximation (although the authors indicated the molecular-fields approximation) [131]. The slitting of the TDS into two peaks is connected with the inhomogeneity of the sites, and not with lateral interactions (see [145]). Sales and Zgrablich [463] studied the joint influence of surface inhomogeneity and the interaction of the nearest and second neighbors by the Monte Carlo method.

Restricted Particle Mobility. Above we dealt with processes under conditions of the equilibrium distribution of the particles on a surface [or local equilibrium when evaluating $D(\theta)$]. When studying systems in which the rate of particle migration is comparable with the rate of their reactions, one must go over to models taking their restricted mobility into account. Kinetic equations relative to θ_{ij} that in an equilibrium distribution obey the algebraic equations (2.45) are a typical feature of models with restricted mobility of the reactants [206, 230, 231, 464]. (In the chaotic and molecular-field approximations there are no functions θ_{ij}, and the equations in these approximations do not depend on the nature of the particle mobility [230].)

A lower particle mobility is attended by a lower intensity of the high-temperature TDS peak for nondissociating molecules, while the valley between the peaks shifts to higher temperatures ($R = 1$, $z = 4$, collision model) [464]. A satisfactory description of the TDS has been obtained for the system CO/W (210) ($\varepsilon_{\mathrm{COCO}} = -12$). (We shall note the incorrectness of the approximation used, namely, the critical temperature is lower than that of the exact solution).

The kinetic equations of surface processes in the transition state model (quasichemical approximation, $R = 1$) with a view to the restricted mobility of the reactants have been constructed by the author elsewhere [210]. The basic types of elementary processes have been treated: $A + Z \rightleftarrows AZ$, $B_2 + 2Z \rightleftarrows 2BZ$, $AZ + BZ \rightarrow 2Z + C$. The kinetic equations for the function θ_{ij} make it possible to consider the influence of the initial distributions of the reactants (at a given composition thereof). Analysis of the TDS [231] has shown that if the particles form islets or macrodomains, the TDS split at low initial coverages $\theta_A(0)$ (with rapid migration, splitting occurs when $\theta_A(0) > 0.6$ [145]). As a result, the kinetics of a process depends on the conditions of formation of the initial coverage. This allows us to qualitatively explain the differences in the TDS for the systems CO/Pd (111) [227] where the initial coverages formed at different temperatures, and O_2/Pt [228, 229] where the initial coverages formed owing to different processes of chemical treatment of the surface. Similar schemes of processes in the collision model have been studied [465, 466] (quasichemical approximation, $R = 1$) for different interaction parameters and initial coverages. For a rapid chemical reaction, the evolution of particle distribution depends on the rate constant of the reactions.

Geldart *et al.* [467] have studied how the finite dimensions of a one-dimensional lattice and the readsorption processes affect the dynamic characteristics of adsorption processes ($R = 1$, collision model, quasichemical approximation) with a view to the restricted mobility of the particles. They found that the first order of desorption is realized only at high temperatures and low coverages. In other cases, it is less than unity, including a zero value.

The models considering the change in the concentration profile of the subsurface region in a multicomponent solid have the same structure of the equations as models describing the restricted mobility of particles on a surface [210]. As a result of the interaction of particles (interaction between atoms of a solid, between adsorbed molecules and atoms of the solid, between adsorbed molecules), the states of the concentration profile and the concentrations of the adsorbed molecules are related to each other, and their evolution in time is described by a common system of kinetic equations. The latter make it possible to describe the influence of the temperature and gas phase, including the reaction one, on the states of the gas-solid interface. In [209] (quasichemical approximation, transition state model, $R = 1$), the author has considered a rough phase interface, in [210] — a smooth one, while in [206] the equations have been generalized for a two-body potential with an arbitrary interaction radius. On the basis of the developed approaches [206, 209, 210], one can obtain the hierarchy of kinetic models of surface processes differing in the scale of the space-time averagings (distributed two- and one-dimensional models, and points models with a restricted mobility of all or part of the components or with an equilibrium distribution of the particles on a surface), but using a common set of interaction parameters and rate constants of the elementary reactions.

Numerical Methods of Studying Kinetics. In addition to analytical models, the kinetics of nonideal reaction systems is studied by the Monte Carlo method and the method of molecular dynamics. These methods are quite popular for analyzing the equilibrium and dynamic characteristics of condensed systems [468, 469].

Apparently, the Monte Carlo method was used for the first time in the kinetics of surface processes when describing desorption in the system C_2N_2/Pt (110) [449] ($R = 1$, collision model). It was found that when considering anisotropy in particle interaction in the directions [001] and [110], except for $\varepsilon_{110} \approx \varepsilon_{001}$, the process can be considered to go on independently on individual chains. A description of the kinetics of dissociative adsorption for the system NO/Co (0001) [392] ($\varepsilon_{NN} = 13.4$, $\varepsilon_{OO} = -26.9$) indicates the weak attraction of the atoms N — O and the formation of islets of nitrogen atoms. The model explains the varying values of the order of the desorption rate of N_2 in the presence of oxygen.

Calculations of $D(\theta)$ by the Monte Carlo method (collision model) at $R = 1$ [421] and $R = 2$ [470] point to a symmetrical peak at $\theta = 1/2$ for the repulsion of the first neighbors. The peak becomes sharply pronounced for attraction of the second neighbors. It is considered that the superstructure $p(2 \times 1)$ forms in the system O/W (110) [246]. Murch [471] for $T_{cr} > T$ (where T_{cr} is the critical temperature of ordering) and $\theta \approx 1/2$ obtained the superstructure $c(2 \times 2)$ (Monte Carlo method, $R = 2$, collision model), and the curve of D vs. θ has a sharp peak at $\theta = 1/2$. The process of formation of the phase $c(2 \times 2)$ at the expense of surface diffusion of the particles has shown [472] (Monte Carlo method, $R = 2$, collision model) that the phase grows at $T < T_{cr}$ by the type of nucleus formation, while the supply of the substance can be described by a diffusion equation. At $T > T_{cr}$, long-range order vanishes, but short-range order remains, which for $\varepsilon_{AA}(2) > 0$ facilitates the formation of islets.

Consideration of the anisotropy of diffusion in the system O/W (110) [458] showed that the motion of particles along a one-dimensional row of atoms occurs much more rapidly than motion across it, and when $T < T_{cr}$ and $\theta \approx 1/2$ the phase $p(2 \times 1)$ forms. The relation $D(\theta)$ can have three peaks.

The data on diffusion in the system O/W (110) [473] were described with the aid of a potential including two-body interactions of three neighbors and three-body interactions of the nearest neighbors [457] (collision model). A similar model was employed when considering the anisotropy of diffusion of O and D atoms on the W (110) face [458]. For O atoms, agreement with experimental results has been obtained, while for D atoms it has not been obtained, which indicates a different nature of the interaction (see [31]).

A comparison of the results obtained by the Monte Carlo method and with the aid of analytic systems of equations ($R = 1$) [392, 421, 423, 474] points to a good coincidence.

Stiles and Metiu [406] used the Monte Carlo method to consider processes of migration and reaction of adsorbed molecules of A and B (collision model, $R = 1$). When $\varepsilon_{AA} > 0$, ε_{AB} and $\varepsilon_{BB} < 0$, islets of A molecules form. The rate of the reaction obeys the law of mass action only at high temperatures, while it changes greatly if the component A experiences a phase transition. A similar problem was considered by Silverberg and Ben-Shul [407] using a combined approach, namely, the Monte Carlo method was employed to describe the slow processes of migration of the A molecules and the reaction AZ + BZ, while the distribution of the mobile B molecules was described by modified equations in the quasichemical approximation.

The method of molecular dynamics is only beginning to be applied to surface processes. To describe them, it is necessary to go over to time intervals of about 10^{-2} s and greater. Here Newton's equations become reformed into Langevin's equations — kinetic equations with a random disturbing function on the right-hand sides. This transition has been considered by Zeizi *et al.* [475] for rarefied systems of adsorbed atoms and molecules. It takes account of the internal degrees of freedom of the adsorbate. A good description of the preexponential factor of the desorption rate for the systems K/W (111), Xe/W (111), and Co/Ni (111) has been obtained. Redondo *et al.* [476] applied the model to the system Xe/W (111) with large coverages. The obtained laws for attracted and repelled particles coincide with all the laws obtained in the lattice gas model [145, 388, 389].

The models treated above were subsequently used many times to describe experimental data. The kinetics of adsorption of oxygen on polycrystalline copper [477] was described by formulas advanced by King and Wells [438]. The kinetics of adsorption processes in the system Hg/Ni {100} [478] was described in a model of a two-dimensional nonideal layer with the use of the van der Waals equation (collision model). Formulas proposed by Goymour and King [381] and by Zhdanov, [403] were used for adsorption processes in the system S/Pt (100) [479]. In both cases, poor agreement with experimental data for conditions of constancy of the interaction parameters was obtained, therefore the approximation $\varepsilon_{SS} = \varepsilon_0 + (\varepsilon_1 - \varepsilon_0)\theta^4$ was employed, where $\varepsilon_0 = -17.3$, $\varepsilon_1 = -24.8$ [381], and $\varepsilon_0 = -23.1$, $\varepsilon_1 = -31.7$ [403]. It is considered that the model must take into account the reorganization of the surface and the simultaneous desorption of the atoms and molecules of S_2.

Inhomogeneous Surface. The TDS of interacting particles at a distance of the first and second neighbors were calculated by the Monte Carlo method (collision model [463]). Lattices with two types of sites were considered with a different way of their discrete and Gaussian distributions. The number of peaks in the splitting of the TDS was found to depend both on the interaction between the particles and on the nature of the lattice inhomogeneity. For the system H/Ru (001), an analysis of data on $D(\theta)$ by the Monte Carlo method showed [480] that adsorbate-adsorbate interaction plays virtually no role, though the data on the TDS point to a strong relation of the effective desorption activation energy on the coverage. It is considered that the main role is played by the nonequivalent arrangements of the H atoms over three Ru atoms depending on whether an Ru atom is present or absent in the second subsurface layer.

Restricted Particle Mobility. The joint influence of repulsion of the first neighbors and slow mobility of the particles explains the "anomalously" sharp peaks of the TDS [481] and leads to a qualitative description of the TDS for the system CO/Ru (001). Account of the migration of the dimer particles (collision model, $R = 1$) for lattices with $z = 2$ and 4 results in the realization of a single peak of the TDS [482]. Evans and Hoffman [483] studied the kinetics of monomolecular desorption of interacting particles for the accurate solution of the problem on a one-dimensional lattice in the absence of particle migration. They investigated the influence of the formation of clusters of particles attracting one another on the order and on the effective desorption activation energy. An unusual scaling behavior of both characteristics of the process was discovered. For particles repelling one another, the splitting of the TDS into three peaks was discovered.

Nonmonolayer Models. Paunov and Bauer [484] considered the process of formation of two Ag layers on Mo (110) (different approximations: molecular-field, quasichemical, the two-dimensional van der Waals model, collision model). They found the interaction energy of the silver atoms in the first layer to be 26.0, and in the second 21.2. Zhdanov [485, 486] showed that the dissolution of adsorbed particles in an unlimited reservoir of a solid reduces the high-temperature peak of the TDS. The interaction of the activated complex with its neighbors is considered for the dissolution step and is not considered for desorption. The results are consistent qualitatively with the data for O_2/Ru (001). The behavior of the TDS for A_2 molecules in the dissolution of B particles was also dealt with. The dissolution of B particles was found to result in splitting of the TDS, and at high temperatures the desorption peak lowers with an increasing initial coverage. It is considered that this behavior of the TDS explains the desorption of N_2 in the system NO/Rh (110)

The notions on the reorganization of the surface layer of platinum have been used for describing the TDS of H_2/Pt (100) [486, 487] (molecular-field approximation, collision model, $R = 1$). Reorganization of the surface shifts the peak of the TDS to higher temperatures with a growth in the initial coverage. Qualitative agreement with experimental data has been obtained when $z\varepsilon_{HH} = -12.6$ and -21 for the initial and reorganized regions of the surface.

We must also note the work of Joyce *et al.* [488]. They studied the possibility of determining the potential of lateral interaction of Xe atoms on a W (110) surface from data on the interaction potential of atoms in the gas phase with a view to the influence on it of the metal surface considered as an ideal conductor. The constructed potentials were employed for describing the TDS of the system Xe/W (110) [443] with a constant value of the preexponential factor and an effective desorption activation energy varying linearly with the coverage. Satisfactory agreement was obtained with the TDS [443] at $z\varepsilon_{XeXe} = 2.8$ (the notation is as in the model with $R = 1$, molecular-field approximation). The values of $z\varepsilon = 7.46$, 4.5, and 7.18 for the systems Xe/W (111), Ar/Ag (111), and Xe/Ag (111), respectively, were also calculated.

We shall note in concluding that the dynamic characteristics of nonideal models of surface processes are only beginning to be studied, but it is already clear that at least two neighbors must be considered, and the approximations must retain the correlation effects. A more accurate description of the internal degrees of freedom of the adsorbed particles and their size will be an important direction of investigations. The consideration of a large number of physical factors will naturally result in more cumbersome mathematical models. Numerical methods will be strong competitors thereof. But in many cases, analytical models will be more convenient for the qualitative analysis of reaction rates, dynamic portraits, critical conditions of changing modes, and for a "utilitarian" description of experimental data. It is very probable that the joint use of analytical models and numerical imitations will be the most fruitful.

APPENDIX 3

Hierarchy of Mathematical Models of Surface Reactions

Mathematical models of processes are closed equations whose solution enables one to determine characteristics of interest for a system being considered. The evolution of the states of a system in time is described by kinetic equations. Their form depends on the mobilities of the system's components and on the space-time scale employed to study the relevant process. The choice of the scale depends on the goals and tasks of investigations. The molecular-kinetic theory makes it possible to formulate the general principles of constructing mathematical models, the ways of closing them, and the transition to a specific form of the right-hand sides of the kinetic equations depending on the selected typical value of the time interval for which the time derivative on the left-hand side of the equations is being determined.

The cluster approach presents any process in the form of a set of elementary processes that are realized on a certain sequence of coupled lattice sites. The interval relating to the most rapid elementary process of the change in the state of occupancy of lattice sites is the minimal typical time interval $\Delta\tau_{\mathrm{m}}$. Recall that kinetic equations (see Chaps. 5 and 7) relate to elementary processes associated with the chemical transformation of particles of one species into another one and with their migration, while the relaxation processes attending them are considered to be rapid. In principle, it is not difficult to negate this assumption and take into account the influence of the kinetics of relaxation processes on the processes of a change in the states of occupancy of lattice sites (see, for example, [489]).

An increase in the time interval $\Delta\tau$ is attended by a transition to the region of characteristic time intervals of slower elementary processes, while the description of rapid processes becomes more approximate. The degree of approximation depends on the ratio $\eta = \Delta\tau/\Delta\tau_{\mathrm{m}}$. With an increase in η, the multifold realization of rapid processes may greatly change the state of a system.

Two types of processes are traditionally singled out [55-58]: (1) when the processes of chemical transformation are the limiting ones (the rate of particle migration is much superior to the rate of chemical transformation), and (2) when the processes of particle migration are the limiting ones (the rate of chemical transformation is much superior to the rate of particle migration). In the first case, an increase in η signifies greater mixing of the particles because of their multifold migrations. This is accompanied by a growth in the region of space whose particles are involved in such mixing. This leads to erosion of the initial fluctuations in the density of particle distribution, and the fluctuations do not virtually affect the processes of chemical transformation. With large differences between the rate constants of migration and of chemical transformation, an equilibrium distribution of the components sets in in the system. In the second case of diffusion-controlled processes, a growth in η retards the processes of particle mixing and increases the role of the initial fluctuations in the density of reactant distribution.

As a result, the initial statement of the problem, including information on the mechanism of the process (a set of elementary steps of chemical transformations and migrations of the particles), particle interactions, typical times of realization of the elementary processes (they are usually characterized by the rate constants), and the employed way of closing the systems of kinetic equations lead to a hierarchy of the mathematical models differing in the dimension of the system of kinetic equations and the form of their right-hand sides. In inhomogeneous systems, the form of the models is also affected by the nature of the inhomogeneities (which can alter the mechanism of the process, the rate constants of a step, and the particle interactions) and by the external fields. The form of the kinetic equations is simplified if we limit ourselves to particular cases of reactant concentrations, e.g. to the region of low concentrations or to an excess of a component.

Let us consider the hierarchy of mathematical models describing a process consisting of the steps: (1) $A + Z \rightleftarrows AZ$; (2) $B_2 + 2Z \rightleftarrows 2BZ$; and (3) $AZ + BZ \rightarrow 2Z + C$ proceeding on a homogeneous surface and distinguished by their accuracy of describing the distribution of the reaction system particles. With a methodological goal in view, we shall supplement the process with the step (4) $BZ + A \rightarrow Z + C$ (for calculations of the reaction of oxidation of CO molecules on platinum catalysts at low pressures, this step must be omitted). The steps of migration of adsorbed A and B particles are written as (5) $jZ_f + vZ_g \rightleftarrows vZ_f + jZ_g$, where j = A, B; f and g are adjacent sites, v is a vacant site (a vacancy). The index α corresponds to the indicated step numbers. We shall consider the interactions of the first and second neighbors in the quasichemical approximation. We shall first construct equations for the distributed (two- and one-dimensional) models, and then for point models. In conclusion, we shall discuss how the form of the systems of equations alters for a great difference in the mobilities of the reactants.

Distributed Models of a Reaction

Two-Dimensional Model. The kinetic equations for the local concentrations of the adsorbed particles, in accordance with the general structure of equations (7.27), have the form

$$\begin{aligned}\dot{\theta}_f^{A} &= U_f^{v}(1) - U_f^{A}(1) - \sum_g [U_{fg}^{AB}(3) + U_{fg}^{Av}(5) - U_{fg}^{vA}(5)] \\ \dot{\theta}_f^{B} &= \sum_g [U_{fg}^{vv}(2) - U_{fg}^{BB}(2) - U_{fg}^{BA}(3) - U_{fg}^{Bv}(5) \\ &\quad + U_{fg}^{vB}(5)] - U_f^{B}(4)\end{aligned} \qquad \text{(Ap. 3.1)}$$

The addends on the right-hand sides of (Ap. 3.1) are the rates of the elementary processes (whose step numbers are indicated in parentheses) the particles participating in an elementary process are defined by superscripts, and the site numbers, by subscripts):

The adsorption-desorption of A molecules by formula (7.11) is expressed as

$$\begin{aligned}U_f^{v}(1) &= K_v(1)\theta_f^{v} \prod_{r=1}^{2} \prod_{g_r} S_{fg}^{v}(r|1) \\ U_f^{A}(1) &= K_A(1)\theta_f^{A} \prod_{r=1}^{2} \prod_{g_r} S_{fg}^{A}(r|1)\end{aligned} \qquad \text{(Ap. 3.2)}$$

Here and below we shall consider in an explicit form that the rate constants and interaction parameters do not depend on the numbers of the lattice sites.

The adsorption-desorption of B_2 molecules, by formula (7.15), is expressed as

$$\begin{aligned}U_{fg}^{vv}(2) &= K_{vv}(2)\theta_{fg}^{vv}(1) \prod_{h=1}^{3} S_{fh}^{v}(2) \prod_{h=4}^{6} S_{gh}^{v}(2) \prod_{h=7}^{10} S_{fgh}^{vv}(2) \\ U_{fg}^{AA}(2) &= K_{AA}e^{-\beta\varepsilon_{AA}(1)}\theta_{fg}^{AA}(1) \prod_{h=1}^{3} S_{fh}^{A}(2) \prod_{h=4}^{6} S_{gh}^{A}(2) \prod_{h=7}^{10} S_{fgh}^{AA}(2)\end{aligned} \qquad \text{(Ap. 3.3)}$$

Figure 7.10 shows neighboring sites h counted from the fixed position of the central sites f and g on which a bimolecular reaction occurs. Sites at which neighboring particles interact either only with a particle at site f or only with one at site g correspond to the numbers h = 1-6. Sites 2 and 5 are first neighbors, and sites 1, 3, 4, and 6 are second ones. At sites 7-10, the neighboring particles interact with both central particles. In the expressions for the rates of reactions, these site numbers unambiguously determine the distances between interacting particles, therefore the indices r and ω_r may be omitted.

Reactions by an adsorption and collision mechanisms:

$$U_{fg}^{AB}(3) = K_{AB}(3)e^{-\beta\varepsilon_{AB}(1)}\theta_{fg}^{AB}(1) \prod_{h=1}^{3} S_{fh}^{A}(3) \prod_{h=4}^{6} S_{gh}^{B}(3)$$

$$\times \prod_{h=7}^{8} S_{fgh}^{AB}(3) \prod_{h=9}^{10} S_{gfh}^{BA}(3) \qquad \text{(Ap. 3.4)}$$

$$U_f^B(4) = K_B(4)\theta_f^B \prod_{r=1}^{2} \prod_{g_r} S_{fg}^{B}(r|4)$$

Migration of adsorbed particles (i = A, B):

$$U_{fg}^{iv}(5) = K_{iv}(5)\theta_{fg}^{iv}(1) \prod_{h=1}^{3} S_{fh}^{i}(5) \prod_{h=4}^{6} S_{gh}^{v}(5) \prod_{h=7}^{8} S_{fgh}^{iv}(5) \prod_{h=9}^{10} S_{gfh}^{vi}(5)$$

$$\text{(Ap. 3.5)}$$

For the reverse direction of a jump of a particle i from site g to a vacant site f, the places of the indices i and v should be exchanged in formula (Ap. 3.5).

The functions $S_{fh}^{i}(r|\alpha)$ and $S_{fgh}^{ij}(\alpha)$ are defined by formulas (7.11) and (7.15), $s = 3$ in them, and the interaction parameters do not contain the numbers of the sites.

The system of equations for the local concentrations is supplemented with equations for the probabilities of the two-body configurations $\theta_{fg}^{ij}(r)$ at a distance of r, where $r = 1$ and 2, that obey normalizing relations (7.30). With a view to these relations for $s = 3$, it is sufficient to write the equations relative to the functions $\theta_{fg}^{ij}(r)$, where ij = AA, AB, BA, and BB. For particles at the distance of first neighbors, the equations have the following form:

for pair "AA"

$$\dot{\theta}_{fg}^{AA}(1) = U_{fg}^{vA}(1|1) - U_{fg}^{\bar{A}A}(1|1) + U_{fg}^{Av}(1|1) - U_{fg}^{A\bar{A}}(1|1)$$

$$- \sum_{h=7,2,8} [U_{hfg}^{\overline{BA}A}(1|3) + U_{hfg}^{\overline{vA}A}(1|5) - U_{hfg}^{\overline{Av}A}(1|5)]$$

$$- \sum_{h=9,5,10} [U_{fgh}^{A\overline{AB}}(1|3) + U_{fgh}^{A\overline{Av}}(1|5) - U_{fgh}^{A\overline{vA}}(1|5)] \qquad \text{(Ap. 3.6)}$$

for pair "BB"

$$\dot{\theta}^{BB}_{fg}(1) = U^{vv}_{fg}(2) - U^{BB}_{fg}(2) - U^{\overline{B}B}_{fg}(4) - U^{B\overline{B}}_{fg}(4)$$

$$+ \sum_{h=7,\ 2,\ 8} [U^{\overline{vv}B}_{hfg}(1|2) - U^{\overline{BB}B}_{hfg}(1|2) - U^{\overline{AB}B}_{hfg}(1|3)$$

$$- U^{\overline{vB}B}_{hfg}(1|5) + U^{\overline{Bv}B}_{hfg}(1|5)] + \sum_{h=9,\ 5,\ 10} [U^{B\overline{vv}}_{fgh}(1|2)$$

$$- U^{B\overline{BB}}_{fgh}(1|2) - U^{B\overline{Bv}}_{fgh}(1|5) + U^{B\overline{vB}}_{fgh}(1|5) - U^{B\overline{BA}}_{fgh}(1|3)] \qquad \text{(Ap. 3.7)}$$

for pair "AB"

$$\dot{\theta}^{AB}_{fg}(1) = U^{vB}_{fg}(1|1) - U^{AB}_{fg}(1|1) - U^{AB}_{fg}(1|4) - U^{AB}_{fg}(3)$$

$$- \sum_{h=7,2,8} [U^{\overline{BA}B}_{hfg}(1|3) + U^{\overline{vA}B}_{hfg}(1|5) - U^{\overline{Av}B}_{hfg}(1|5)]$$

$$+ \sum_{h=9,5,10} [U^{A\overline{vv}}_{fgh}(1|2) - U^{A\overline{BB}}_{fgh}(1|2) - U^{A\overline{BA}}_{fgh}(1|3)$$

$$- U^{A\overline{Bv}}_{fgh}(1|5) + U^{A\overline{vB}}_{fgh}(1|5)] \qquad \text{(Ap. 3.8)}$$

The equation for the pair "BA" is written by changing the places of the superscripts A and B in (Ap. 3.8).

When constructing the kinetic equations for the functions θ^{ij}_{fg} (2) (sites f and $g = 9$ in Fig. 7.10), one must have in view that the state of occupancy of sites f and g cannot change simultaneously because bimolecular reactions occur only on the nearest sites. This leads to the following distinctions in the system of equations for the second neighbors from the system (Ap. 3.6)-(Ap. 3.8): (1) the addends U^{ij}_{fg} (α) are absent in Eqs. (Ap. 3.7) and (Ap. 3.8); and (2) summation over h is performed over all the nearest neighbors of sites f and g.

For instance, for the pair $\theta^{AB}_{fg}(2)$, the equation will be written as

$$\dot{\theta}^{AB}_{fg}(2) = U^{\overline{v}B}_{fg}(2|1) - U^{\overline{A}B}_{fg}(2|1) - U^{A\overline{B}}_{fg}(2|4) - \sum_{h} [U^{\overline{BA}B}_{hfg}(2|3)$$

$$+ U^{\overline{vA}B}_{hfg}(2|5) - U^{\overline{Av}B}_{hfg}(2|5)] + \sum_{h} [U^{A\overline{vv}}_{fgh}(2|2)$$

$$- U^{A\overline{BB}}_{fgh}(2|2) - U^{A\overline{BA}}_{fgh}(2|3) - U^{A\overline{Bv}}_{fgh}(2|5) + U^{A\overline{vB}}_{fgh}(2|5)] \qquad \text{(Ap. 3.9)}$$

The addends on the right-hand sides of Eqs. (Ap. 3.6)-(Ap. 3.9) are expressed in terms of the rates of the elementary processes in accordance with formulas (7.12) and (7.16). For one-site processes $\alpha = 1$ and 4 no explanations are needed. For two-site processes depending on the distance between sites h and g, we have

$$U_{hfg}^{ijn}(r|\alpha) = U_{hf}^{ij}(\alpha) \times \begin{cases} \psi_{fg}^{jn}(r|\alpha), & r_{in} > 2,\ r = 1;\ 2 \\ \psi_{fhg}^{jin}(2|\alpha), & r_{in} = 2,\ r = 1 \\ \psi_{hfg}^{ijn}(1|\alpha), & r_{in} = 1,\ r = 2 \end{cases} \qquad \text{(Ap. 3.10)}$$

The site $h = 2$ (see Fig. 7.10) corresponds to the distance $r_{in} > 2$, and the sites $h = 7$ and 8 to the distance $r_{in} = 2$ when $r = 1$; when $r = 2$, the sites $h = 2$ and 3 correspond to the distance $r_{in} > 2$, and sites $h = 8$ and 11 to $r_{in} = 1$. The expressions $U_{fgh}^{ijn}(r|\alpha)$ are written in a similar way when the particles j and n react.

The constructed system of equations is a closed one. It is solved with the preset initial conditions $\theta_f^i(\tau = 0)$, $\theta_{fg}^{ij}(1, \tau = 0)$, $\theta_{fg}^{ij}(2, \tau = 0)$. The system of equations makes it possible to describe arbitrary distributions of particles on a surface and their evolution in time. The only shortcoming is the large dimension. The minimal fragment of a lattice on which a process with cyclic boundary conditions should be described is 4 × 4. It is therefore natural to raise the question of approximating the description of particle distribution to lower the dimension of the system of equations without "losing the effects being described. In this connection, we shall consider simpler models, namely, one-dimensional and possible point models.

One-Dimensional Model of a Reaction. As in Sec. 5.3, when going over to a one-dimensional model, we shall take into account the difference in the occupancies of sites of various sublattices within the confines of each layer into which the two-dimensional surface is divided. If we consider the superstructure $c(2 \times 2)$, the local density of particles i at sites of type σ ($\sigma = \delta, \gamma$) of layer k is determined by formulas (5.22). The two-body functions $\theta_{k\sigma,m\lambda}^{ij}$ showing the probability of finding particles i and j at sites of type σ and λ in layers k and m at a distance of r are obtained from the two-body functions of the two-dimensional model as follows:

$$\theta_{k\sigma,\ m\lambda}^{ij}(r) = 2 \sum_{l=0}^{L/2} \sum_{h} \theta_{(k2l),\ (mh)}^{ij}(r)/(\tilde{z}_{\sigma\lambda}(r)L) \qquad \text{(Ap. 3.11)}$$

If $r = 1$, we have $\sigma \neq \lambda$, and if $r = 2$, we have $\sigma = \lambda$. The second sum over h is taken over adjacent sites of type λ at a distance r from site σ in layer k. Then number of such sites, designated by $\tilde{z}_{\sigma\lambda}(r)$, is $\tilde{z}_{\sigma\lambda}(1) = 1$ ($m = k \pm 1$) and 2 ($m = k$); $\tilde{z}_{\sigma\sigma}(2) = 2$ ($m = k \pm 1$); when $m = k$, second neighbors are absent. The transition to the indicated description of particle distribution cuts the number of equations for local densities $L/2$ times because the state of the sites of each row is set by four functions: $\theta^{A}_{k\delta}, \theta^{A}_{k\gamma}, \theta^{B}_{k\delta}, \theta^{B}_{k\gamma}, \left(\sum_{i=A}^{v} \theta^{i}_{k\sigma} = 1\right)$ The structure of kinetic equations (Ap. 3.1) for them does not change, for example for i = A

$$\dot{\theta}^{A}_{k\delta} = U^{v}_{k\delta}(1) - U^{A}_{k\delta}(1) - \sum_{m} \tilde{z}_{km}(1)[U^{AB}_{k\delta m\gamma}(3) + U^{Av}_{k\delta m\gamma}(5) - U^{vA}_{k\delta m\gamma}(5)] \quad \text{(Ap. 3.12)}$$

Instead of the site number f, the layer number k and the site type δ are used, and instead of g — layer number m and site type γ, where $m = k, k \pm 1$.

The same substitution of the subscripts has been performed in expressions for the rates of elementary reactions (7.11) and (7.15): $S^{i}_{fh}(\alpha) \rightarrow S^{i}_{k\sigma,\ m\lambda}(\alpha)$; if f and h are first neighbors, we have $\lambda \neq \sigma$, and if they are second neighbors, we have $\sigma = \lambda$; $S^{ij}_{fgh}(\alpha) \rightarrow S^{ij}_{k\sigma,\ m\lambda,\ n\lambda}(\alpha)$, because sites f and g can only be first neighbors, while on a lattice with $z = 4$, sites g and h are second neighbors. Hence, the rates of adsorption of A and the reactions of step 3 will be rewritten in the form

$$U^{v}_{k\sigma}(1) = K_{v}(1)\theta^{v}_{k\sigma} \prod_{m=k+1}^{k-1} [S^{v}_{k\sigma,\ m\lambda}(1|1)]^{z_{km}(1)}[S^{v}_{k\sigma,\ m\sigma}(2|1)]^{z_{km}(2)}$$

$$U^{AB}_{k\delta,\ k+1\gamma}(3) = K_{AB}(3)e^{-\beta\varepsilon_{AB}(1)}\theta^{AB}_{k\delta,\ k+1\gamma}(1)S^{A}_{k\delta,\ k-1\gamma}(1|3) \times [S^{A}_{k\delta,\ k-1\delta}(2|3)]^{2}[S^{AB}_{k\delta,\ k+1\gamma,\ k\gamma}(3)S^{BA}_{k+1\gamma,\ k\delta,\ k+1\delta}(3) \times S^{B}_{k+1\gamma,\ k+2\gamma}(2|3)]^{2}S^{B}_{k+1\gamma,\ k+2\delta}(1|3) \quad \text{(Ap. 3.13)}$$

The remaining addends of the equations for $\theta^{i}_{k\sigma}$ are written similarly.

The equations for two-body probabilities are the next ones of the system. Two cases are possible for the nearest neighbors [Eqs. (Ap. 3.6)-(Ap. 3.8)], namely, a pair of sites is in row k, and a pair of sites is between rows k and $k + 1$. In the first case, if $i = j$ = A or B, we have one equation each for $\theta^{ii}_{k\delta k\gamma}(1)$, while if i = A and j = B, we need two equations for each layer relative to $\theta^{AB}_{k\delta k\gamma}(1)$ and $\theta^{AB}_{k\gamma k\delta}(1)$. Owing to the symmetry of the equations relative to the simultaneous interchange of the superscripts and subscripts, the equations for the pair "BA" are not needed. In the second case, the number of required equations is doubled because neighboring sites in adjacent layers relate to pairs of the type of $\delta\gamma$ and $\gamma\delta$. As a whole, the number of equations for the nearest pairs is cut $L/3$ times. As an example, we shall write the equations for the functions $\theta^{AB}_{k\delta m\gamma}(1)$, where $m = k$ or $k + 1$:

$$\dot{\theta}^{AB}_{k\delta k\gamma}(1) = U^{\bar{v}B}_{k\delta k\gamma}(1|1) - U^{\bar{A}B}_{k\delta k\gamma}(1|1) - U^{A\bar{B}}_{k\delta k\gamma}(1|4) - U^{AB}_{k\delta k\gamma}(3)$$
$$- \sum_{m=k-1}^{k+1} (z_{km}(1) - \Delta_{km})[U^{\overline{BA}B}_{I}(1|3) + U^{\overline{vA}B}_{I}(1|5)$$
$$- U^{\overline{Av}B}_{I}(1|5) - U^{A\overline{vv}}_{II}(1|2) + U^{A\overline{BB}}_{II}(1|2) + U^{A\overline{BA}}_{II}(1|3)$$
$$+ U^{A\overline{Bv}}_{II}(1|5) - U^{A\overline{vB}}_{II}(1|5)] \qquad \text{(Ap. 3.14)}$$

$$\dot{\theta}^{AB}_{k\delta,\ k+1\gamma}(1) = U^{\bar{v}B}_{k\delta,\ k+1\gamma}(1|1) - U^{\bar{A}B}_{k\delta,\ k+1\gamma}(1|1) - U^{A\bar{B}}_{k\delta,\ k+1\gamma}(1|4)$$
$$- U^{AB}_{k\delta,\ k+1\gamma}(3) - \sum_{m=k-1}^{k+1} (z_{km}(1) - \Delta_{k+1m})[U^{\overline{BA}B}_{III}(1|3)$$
$$+ U^{\overline{vA}B}_{III}(1|5) - U^{\overline{Av}B}_{III}(1|5)] + \sum_{m=k}^{k+2} (z_{k+1m}(1)$$
$$- \Delta_{mk})[U^{A\overline{vv}}_{IV}(1|2) - U^{A\overline{BB}}_{IV}(1|2) - U^{A\overline{BA}}_{IV}(1|3)$$
$$- U^{A\overline{Bv}}_{IV}(1|5) + U^{A\overline{vB}}_{IV}(1|5)]$$

To simplify the writing of the subscripts, the following abbreviated notation has been introduced: I $= m\gamma, k\delta, k\gamma$; II $= k\delta, k\gamma, m\delta$; III $= m\gamma, k\delta, k+1\gamma$; IV $= k\delta, k+1\gamma, m\delta$. The addends of equations of the type of (Ap. 3.14) are expressed in terms of the rates of elementary processes of the type of (Ap. 3.13) in exactly the same way as indicated above for a two-dimensional model.

The number of equations for pairs at a distance of the second neighbors diminishes $L/2$ times because there are such pairs only between adjacent layers. Equation (Ap. 3.9) will be written for them as follows:

$$\dot{\theta}^{AB}_{k\delta,\ k+1\delta}(2) = U^{\bar{v}B}_{k\delta,\ k+1\delta}(2|1) - U^{\bar{A}B}_{k\delta,\ k+1\delta}(2|1) - U^{A\bar{B}}_{k\delta,\ k+1\delta}(2|4)$$
$$- \sum_{m=k-1}^{k+1} z_{km}(1)[U^{\overline{BA}B}_{V}(2|3) + U^{\overline{vA}B}_{V}(2|5)$$
$$- U^{\overline{Av}B}_{V}(2|5)] + \sum_{m=k}^{k+2} z_{k+1m}(1)[U^{A\overline{vv}}_{VI}(2|2)$$
$$- U^{A\overline{BB}}_{VI}(2|2) - U^{A\overline{BA}}_{VI}(2|3) - U^{A\overline{Bv}}_{VI}(2|5) + U^{A\overline{vB}}_{VI}(2|5)]$$
$$\text{(Ap. 3.15)}$$

where V $= m\gamma, k\delta, k+1\delta$; VI $= k\delta, k+1\delta, m\gamma$.

The system of equations is solved with the given initial conditions $\theta^{i}_{k\sigma}(\tau = 0)$, $\theta^{ij}_{k\sigma,\ m\lambda}(1, \tau = 0)$, $\theta^{ij}_{k\sigma,\ m\sigma}(2, \tau = 0)$.

If ordering of the particles is disregarded, the dimension of the system decreases L times in comparison with that of a two-dimensional model. In this case, according to definition (5.21), the indices δ and γ indicating the type of sublattice must be omitted in the equations of the system and the expressions for the rates of the elementary reactions (Ap. 3.12)-(Ap. 3.15).

Point-Like Models of a Reaction

In point-like models, the averaging procedure is conducted over all the sites of a lattice, therefore diffusion flows are absent in the models. But the effects of particle mobility can be taken into account via the processs of their local redistribution that change the short-range order. This leads to a model considering the restricted mobility of all the reactants, and to a model with rapid surface mobility of the reactants.

Restricted Mobility of the Reactants. We shall characterize the average concentration of particles and the average fraction of their pairs by the following relations:

$$\theta_\sigma^i = \sum_{k=1}^{L} \theta_{k\sigma}^i/L, \; \theta_{\sigma\lambda}^{ij}(r) = \sum_{k=1}^{L} \sum_{m} \theta_{k\sigma,\, m\lambda}^{ij}(r)/(z_{\sigma\lambda}(r)L) \quad \text{(Ap. 3.16)}$$

that take into account the ordered distribution of the particles on a surface. For the mean values of the functions (Ap. 3.12), the kinetic equations give only relaxation evolution. The concentrations of the particles obey the equations ($\sigma = \delta, \gamma$)

$$\begin{aligned} \dot\theta_\sigma^{\mathrm{A}} &= U_\sigma^{\mathrm{v}}(1) - U_\sigma^{\mathrm{A}}(1) - z_{\sigma\lambda}(1)[U_{\sigma\lambda}^{\mathrm{AB}}(3) + U_{\sigma\lambda}^{\mathrm{Av}}(5) - U_{\sigma\lambda}^{\mathrm{vA}}(5)] \\ \dot\theta_\sigma^{\mathrm{B}} &= z_{\sigma\lambda}(1)[U_{\sigma\lambda}^{\mathrm{vv}}(2) - U_{\sigma\lambda}^{\mathrm{BB}}(2) - U_{\sigma\lambda}^{\mathrm{BA}}(3) - U_{\sigma\lambda}^{\mathrm{Bv}}(5) \\ &\quad + U_{\sigma\lambda}^{\mathrm{vB}}(5)] - U_\sigma^{\mathrm{B}}(4) \end{aligned} \quad \text{(Ap. 3.17)}$$

where $z_{\sigma\lambda}(r)$ is the number of neighbors of type λ that are neighbors of a site of type σ at a distance of r; for a square lattice, we have $z_{\delta\gamma}(1) = z_{\gamma\delta}(1) = 4$, and $z_{\delta\delta}(2) = z_{\gamma\gamma}(2) = 4$. The rates of elementary processes in the general form will be written as follows:

$$\begin{aligned} U_\sigma^i(\alpha) &= K_i(\alpha)\theta_\sigma^i \prod_{r=1}^{2} \prod_{\lambda} [S_{\sigma\lambda}^i(r|\alpha)]^{z_{\sigma\lambda}(r)} \\ U_{\sigma\lambda}^{ij}(\alpha) &= K_{ij}(\alpha)\theta_{\sigma\lambda}^{ij}(1)e^{-\beta\varepsilon_{ij}(1)}S_{\sigma\lambda}^i(1|\alpha)S_{\lambda\sigma}^j(1|\alpha) \\ &\quad \times [S_{\sigma\sigma}^i(2|\alpha)S_{\lambda\lambda}^j(2|\alpha)S_{\sigma\lambda\lambda}^{ij}(\alpha)S_{\lambda\sigma\sigma}^{ji}(\alpha)]^2 \end{aligned} \quad \text{(Ap. 3.18)}$$

where the functions $S^{i}_{\sigma\lambda}(r|\alpha)$ and $S^{ij}_{\sigma\lambda\lambda}(\alpha)$ have the form of (7.11) and (7.15) in which the site numbers have been replaced with the indices of the site types. As previously, the type of a pair of sites is unambiguously related to the distance between them.

The number of equations for two-body probabilities is also sharply reduced. For the first neighbors, four equations relative to $\theta^{ij}_{\delta\gamma}(1)$ are needed, ij = AA, AB, BA, and BB, while for the second neighbors three equations relative to $\theta^{ij}_{\sigma\sigma}(2)$ are needed, ij = AA, AB, and BB. The normalizing relations have the form

$$\sum_{j=\mathrm{A}}^{\mathrm{v}} \theta^{ij}_{\sigma\lambda}(r) = \theta^{i}_{\sigma}, \quad \sum_{i=\mathrm{A}}^{\mathrm{v}} \theta^{ij}_{\sigma\lambda}(r) = \theta^{i}_{\lambda}, \quad \sum_{i=\mathrm{A}}^{\mathrm{v}} \theta^{i}_{\sigma} = 1 \qquad \text{(Ap. 3.19)}$$

The kinetic equations for the functions $\theta^{ij}_{\sigma\lambda}(2)$ are similar to the system (Ap. 3.6)-(Ap. 3.9) in which the site numbers have been replaced with the indices of the site types; the number h has been replaced by γ if h is the nearest site to a type δ site, and by δ if h is the nearest site to a type γ site.

If for a square lattice we introduce the notation

$$\begin{aligned} W^{\overline{ij}k}_{\delta\gamma\delta}(1|\alpha) &= U^{\overline{ij}k}_{\delta\gamma\delta}(\omega_1 = 1|\alpha) + 2U^{\overline{ij}k}_{\delta\gamma\delta}(\omega_1 = 2|\alpha) \\ W^{\overline{ij}k}_{\sigma\sigma\lambda}(2|\alpha) &= 2[U^{\overline{ij}k}_{\sigma\sigma\lambda}(\omega_2 = 1|\alpha) + U^{\overline{ij}k}_{\sigma\sigma\lambda}(\omega_2 = 2|\alpha)] \end{aligned} \qquad \text{(Ap. 3.20)}$$

where $\omega_1 = 1$ corresponds to a linear arrangement of the sites, when $\omega_1 = 2$ the lines joining the sites form a right angle, while when $\omega_2 = 1$ and ω_2 they form an obtuse and acute angles, Eqs. (Ap. 3.8) and (Ap. 3.9), for example, will be rewritten as follows:

$$\begin{aligned} \dot{\theta}^{\mathrm{AB}}_{\delta\gamma}(1) = {} & U^{\bar{\mathrm{v}}\mathrm{B}}_{\delta\gamma}(1|1) - U^{\bar{\mathrm{A}}\mathrm{B}}_{\delta\gamma}(1|1) - U^{\mathrm{A}\bar{\mathrm{B}}}_{\delta\gamma}(1|4) - U^{\mathrm{AB}}_{\delta\gamma}(3) \\ & - [W^{\overline{\mathrm{BA}}\mathrm{B}}_{\gamma\delta\gamma}(1|3) + W^{\overline{\mathrm{vA}}\mathrm{B}}_{\gamma\delta\gamma}(1|5) - W^{\overline{\mathrm{Av}}\mathrm{B}}_{\gamma\delta\gamma}(1|5)] \\ & - W^{\mathrm{A}\overline{\mathrm{BB}}}_{\delta\gamma\delta}(1|2) - W^{\mathrm{A}\overline{\mathrm{vv}}}_{\delta\gamma\delta}(1|2) + W^{\mathrm{A}\overline{\mathrm{BA}}}_{\delta\gamma\delta}(1|3) \\ & + W^{\mathrm{A}\overline{\mathrm{Bv}}}_{\delta\gamma\delta}(1|5) - W^{\mathrm{A}\overline{\mathrm{vB}}}_{\delta\gamma\delta}(1|5)] \end{aligned} \qquad \text{(Ap. 3.21)}$$

$$\begin{aligned} \dot{\theta}^{\mathrm{AB}}_{\sigma\sigma}(2) = {} & U^{\bar{\mathrm{v}}\mathrm{B}}_{\sigma\sigma}(2|1) - U^{\bar{\mathrm{A}}\mathrm{B}}_{\sigma\sigma}(2|1) - U^{\mathrm{A}\bar{\mathrm{B}}}_{\sigma\sigma}(2|4) - [W^{\overline{\mathrm{BA}}\mathrm{B}}_{\lambda\sigma\sigma}(2|3) \\ & + W^{\overline{\mathrm{vA}}\mathrm{B}}_{\lambda\sigma\sigma}(2|5) - W^{\overline{\mathrm{Av}}\mathrm{B}}_{\lambda\sigma\sigma}(2|5) - W^{\mathrm{A}\overline{\mathrm{vv}}}_{\sigma\sigma\lambda}(2|2) \\ & + W^{\mathrm{A}\overline{\mathrm{BB}}}_{\sigma\sigma\lambda}(2|2) + W^{\mathrm{A}\overline{\mathrm{BA}}}_{\sigma\sigma\lambda}(2|3) + W^{\mathrm{A}\overline{\mathrm{Bv}}}_{\sigma\sigma\lambda}(2|5) - W^{\mathrm{A}\overline{\mathrm{vB}}}_{\sigma\sigma\lambda}(2|5)] \end{aligned}$$

The addends of equations of type (Ap. 3.21) are expressed in terms of the rates of elementary processes (Ap. 3.18) by formulas similar to (7.12) and

(7.16). The system of equations is solved with the given initial conditions for $\theta^i_\sigma(\tau = 0)$, $\theta^{ij}_{\delta\gamma}(1,\ \tau = 0)$, and $\theta^{ij}_{\sigma\sigma}(2,\ \tau = 0)$.

When going over to a model without ordering of the particles

$$\theta_i = \sum_{k=1}^{L} \theta^i_k/L,\ \theta_{ij}(r) = \sum_{k=1}^{L} \sum_{q} \theta^{ij}_{kq}(r)/(z(r)L) \qquad \text{(Ap. 3.22)}$$

system (Ap. 3.17) is simplified: singling out of sites δ and γ is absent, and the migration contributions between different sublattices are compensated. The system of equations acquires the form

$$\begin{aligned} \dot{\theta}_{\mathrm{A}} &= U^{\mathrm{v}}(1) - U^{\mathrm{A}}(1) - z(1)U^{\mathrm{AB}}(3) \\ \dot{\theta}_{\mathrm{B}} &= z(1)[U^{\mathrm{vv}}(2) - U^{\mathrm{BB}}(2) - U^{\mathrm{AB}}(3)] - U^{\mathrm{B}}(4) \end{aligned} \qquad (3.23)$$

The probabilities of two-body configurations are determined by solving the kinetic equations relative to $\theta_{ij}(r)$, where ij = AA, AB, or BB, and $r = 1$ and 2. As previously, we shall limit ourselves to the equations for $\theta_{\mathrm{AB}}(r)$:

$$\begin{aligned} \dot{\theta}_{\mathrm{AB}}(1) = {} & U^{\bar{\mathrm{v}}\mathrm{B}}(1|1) - U^{\bar{\mathrm{A}}\mathrm{B}}(1|1) - U^{\mathrm{AB}}(3) - U^{\mathrm{A}\bar{\mathrm{B}}}(4) \\ & + W^{\mathrm{A}\overline{\mathrm{vv}}}(1|2) - W^{\mathrm{A}\overline{\mathrm{BB}}}(1|2) - W^{\overline{\mathrm{BA}}\mathrm{B}}(1|3) \\ & - W^{\mathrm{A}\overline{\mathrm{BA}}}(1|3) - W^{\overline{\mathrm{vA}}\mathrm{B}}(1|5) + W^{\overline{\mathrm{Av}}\mathrm{B}}(1|5) \\ & - W^{\mathrm{A}\overline{\mathrm{Bv}}}(1|5) + W^{\mathrm{A}\overline{\mathrm{vB}}}(1|5) \end{aligned} \qquad \text{(Ap. 3.24)}$$

$$\begin{aligned} \dot{\theta}_{\mathrm{AB}}(2) = {} & U^{\bar{\mathrm{v}}\mathrm{B}}(2|1) - U^{\bar{\mathrm{A}}\mathrm{B}}(2|1) - U^{\mathrm{A}\bar{\mathrm{B}}}(2|4) + W^{\mathrm{A}\overline{\mathrm{vv}}}(2|2) \\ & - W^{\mathrm{A}\overline{\mathrm{BB}}}(2|2) - W^{\overline{\mathrm{BA}}\mathrm{B}}(2|3) - W^{\mathrm{A}\overline{\mathrm{BA}}}(2|3) \\ & - W^{\overline{\mathrm{vA}}\mathrm{B}}(2|5) + W^{\overline{\mathrm{Av}}\mathrm{B}}(2|5) - W^{\mathrm{A}\overline{\mathrm{Bv}}}(2|5) + W^{\mathrm{A}\overline{\mathrm{vB}}}(2|5) \end{aligned}$$

where the addends on the right-hand sides are determined by formulas (7.12), (7.16), and (Ap. 3.20) in which the subscripts of the site numbers or types have been omitted. The system of equations being considered is solved with the given initial conditions $\theta_i(\tau = 0)$, $\theta_{ij}(1,\ \tau = 0)$, and $\theta_{ij}(2,\ \tau = 0)$.

Rapid Mobility of Reactants. The kinetic system of equations for the concentrations of adsorbed particles coincides with system (Ap. 3.17) for ordering of the particles and with system (Ap. 3.23) for their disordered distribution. (The addends of the equations have been indicated by the author [206, 249].) Here the functions $\theta_{ij}(r)$ are determined from the solution of systems (6.27) or (6.11), respectively. The kinetic equations are solved with given initial conditions θ_i $(\tau = 0)$, i = A or B.

Different Mobilities of Reactants

The equations constructed above in the general case hold for any mobilities of the reactants. A priori information on great differences in the mobilities of reactants makes it possible to lower the dimension of the system of kinetic equations. Let us consider the case when the step of migration of one of the components, e.g. A, is rapid, while the other steps have commensurable characteristic times. In the kinetic equation for $\theta_{fg}^{Ai}(r)$, where i = A or B, and r = 1 or 2 — (Ap. 3.6), (Ap. 3.8), and (Ap. 3.9), the main contribution is made by the addends corresponding to the migration of the A particles, while the remaining addends can be disregarded. As a result, the following four algebraic equations are obtained for the indicated two-body functions:

$$\theta_{fg}^{vA}(r)\theta_{fg}^{AB}(r) = \theta_{fg}^{AA}(r)\theta_{fg}^{vB}(r)\exp\beta[\varepsilon_{vA}(r) + \varepsilon_{AB}(r) - \varepsilon_{AA}(r) - \varepsilon_{vB}(r)] \quad \text{(Ap. 3.25)}$$

$$\theta_{fg}^{vv}(r)\theta_{fg}^{AA}(r) = \theta_{fg}^{Av}(r)\theta_{fg}^{vA}(r)\exp\beta[\varepsilon_{AA}(r) + \varepsilon_{vv}(r) - 2\varepsilon_{Av}(r)]$$

that are equations of the equilibrium distribution of the particles in the quasichemical approximation (6.11), but the functions $\theta_{fg}^{BB}(r)$ do not obey this distribution. For this reason, the local concentrations of the B components will vary and, consequently, the concentrations of the A and v components will also vary. The indicated change in the structure of the system of equations relates to all the levels of the hierarchy. For instance, in a point model with restricted mobility of the B component, the kinetic equation for the function $\theta_{BB}(1)$ will be written as follows:

$$\begin{aligned}\dot{\theta}_{BB}(1) = {} & U^{vv}(1|2) - U^{BB}(1|2) - 2U^{\bar{B}B}(1|4) - 2W^{\overline{AB}B}(1|3) \\ & - 2W^{\overline{BB}B}(1|2) + 2W^{\overline{vv}B}(1|2) - 2W^{\overline{vB}B}(1|5) \\ & + 2W^{\overline{Bv}B}(1|5)\end{aligned} \quad \text{(Ap. 3.26)}$$

where the functions $W^{\overline{ij}k}(1|\alpha)$ have been determined in (Ap. 3.20). The system of kinetic equations in a model with a restricted mobility of the B components is solved at the given initial conditions $\theta_i(\tau = 0)$, i = A or B, $\theta_{BB}(r, \tau = 0)$, r = 1 or 2.

APPENDIX 4

Designations of Particle Configurations and Their Probabilities

$\theta^{i_1 \ldots i_m}_{f_1 \ldots f_m}(\tau | q_1 \ldots q_m)$ is a distribution function characterizing the probability of finding a particle of species i_1 at a site of type q_1 with the number f_1, a particle of species i_2 at a site of type q_2 with the number f_2, etc., up to the m-th particle, inclusive, at the instant τ; on a homogeneous lattice all the sites are equivalent, therefore the indices of the site types are absent — $\theta^{i_1 \ldots i_m}_{f_1 \ldots f_m}(\tau)$, $1 \leqslant i_n \leqslant s$, $1 \leqslant n \leqslant m$, where s is the number of particle species (Sec. 1.3).

Homogeneous Lattice, Cluster with $R = 1$. $\theta^{ij_1 \ldots j_z}_{fg_1 \ldots g_z}$ is the probability of finding a particle i at the central site of cluster f and particles j_n at an adjacent site g_n, $1 \leqslant n \leqslant z$, where z is the number of nearest neighbors (see Sec. 2.1).

Configuration $i\{[n]\}$: the central site of a cluster accommodates a particle i, while its first coordination sphere accommodates n_1 particles of species 1, n_2 particles of species 2, and so on up to n_s particles of species s; $\sum_{i=1}^{s} n_i = z$. The braces $\{[n]\}$ signify $[n_1 \ldots n_{s-1}]$, when $s = 2$ the braces are absent (see Sec. 2.1).

Configuration $i\{n\}$: the central site of a cluster accommodates a particle i, while its first coordination sphere accommodates at least n_1 particles of species 1, at least n_2 particles of species 2, etc., at least n_{s-1} particles of species $s - 1$ (the particles of species s are considered as a supplement up to the complete occupancy of all the neighboring sites); the braces have the same meaning as for $\{[n]\}$ (see Sec. 2.1).

$\theta^{i}_{\{[n]\}}$ is the probability of finding the particle configuration $i\{[n]\}$ on a cluster (see Sec. 2.1).

$\theta^{i}_{\{n\}}$ is the probability of finding the particle configuration $i\{n\}$ on a cluster (see Sec. 2.1).

Homogeneous Lattice, Cluster with $R > 1$. $\theta^{i(j_1 \ldots j_z)R}_{f(g_1 \ldots g_z)R}$ is the probability of finding a particle i at the central site of cluster f and particles j_n at site g_n, $1 \leqslant n \leqslant z(r)$, $1 \leqslant r \leqslant R$, where R is the radius of interaction, and $z(r)$ is the number of sites in the r-th coordination sphere of site f (see Sec. 6.1).

Configuration $i\{[n]\}R$: the central site of a cluster accommodates a particle i, while its r-th coordination sphere accommodates n_r^j particles j, $1 \leqslant j \leqslant s$, $\sum_{j=1}^{s} n_r^j = z(r)$. The braces signify a complete list of all the n_r^j's, $1 \leqslant j \leqslant s - 1$, $1 \leqslant r \leqslant R$, when $s = 2$ the braces are absent (see Sec. 6.1).

Configuration $i\{n\}R$: the central site of a cluster accommodates a particle i, while its r-th coordination sphere accommodates at least n_r^j particles of species j, $i \leqslant j \leqslant s - 1$, $1 \leqslant r \leqslant R$; the braces have the same meaning as for $i\{[n]\}R$ (see Sec. 6.1).

$\theta^i_{\{[n]\}R}$ is the probability of finding the particle configuration $i\{[n]\}R$ on a cluster (see Sec. 6.1).

$\theta^i_{\{n\}R}$ is the probability of finding the particle configuration $i\{n\}R$ on a cluster (see Sec. 6.1).

Cluster with Two Central Sites. Configuration $ij\{[m]\}R$: one of the central sites accommodates a particle i, and the other — a particle j, while the group of equivalent sites $\varkappa_{\omega r}$ contains $m^j_{\omega r}$ particles j, $1 \leqslant j \leqslant s$, $\sum_{j=1}^{s} m^j_{\omega r} = \varkappa_{\omega r}$, $1 \leqslant \omega_r \leqslant \pi_r$, $1 \leqslant r \leqslant R$. The braces signify a complete list of all the $m^j_{\omega r}$'s, $1 \leqslant j \leqslant s - 1$ for all r's and ω_r's (see Sec. 7.1).

$\theta^{ij}_{\{[m]\}R}$ is the probability of finding the particle configuration $ij\{[m]\}R$ on a cluster with two central sites (see Sec. 7.1).

Inhomogeneous Lattice, Cluster with $R = 1$. Configuration of sites $q\{p\}$: the center of a cluster accommodates a site of type q, while its first coordination sphere accommodates sites of type p_n, $1 \leqslant n \leqslant z_q$, where z_q is the number of nearest neighbors of a type q site, $1 \leqslant q \leqslant t$, here t is the number of types of lattice sites (the set of clusters M_1 (see Sec. 3.1).

$f(q\{p\})$ is the probability of finding a cluster of type $q\{p\}$ on a lattice, $d(q\{p\})$ is the conditional probability of finding a cluster with a coordination sphere $\{p\}$ around a type q site on a lattice (Sec. 3.1).

Configuration of sites $q\{[m]\}$: the center of a cluster accommodates a site of type q, while its first coordination sphere accommodates m_{qp} sites of type p, $1 \leqslant p \leqslant t$, $\sum_{p=1}^{t} m_{qp} = z_q$ (the set of clusters M_2).

$f(q\{[m]\})$ is the probability of finding a cluster of type $q\{[m]\}$ on a lattice, $d(q\{[m]\})$ is the conditional probability of finding a cluster with a coordination sphere $\{[m]\}$ around a type q site (see Sec. 3.1).

$\theta^{i\{n\}}_{q\{[m]\}}$ is the probability of finding the particle configuration $i\{[n]\}$ on a type $q\{[m]\}$ cluster; the center of the configuration accommodates a particle i, while its first coordination sphere accommodates the following set of particles: $\{[n]\} = [\bar{n}_{q1} \ldots \bar{n}_{qt}]$, where $\bar{n}_{qp} = n^1_{qp}, \ldots, n^s_{qp}$, here $\sum_{j=1}^{s} n^j_{qp} = m_{qp}$; the braces signify a complete list of the values of all the n^j_{qp}'s, $1 \leqslant j \leqslant s$, $1 \leqslant p \leqslant t$ (Sec. 3.1).

$\theta^{i\{n\}}_{q\{[m]\}}$ is the probability of finding the particle configuration $i\{n\}$ on a type $q\{[m]\}$ cluster; the center of the configuration accommodates a particle i, while its first coordination sphere accommodates the following set of particles: $\{n\} = \bar{n}_{q1} \ldots \bar{n}_{qt}$, where $\bar{n}_{qp} = n^1_{qp} \ldots n^{s-1}_{qp}$, i.e. m_{qp} sites of type p contain at least n^j_{qp} particles j, $1 \leqslant j \leqslant s - 1$ (see Sec. 3.1).

Cluster with Two Central Sites. Configurations of sites of type $qp\{[\lambda\sigma]\}$: one of the central sites of a cluster accommodates a site of type q, and the other a site of type p. The types of the neighboring sites $\{[\lambda\sigma]\}$ around the type q and p sites are given by the relevant sets of indices of the site types (the set of clusters M_1) or by the numbers of the site types (the set of clusters M_2) (see Sec. 4.5).

$f(qp\{[\lambda\sigma]\})$ is the probability of finding a cluster of type $qp\{[\lambda\sigma]\}$ on a lattice, $d(qp\{[\lambda\sigma]\})$ is the conditional probability of finding a cluster with a coordination sphere of type $\{[\lambda\sigma]\}$ around a pair of central sites of type qp (see Sec. 4.5).

Inhomogeneous Lattice, Cluster with $R > 1$. Configuration of sites $q\{[m]\}$: the center of a cluster accommodates a site of type q, while its r-th coordination sphere accommodates $m_{qp}(r)$ sites of type p (the set of clusters M_2); the braces signify a complete list of all the site types, $1 \leqslant p \leqslant t$, $1 \leqslant r \leqslant R$ (see Sec. 6.3).

$f(q\{[m]\}R)$ is the probability of finding a cluster of type $q\{[m]\}R$ on a lattice, $d(q\{[m]\}R)$ is the conditional probability of finding R coordination spheres of type $\{[m]\}R$ around a central type q site on a lattice (see Sec. 6.3).

$\theta^{i\{[n]\}R}_{q\{[m]\}R}$ is the probability of finding the particle configuration $i\{[n]\}R$ on a type $q\{[m]\}R$ cluster; the center of the configuration accommodates a particle i, while $m_{qp}(r)$ sites of type p at a distance of r from a type q site accommodate $n^j_{qp}(r)$ particles j, $1 \leqslant j \leqslant s$, $\sum_{j=1}^{s} n^j_{qp}(r) = m_{qp}(r)$. The braces signify a complete list of all values of $n^j_{qp}(r)$, $1 \leqslant j \leqslant s$, $1 \leqslant p \leqslant t$ (see Sec. 6.3).

$\theta^{i\{n\}R}_{q\{[m]\}R}$ is the probability of finding the particle configuration $i\{n\}R$ on a type $q\{[m]\}R$ cluster; the center of the configuration accommodates a particle i, while $m_{qp}(r)$ sites of type p at a distance of r from a type q site accommodate at least $n^j_{qp}(r)$ particles j, $1 \leqslant j \leqslant (s-1)$ (see Sec. 6.3).

Clusters with two central sites are described in a similar way. For instance, $d(qp\{[m, h]\}R)$ is the conditional probability of finding $m_{qp\xi}(\omega_r)$ sites of type ξ around the central sites of type qp in the group of equivalent sites $\varkappa_{\omega r}(qp)$, $1 \leqslant \xi \leqslant t$, $\sum_{\xi=1}^{t} m_{qp\xi}(\omega_r) = \varkappa_{\omega r}(qp)$ (see Sec. 7.2).

REFERENCES

1. Gibbs, J. W. *Collected Works*, Vols. 1 and 2. New York: Longmans, Green (1928).
2. Rusanov, A. I. *Fazovye ravnovesiya i poverkhnostnye yavleniya* (Phase Equilibria and Surface Phenomena). Leningrad: Khimiya (1967).
3. Prigogine, I. and Defay, R. *Chemical Thermodynamics.* Trans. by D. H. Everett. London: Longmans Green (1954).
4. Kireev, V. A. *Physical Chemistry*, 4th ed. Moscow: Mir Publishers (1979).
5. Landau, L. D. and Lifshits, E. M. *Statisticheskaya fizika* (Statistical Physics), 2nd ed., Vol. 5. Moscow: Nauka (1964).
6. Henderson, D. *J. Chem. Phys.*, **37**: 632-635 (1962).
7. Sanchez, I. C. and Lakombe, R. H. *J. Phys. Chem.*, **80**: 2568 (1976).
8. Batalin, O. Yu., Tovbin, Yu. K., and Fedyanin, V. K. *Zh. Fiz. Khimii*, **53**: 3020-3023 (1980).
9. Zhdanov, G. S. *Fizika tverdogo tela* (Solid State Physics). Moscow State University Press (1961).
10. Wert, C. and Thomson, R. *Physics of Solids.* New York (1964).
11. Kittel, C. *Introduction to Solid State Physics*, 4th ed. New York: Wiley (1976).
12. Krivoglaz, A. M. and Smirnov, A. A. *Teoriya uporyadochivayushchikhsya splavov* (The Theory of Ordering Alloys). Moscow: Fizmatgiz (1958).
13. Khachaturian, A. G. *Teoriya fazovykh prevrashchenii i struktura tverdykh rastvorov* (The Theory of Phase Transformations and Structure of Solid Solutions). Moscow: Nauka (1974).
14. Iveronova, V. I. and Katsnelson, A. A. *Blizhnii poryadok v tverdykh rastvorakh* (Short-range Order in Solid Solutions). Moscow: Nauka (1977).
15. Gufan, Yu. M. *Strukturnye fazovye perekhody* (Structural Phase Transitions). Moscow: Nauka (1982).
16. Santen, R. A. and von Sachtler, W. M. H. *J. Catal.*, **37**: 202 (1974).
17. Williams, F. L. and Nason, D. *Surface Sci.*, **45**: 377-408 (1974).
18. Kymar, V., Kymar, D., and Joshi, S. K. *Phys. Rev. B — Solid State*, **19**: 1954-1962 (1979).
19. Moran-Lopez, J. L. and Falicov, L. M. *Phys. Rev. B — Solid State*, **18**: 2542-2548 (1978).
20. Moran-Lopez, J. L. and Falicov, L. M. *Phys. Rev. B — Solid State*, **18**: 2549-2554 (1978).
21. Farnsworth, H. E. In: E. A. Flood (ed.). *The Solid-Gas Interface*, Vol. 1, pp. 431-446. New York (1967).
22. Laudise, R. *The Growth of Single Crystals.* Englewoof Cliffs, N. J. (1970).
23. De Boer, J. H. *The Dynamic Character of Adsorption.* London (1968).
24. Crowell, A. D. In: E. A. Flood (ed.). *The Solid-Gas Interface*, Vol. 1, pp. 175-202. New York (1967).
25. Volkenshtein, F. F. *Fizikokhimiya poverkhnosti poluprovodnikov* (The Physicochemisty of the Surface of Semiconductors). Moscow: Nauka (1973).
26. Kiselev, V. F. and Krylov, O. V. *Adsorbtsionnye protsessy na poverkhnosti poluprovodnikov i dielektrikov* (Adsorption Processes on the Surface of Semiconductors and Insulators). Moscow: Nauka (1978).

27. Adamson, A. *The Physical Chemistry of Surfaces.* New York (1976).
28. Morrison, S. R. *The Chemical Physics of Surfaces.* New York: Plenum Press (1977).
29. Krylov, O. V. and Kiselev, V. F. *Adsorbtsiya i kataliz na perekhodnykh metallakh i ikh oksidakh* (Adsorption and Catalysis on Transition Metals and Their Oxides). Moscow: Khimiya (1981).
30. Goldansky, V. I., Trakhtenberg, L. I., and Flerov, V. N. *Tunnel'nye yavleniya v khimicheskoi fizike* (Tunnel Phenomena in Chemical Physics). Moscow: Nauka (1986).
31. Tringides, M. and Gomer, R. *Surface Sci.*, **166**: 440-457 (1986).
32. Shub, B. R. Geterogennaya relaksatsiya vnutrennei energii molekul i neravnovesnye protsessy na poverkhnostyakh tverdykh tel (Heterogeneous Relaxation of the Internal Energy of Molecules and Nonequilibrium Processes on Solid Surfaces). *Synopsis of Doctor's Thesis.* Moscow (1983).
33. Roginsky, S. Z. Adsorbtsiya i kataliz na neodnorodnykh poverkhnostyakh (Adsorption and Catalysis on Inhomogeneous Surfaces). Moscow: Acad. of Sci., USSR (1948).
34. Danning, M. J. In: E. A. Flood (ed.). *The Solid-Gas Interface*, Vol. 1, pp. 271-306. New York (1967).
35. Leygraf, C. and Hultquist, G. *Surface Sci.*, **61**: 60-84 (1976).
36. Franken, P. E. C. and Ponec, V. *J. Catal.*, **35**: 417-426 (1974).
37. Kugler, E. L. and Boudart, M. *J. Catal.*, **59**: 201-210 (1979).
38. Wang, T. and Schmidt, L. D. *J. Catal.*, **71**: 411-422 (1981).
39. Sedlacek, J., Hilaire, L., Legare, P., *et al. Surface Sci.*, **115**: 541-552 (1982).
40. Ertl, G., Norton, P. R. and Rustig, *J. Phys. Rev. Lett.*, **49**: 177-180 (1982).
41. Griffiths, K., Jackman, T. E., Davies, J. A., *et al. Surface Sci.*, **138**: 113-124 (1984).
42. Delmon, B. *Introduction a la cinetique hétérogène.* Paris (1969).
43. Rozovsky, A. Ya. *Kinetika topokhimicheskikh reaktsii* (The Kinetics of Topochemical Reactions). Moscow: Khimiya (1974)
44. Barret, P. *Cinetique hétérogène.* Paris (1973).
45. Fast, J. D. *Interaction of Metals and Gases.* Vol. 2. Kinetics and Mechanisms. London: Macmillan (1971).
46. Geld, P. V., Ryabov, R. A., and Mokhracheva, L. P. *Vodorod i fizicheskie svoistva metallov i splavov: Gidridy perekhodnykh metallov* (Hydrogen and the Physical Properties of Metals and Alloys: Hydrides of the Transition Metals). Moscow: Nauka (1985).
47. Andrievsky, R. A. and Umansky, Ya. S. *Fazy vnedreniya* (Interstitial Phases). Moscow: Nauka (1977).
48. Smirnov, A. A. *Teoriya splavov vnedreniya* (The Theory of Interstitial Alloys). Moscow: Nauka (1979).
49. Manning, J. H. *Diffusion Kinetics for Atoms in Crystals.* Princeton, N. J. (1968).
50. Bokshetein, B. S., Bokshtein, S. Z., and Zhykhovitsky, A. A. *Termodinamika i kinetika diffuzii v tverdykh telakh* (The Thermodynamics and Kinetics of Diffusion in Solids). Moscow: Metallurgiya (1974).
51. Bokshtein, B. S. *Diffuziya v metallakh* (Diffusion in Metals). Moscow: Metallurgiya (1978).

52. Geguzin, Ya. E. *Diffuzionnaya zona* (The Diffusion Zone). Moscow: Nauka (1979).
53. Kozhushner, M. A., Prostnev, A. S., and Shub, B. R. *Dokl. AN SSSR*, **279**, 6: 1401-1404 (1984).
54. Prostnev, A. S., Kozhushner, M. A., and Shub, B. R. *Khim. Fizika*, **5**, 85-94 (1986).
55. Emanuel, N. M. and Knorre, D. T. *Kurs khimicheskoi kinetiki* (A Course in Chemical Kinetics), 4th ed. Moscow: Vysshaya Shkola (1984).
56. Eremin, E. N. *Osnovy khimicheskoi kinetiki* (Fundamentals of Chemical Kinetics). Moscow: Vysshaya Shkola (1976).
57. Frank-Kamenetsky, D. A. *Diffuziya i teploperedacha v khimicheskoi kinetike* (Diffusion and Heat Transfer in Chemical Kinetics). Moscow: Acad. of Sci., USSR (1967).
58. Ovchinnikov, A. A., Timashev, S. F., and Belyi, A. A. *Kinetika diffuzionno-kontroliruemykh khimicheskikh protsessov* (Kinetics of Diffusion-Controlled Chemical Processes). Moscow: Khimiya (1986).
59. Slater, J. C. *The Self-Consistent Field for Molecules and Solids.* New York (1975).
60. Kaplan, I. G. *Vvedenie v teoriyu mezhmoleculyarnykh vzaimodeistvii* (Introduction to the Theory of Intermolecular Interactions). Moscow: Nauka (1982).
61. Animalu, A *Intermediate Quantum Theory of Crystalline Solids.* Englewood Cliffs, N. J.: Prentice-Hall (1977).
62. Harrison, W. A. *Solid State Theory.* New York (1970).
63. Tsidilkovsky, I. M. *Zonnaya struktura poluprovodnikov* (The Band Structure of Semiconductors). Moscow: Nauka (1978).
64. Ehrenreich, H. and Schwartz, L. *The Electronic Structure of Alloys.* New York (1976).
65. Harrison, W. A. *Pseudopotentials in the Theory of Metals.* New York (1966).
66. Panin, V. E., Khon, Yu. A., Naumov, I. I., Psakhye, S. G., Landa, A. I., and Chulov, E. V. *Teoriya faz v splavakh* (The Theory of Phases in Alloys). Moscow: Nauka (1984).
67. Gubanov, V. A., Ivanovsky, A. L., and Ryzhkov, M. V. *Kvantovaya khimiya v materialovedenii* (Quantum Chemistry in the Science of Materials). Moscow: Nauka (1987).
68. Bolshov, L. A., Napartovich, A. P., Naumovets, A. G., and Fedorus, A. G. *Uspekhi Fiz. Nauk.*, **122**: 125-158 (1977).
69. Smith, J. (ed.). *Theory of Chemisorption.* West Berlin: Springer (1980).
70. Dunken, H. and Lygin, V. *Quantenchemie der Adsorption an Festkörperoberflächen.* Leipzig (1978).
71. Zhidomirov, G. M. and Mikheikin, I. D. *Klasternoe priblizhenie v kvantovo-khimicheskikh issledovaniyakh khemosorbtsii i poverkhnostnykh struktur* (The Cluster Approximation in the Quantum-Chemical Studies of Chemisorption and Surface Structures). Moscow: VINITI (1984).
72. Ertl, G. *Surface Sci.*, **152/153**: 328-337 (1985).
73. Benson, G. C. and Yun, K. S. In: E. A. Flood (ed.). *The Solid-Gas Interface*, Vol. 1, pp. 203-270. New York (1967).
74. Smorjai, G. A. *Ann. Rev. Phys. Chem.*, **19**: 251-269 (1968).

75. Steele, W. A. In: E. A. Flood (ed.). *The Solid-Gas Interface*, Vol. 1, pp. 307-370. New York (1967).
76. Hirschfelder, J., Curtiss, C., and Bird, R. *Molecular Theory of Gases and Liquids.* New York (1954).
77. Hill, T. *Statistical Mechanics.* New York (1956).
78. Barker, J. A. *Lattice Theories of the Liquid State.* Oxford: Pergamon Press (1963).
79. Girifalco, L. A. *Statistical Physics of Materials.* New York (1973).
80. Chebotin, V. I. *Fizicheskaya khimiya tverdogo tela* (The Physicochemistry of a Solid). Moscow: Khimiya (1982).
81. Ono, S. and Kondo, S. *Molecular Theory of Surface Tension in Liquids.* West Berlin (1960).
82. Smirnova, N. A. *Kolloid. Zhurn.*, **41**: 1152-1156 (1979).
83. Smirnova, N. A. *Kolloid. Zhurn.*, **39**: 92-99 (1977).
84. Gurov, K. P., Kartaskhin, B. A., and Ugaste, Yu. E. *Vzaimnaya diffuziya v mnogofaznykh metallicheskikh sistemakh* (Interdiffusion in Multiphase Metal Systems). Moscow: Nauka (1981).
85. Kiperman, S. L. *Vvedenie v kinetiku geterogennykh kataliticheskikh reaktsii* (Introduction into the Kinetics of Heterogeneous Catalytic Reactions). Moscow: Nauka (1964).
86. Balandin, A. A. *Mul'tipletnaya teoriya kataliza* (The Multiplet Theory of Catalysis), Parts 1 and 2. Moscow State University Press (1963-1964).
87. Snagovsky, Yu. S. and Ostrovsky, G. M. *Modelirovanie kinetiki geterogennykh kataliticheskikh protsessov* (Modeling the Kinetics of Heterogeneous Catalytic Processes). Moscow: Khimiya (1976).
88. Kiperman, S. L. *Osnovy khimicheskoi kinetiki v geterogennom katalize* (Fundamentals of Chemical Kinetics in Heterogeneous Catalysis). Moscow: Khimiya (1979).
89. Rozovsky, A. Ya. *Geterogennye khimicheskie reaktsii: kinetika i makrokinetika* (Heterogeneous Chemical Reactions: Kinetics and Macrokinetics). Moscow: Nauka (1980).
90. Yablonsky, G. S., Bykov, V. I., and Gorban, A. N. *Kineticheskie modeli kataliticheskikh reaktsii* (Kinetic Models of Catalytic Reactions). Novosibirsk: Nauka (1983).
91. Smirnov, A. A. *Molekulyarno-kineticheskaya teoriya metallov* (The Molecular-Kinetic Theory of Metals). Moscow (1966).
92. Huang, K. *Statistical Mechanics.* New York (1963).
93. Shneider, V. E. *Zh. Eksper. i Teor. Fiziki*, **66**: 638-646 (1974).
94. Baxter, R. J. *Exactly Solved Models in Statistical Mechanics.* London: Academic Press (1982).
95. Anderson, R. (ed.) *Experimental Methods in Catalytic Research.* New York (1968).
96. Tayadevaiah, T. S. and Vanselow, R. (eds.) *Surface Science: Recent Progress and Perspectives.* Cleveland: CRC Press, inc. (1974).
97. Roberts, M. and McKee, C. *Chemistry of the Metal-Gas Interface.* Oxford University Press (1978).
98. Golze, M., Grunze, M., and Unertl, W. *Progr. Surface Sci.*, **22**: 101-180 (1986).

99. Zhukhovitsky, A. A. and Shvartsman, A. A. *Fizicheskaya Khimiya* (Physical Chemistry). Moscow: Metallurgiya (1968).
100. Moelwyn-Hughes, E. A. *Physical Chemistry.* New York (1961).
101. Bogolyubov, N. N. *Problemy dinamicheskoi teorii v statisticheskoi fizike* (Problems of the Dynamic Theory in Statistical Physics). Moscow: OGIZ (1946).
102. Bogolyubov, N. N. (ed.) *Statisticheskaya fizika i kvantovaya teoriya polya* (Statistical Physics and the Quantum Theory of a Field). Moscow: Nauka (1973).
103. Bazarov, I. P. *Statisticheskaya teoriya kristallicheskogo sostoyaniya* (The Statistical Theory of the Crystal State). Moscow State University Press (1972).
104. Bazarov, I. P. and Nikolaev, P. N. *Korrelyatsionnaya teoriya kristalla* (The Correlation Theory of a Crystal). Moscow State University Press ((1981).
105. Fisher, I. Z. *Statisticheskaya teoriya zhidkostei* (The Statistical Theory of Liquids). Moscow: Fizmatgiz (1961).
106. Croxton, C. A. *Liquid State Physics. A Statistical Mechanical Introduction.* Cambridge (1974).
107. Yukhnovsky, I. R. and Golovko, M. F. *Statisticheskaya teoriya klassicheskikh ravnovesnykh sistem* (The Statistical Theory of Classical Equilibrium Systems). Kiev: Naukova Dumka (1980).
108. Tyablikov, S. V. and Fedyanin, V. K. *Fizika Metallov i Metallovedenie*, **23**: 193-199 (1967).
109. Fedyanin, V. K. *Metod korrelyatsionnykh funktsii v modeli Izinga* (The Method of Correlation Functions in the Ising Model). Tartu University Press (1969).
110. Fedyanin, V. K. In: N. N. Bogolyubov (ed.). *Statisticheskaya fizika i kvantovaya teoriya polya* (Statistical Physics and the Quantum Theory of a Field), pp. 241-261. Moscow: Nauka (1973).
111. Kuzmin, V. L. *Teoret. i Mat. Fizika*, **28**: 389-397 (1976).
112. Rott, L. A. *Statisticheskaya teoriya molekulyarnykh sistem: metod korrelyativnykh funktsii uslovnykh raspredelenii* (The Statistical Theory of Molecular Systems: the Method of Correlative Functions of Conditional Distributions). Moscow: Nauka (1979).
113. Shulepov, Yu. V. and Aksenenko, E. V. *Reshetochnyi gaz: vvedenie v teoriyu i izbrannye prilozheniya* (A Lattice Gas: Introduction to Theory and Selected Applications). Kiev: Naukova Dumka (1981).
114. Balagurov, B. Ya., Vaks, V. G., and Zaitsev, R. O. *Fiz. Tv. Tela*, **16**: 2302-2309 (1974).
115. Vaks, V. G. and Zein, N. E. *Zh. Eksper. i Teor. Fiziki*, **67**: 1082-1100 (1974).
116. Tovbin, Yu. K. *Kinetika i Kataliz*, **19**: 1453-1462 (1978).
117. Gnedenko, B. V. and Khinchin, A. Ya. *Elementarnoe vvedenie v teoriyu veroyatnostei* (Elementary Introduction to the Theory of Probability), 8th ed., revised. Moscow: Nauka (1976).
118. Rumshisky, L. Z. *Elementy teorii veroyatnostei* (Elements of the Theory of Probability), 5th ed., revised. Moscow: Nauka (1976).
119. Feller, W. *An Introduction to Probability Theory and Its Applications.* Vol. 1, 3rd ed., New York: Wiley (1968).
120. Chapman, S. and Cowling, T. *The Mathematical Theory of Non-Uniform Gases.* Cambridge (1952).

121. Cercignani, C. *Theory and Application of the Boltzmann Equation.* London (1975).
122. Koga, T. *Introduction to Kinetic Theory. Stochastic Processes in Gaseous Systems.* Oxford: Pergamon Press (1970).
123. Borovsky, I. B., Gurov, K. P., Marchukova, I. D., and Uchaste, E. Yu. *Protsessy vzcimnoi diffuzii v splavakh* (Processes of Interdiffusion in Alloys). Moscow: Nauka (1973).
124. Bird, R. B., Stewart, W. E., and Lightfoot, E. N. *Transport Phenomena*, 5th ed. New York: Wiley (1960).
125. Tikhonov, A. I. and Samarsky, A. A. *Uravneniya matematicheskoi fiziki* (Equations of Mathematical Physics), 3rd ed., revised. Moscow: Nauka (1966).
126. Klimontovich, Yu. L. *Statisticheskaya fizika* (Statistical Physics). Moscow: Nauka (1982).
127. Tovbin, Yu. K. In: *Materialy II Vsesoyuz. konf. po poverkhnost. yavleniyam v zhidkostyakh* (Materials of the 2nd All-Union Conference on Surface Phenomena in Liquids, Leningrad, February. 1978), p. 168. Leningrad State University Press (1978).
128. Balescu, R. *Equilibrium and Non-Equilibrium Statistical Mechanics.* New York (1975).
129. Mayer, J. and Goeppert-Mayer, M. *Statistical Mechanics.* New York (1977).
130. Resibois, P. and de Leener, M. *Classical Kinetic Theory of Fluids.* New York (1977).
131. Cordoba, A. and Luque, J. J. *Phys. Rev. B — Solid State*, **26**: 4028-4034 (1982).
132. Tovbin, Yu. K. *Klasternyi podkhod k teorii molekulyarnykh protsessov v kondensirovannoi faze* (The Cluster Approach to the Theory of Molecular Processes in a Condensed Phase). *Doctor's Thesis.* Moscow (1985).
133. Eyring, H. *J. Chem. Phys.*, **3**: 107-115 (1935).
134. Glasstone, S., Laidler, K., and Eyring, H. *The Theory of Rate Processes.* New York (1941).
135. Tovbin, Yu. K. *Zh. Fiz. Khimii*, **56**: 1698-1702 (1982).
136. Temkin, M. I. *Zh. Fiz. Khimii*, **15**: 296-332 (1941).
137. Temkin, M. I. *Zh. Fiz. Khimii*, **11**: 169-189 (1938).
138. Snagovsky, Yu. S. *Kinetika i Kataliz*, **17**: 92-101 (1976).
139. Snagovsky, Yu. S. *Kinetika i Kataliz*, **19**: 932-941 (1978).
140. Tovbin, Yu. K. *Kinetika i Kataliz*, **21**: 1165-1173 (1980).
141. Tovbin, Yu. K. *Zh. Fiz. Khimii*, **56**: 273-283 (1981).
142. Bethe, H. A. *Proc. Roy. Soc. London A*, **150**: 552-575 (1935).
143. Peierls, R. *Proc. Roy. Soc. London A*, **154**: 207-222 (1936).
144. Guggenheim, E. A. *Proc. Roy. Soc. London A*, **148**: 304-312 (1935).
145. Tovbin, Yu. K. *Kinetika i Kataliz*, **20**: 1226-1234 (1979).
146. Bragg, W. L. and Williams, E. I. *Proc. Roy. Soc. London A*, **145**: 699-730 (1934).
147. Wang, J.-S. *Proc. Roy. Soc. London A*, **161**: 127-140 (1937).
148. Zagrebnov, V. A. and Fedyanin, V. K. *Teoret. i Mat. Fizika*, **10**: 124-139 (1972).
149. Tovbin, Yu. K. and Fedyanin, V. K. *Kinetika i Kataliz*, **15**: 524-526 (1974).
150. Trapnell, B. *Chemisorption.* London (1955).
151. Hill, T. L. *J. Chem. Phys.*, **17**: 762-771 (1949).
152. Kramers, H. A. and Wannier, G. H. *Phys. Rev.*, **60**: 252-262 (1941).

153. Domb, C. *Adv. Phys.*, **9**: 149-361 (1960).
154. Stanley, H. *Introduction to Phase Transitions and Critical Phenomena.* Oxford (1971).
155. Pace, E. In: E. A. Flood (ed.). *The Solid-Gas Interface*, Vol. 1, p. 105. New York (1967).
156. Tovbin, Yu. K. and Fedyanin, V. K. *Kinetika i Kataliz*, **13**: 1085-1087 (1972).
157. Tovbin, Yu. K. and Fedyanin, V. K. *Zh. Fiz. Khimii*, **47**: 2886-2888 (1973).
158. Tovbin, Yu. K. and Fedyanin, V. K. *Fiz. Tv. Tela*, **17**: 1511-1513 (1975).
159. Rushbrooke, G. S. *Proc. Roy. Soc. London A*, **166**: 296-315 (1938).
160. Wang, J.-S. *Proc. Cambridge Philos. Soc.*, **34**: 247-252 (1938).
161. Tovbin, Yu. K. and Fedyanin, V. K. *Zh. Fiz. Khimii.*, **50**: 209 (1976).
162. Chretilin, M., *et al.* (eds.). *Statistical Physics. Phase Transitions and Superfluidity*, Vols. 1 and 2. New York: Gordon and Breach Science Publishers (1968).
163. Guggenheim, E. A. *Mixture.* Oxford University Press (1952).
164. Dziembaj, R. and Wedler, G. *Surface Sci.*, **134**: 283-296 (1983).
165. Kiskinova, M., Pirug, G., and Bonzel, H. P. *Surface Sci.*, **133**: 321-343 (1983).
166. Stuve, E. M. and Madix, R. J. *Surface Sci.*, **146**: 155-178 (1984).
167. Angel, T. and Ertl, G. *Adv. Catal.*, **28**: 1-78 (1979).
168. Barrer, R. M. and Klinowski, J. *J. Chem. Soc., Faraday Trans., Part II*, **74**: 904-917 (1978).
169. Okunev, B. N., Kaminsky, V. A., and Tovbin, Yu. K. *Kolloid. Zhurn.*, **47**: 1110-1116 (1985).
170. Somorjai, G. Z. *Chemistry in Two Dimension Surfaces.* Ithaca: Cornell University Press (1981).
171. Tovbin, Yu. K. *Kinetika i Kataliz*, **24**: 308-316 (1983).
172. Tovbin, Yu. K. *Poverkhnost'. Fizika, Khimiya, Mekhanika*, **9**: 15-24 (1982).
173. Ryazanov, M. I. and Tilinin, N. S. *Issledovanie poverkhnosti po obratnomu rasseyaniyu chastits* (Studying a Surface by the Back-scattering of Particles). Moscow: Energoatomizdat (1985).
174. Tovbin, Yu. K. *Kolloid. Zhurn.*, **45**: 707-714 (1983).
175. Melikhov, I. V. and Merkulova, M. S. *Sokristallizatsiya* (Cocrystallization). Moscow: Khimiya (1975).
176. Fujinaga, Y. *Surface Sci.*, **64**: 751-770 (1977).
177. Tovbin, Yu. K. *Kinetika i Kataliz*, **24**: 317-325 (1983).
178. Tovbin, Yu. K. *Dokl. AN SSSR*, **235**, 3: 641-643 (1977).
179. Tovbin, Yu. K. *Zh. Fiz. Khimii*, **56**: 686-690 (1982).
180. Tovbin, Yu. K., Chelnokova, O. V., and Cherkasov, A. V. *Zh. Fiz. Khimii*, **62**: 1598-1603 (1988).
181. Tovbin, Yu. K. *Dokl. AN SSSR*, **260**, 3: 679-682 (1981).
182. Avgul, N. N., Kiselev, A. V., and Poshkus, D. P. *Adsorbtsiya gazov i parov na odnorodnykh poverkhnostyakh* (Adsorption of Gases and Vapor on Homogeneous Surfaces). Moscow: Khimiya (1975).
183. Matecki, M., Thomy, A., and Duval, X. *Surface Sci.*, **75**: 142 (1978).
184. Tovbin, Yu. K. *Zh. Fiz. Khimii*, **61**: 3380-3384 (1987).
185. Tovbin, Yu. K. *Zh. Fiz. Khimii*, **56**: 691-695 (1982).
186. Klyachko, A. L. *Kinetika i Kataliz*, **13**: 1218-1223 (1978).
187. Tovbin, Yu. K. and Chalnokova, O. V. *Zh. Fiz. Khimii*, **62**: 714-720 (1988).

188. Cartner, D. G. and Somorjai, G. A. *Chem. Rev.*, **79**: 233-252 (1979).
189. Roelofs, L. D. and Estrup, P. J. *Surface Sci.*, **125**: 51-73 (1983).
190. Mozolkov, A. E. and Fedyanin, V. K. *Difraktsiya medlennykh elektronov poverkhnost'yu* (Diffraction of Low-Energy Electrons by a Surface). Moscow: Energoizdat (1982).
191. Ohtani, H., Kao, C.-T., Van Hove, M. A., and Somorjai, G. A. *Progr. Surface Sci.*, **22**: 155-316 (1987).
192. Tovbin, Yu. K. *Kinetika i Kataliz*, **28**: 1148-1157 (1987).
193. Tovbin, Yu. K. *Teor. i Eksper. Khimiya*, **18**: 417-431 (1982).
194. Roginsky, S. Z. and Todes, O. M. *Probl. Kinetiki i Kataliza*, **7**: 339-359 (1949).
195. Tovbin, Yu. K. *Poverkhnost'. Fizika, Khimiya, Mekhanika*, **6**: 33-45 (1985).
196. Burton, W., Kabrera, G. H., and Frank, F. *Philos. Trans. Roy. Soc. London A*, **243**: 299-358 (1951).
197. Voronkov, V. V. *Kristallografiya*, **13**: 19-25 (1986).
198. Pfeiffer, H. *Phys. Status Solidi* (a). **7**: 239-249 (1982).
199. Kuzovkov, V. N. and Cherepanova, T. A. *Izv. AN Latv. SSR. Ser. Fiz. i Tekh. Nauk*, **2**: 122-125 (1983).
200. Frenkel, Ya. I. *Kineticheskaya Teoriya Zhidkostei* (The Kinetic Theory of Liquids). Leningrad: Nauka (1975).
201. Vyatkin, A. F., Tseitlin, E. M., and Antropov, A. S. *Poverkhnost'. Fizika, Khimiya, Mekhanika*, **3**: 123-129 (1983).
202. Temkin, M. I. *Zh. Fiz. Khimii*, **20**: 105-110 (1946).
203. Kozheurov, V. A. *Termodinamika metallurgicheskikh shlakov* (Thermodynamics of Metallurgical Slags). Sverdlovsk: Metallurgizdat (1955).
204. Tretyakov, Yu. D. *Khimiya nestekhiometricheskikh okislov* (The Chemistry of Nonstoichiometric Oxides). Moscow State University Press (1974).
205. Kofstad, P. *Nonstoichiometry, Diffusion, and Electrical Conductivity in Binary Metal Oxides.* New York (1972).
206. Tovbin, Yu. K. *Dokl. AN SSSR*, **277**, 4: 917-921 (1984).
207. Kawasaki, K. In: *Phase Transitions and Critical Phenomena*, Vol. 2, pp. 443-501. New York: Academic Press (1972).
208. Saito, Y. and Muller-Krumbhaar, H. *J. Chem. Phys.*, **70**: 1078 (1979).
209. Tovbin, Yu. K. *Dokl. AN SSSR*., **267**, 6: 1415-1419 (1982).
210. Tovbin, Yu. K. *Kinetika i Kataliz*, **25**: 26-36 (1984).
211. Nikitin, E. E. *Teoriya elementarnykh atomno-molekulyarnykh protsessov v gazakh* (Theory of Elementary Atomic and Molecular Processes in Gases). Moscow: Khimiya (1970).
212. Tovbin, Yu. K. *Zh. Fiz. Khimii*, **55**: 284-304 (1981).
213. Tovbin, Yu. K. and Fedyanin, V. K. *Kinetika i Kataliz*, **19**: 1202-1207 (1978).
214. Kwan, T. J. *J. Phys, Chem.*, **60**: 1033-1037 (1956).
215. Low, M. I. *Chem. Rev.*, 3: 267-297 (1960).
216. Tovbin, Yu. K. and Fedyanin, V. K. *Fiz. Tv. Tela.*, **22**: 1599-1605 (1980).
217. Elenin, G. G., Sinilo, A. V., and Tovbin, Yu. K. *Kinetika i Kataliz*, **27**: 960-965 (1986).
218. Tataurov, V. A. and Ivanov, V. P. In: *Materialy Vsesoyuz. konf. "Nestatsionarnye protsessy v katalize"* (Materials of All-Union Conference "Nonstationary

Processes in Catalysis"), Vol. 2, pp. 170-175. Novosibirsk: IK SO AN SSSR (1979).

219. Surovtsev, S. Yu. and Tovbin, Yu. K. *Poverkhnost'. Fizika, Khimiya, Mekhanika*, 5: 22-28 (1985).
220. Steininger, H., Lehwald, S., and Ibach, H. *Surface Sci.*, **123**: 264-282 (1982).
221. Steininger, H., Lehwald, S., and Ibach, H. *Surface Sci.*, **123**: 1-17 (1982).
222. Pacia, N., Cassuto, A., Pentenero, A., *et al. J. Catal.*, **41**: 455-465 (1976).
223. Gland, J. L. and Kollin, E. B. *J. Chem. Phys.*, **78**: 963-974 (1983).
224. Cox, M. P., Ertl, G., Imbihl, R., *et al. Surface Sci.*, **134**: L517-L523 (1983).
225. Tovbin, Yu. K. and Cherkasov, A. V. *Teor. i Eksper. Khimiya*, **20**: 507-512 (1984).
226. Tovbin, Yu. K. *Dokl. AN SSSR*, **262**, **4**: 929-931 (1982).
227. Kiskinova, M. P. and Bliznakov, G. M. *Surface Sci.*, **123**: 61 (1982).
228. Shmachkov, V. A., Malakhov, V. F., and Kolchin, A. M. *Probl. Kinetiki i Kataliza*, **18**: 170-181 (1978).
229. Malakhov, V. F., Shmachkov, V. A., Vasilyev, V. Yu., and Kolchin, A. M. In: *Materialy Vsesouyz. konf. "Nestatsionarnye protsessy v katalize"* (Materials of All-Union Conference "Nonstationary Processes in Catalysis"), Vol. 1, pp. 76-82. Novosibirsk: IK SO AN SSSR (1979).
230. Tovbin, Yu. K. and Murovtsev, A. N. *Zh. Fiz. Khimii*, **58**: 2286-2290 (1984).
231. Tovbin, Yu. K. *Kinetika i Kataliz*, **27**: 655-662 (1986).
232. Zakharov, A. Yu. and Markachev, Yu. E. *Sovremennye pakety i biblioteki programm matematicheskogo obespecheniya* (Modern Program Software Packages and Libraries): Preprint of M. V. Keldysh Inst. of Appl. Math., Acad. of Sci., USSR. Moscow (1987).
233. Polak, L. S., Goldenberg, M. Ya., and Levitsky, A. A. *Vychislitel'nye metody v khimicheskoi kinetike* (Calculation Methods in Chemical Kinetics). Moscow: Nauka (1984).
234. Tewarson, R. *Sparse Matrices. New York (1973).*
235. Tovbin, Yu. K. and Filimonova, O. A. In: *Modelirovanie rosta kristallov*: *Tez. dokl. II Vsesoyuz. konf.* (Modeling the Growth of Crystals: Abstracts of Reports to 2nd All-Union Conf.). Riga: Latvian University (1987).
236. *Metally i splavy kak membrannye katalizatory*: *Sb. nauch. st.* (Metals and Alloys as Membrane Catalysts: Collected Scientific Articles). Moscow: Nauka (1981).
237. Batist, PH. A., Kapteijns, C. J., Lippens, B. C., *et al. J. Catal.*, **7**: 33-49 (1967).
238. Gelbshtein, A. I., Mishchenko, Yu. A., and Goldshtein, N. D. *Dokl. AN SSSR*, **205**, 2: 374-376 (1972).
239. *Membrannye katalizatory, pronitsaemye dlya vodoroda i kisloroda*: *Sb. nauch. st.* (Membrane Catalysts Permeable to Hydrogen and Oxygen: Collected Scientific Articles). Moscow: Nauka (1985).
240. Tovbin, Yu. K. In: *Tez. dokl. I Vsesoyuz. konf. "Dinamika protsessov i apparatov khimicheskoi tekhnologii"* (Abstracts of Reports to 1st All-Union Conference: "Dynamics of Processes and Apparatus of Chemical Technology"), p. 27. Voronezh Ped. Inst. (1982).
241. Kaminsky, V. A. and Tovbin, Yu. K. *Kinetika i Kataliz*, **24**: 218-224 (1983).
242. Tovbin, Yu. K. and Fedyanin, V. K. *Zh. Fiz. Khimii*, **54**: 3132-3136 (1980).
243. Muravyev, V. A. and Panteleev, V. A. *Fiz. Tv. Tela*, **20**: 3245-3250 (1978).

244. Sankey, D. F. and Fedders, P. A. *Phys. Rev. B — Solid State*, **15**: 3586-3591 (1977).
245. Gokhman, E. M., Tovbin, Yu. K., and Fedyanin, V. K. In: *Fizika mezhfaznykh yavlenii: Sb. nauch. st.* (Physics of Interphase Phenomena: Collected Scientific Articles), pp. 145-153. Nal'chik: KBGU (1981).
246. Butz, R. and Wagner, H. *Surface Sci.*, **63**: 448-459 (1977).
247. Nartikoev, R. K. Poverkhnostnaya diffuziya adsorbirovannykh gazov na metallakh (Surface Diffusion of Adsorbed Gases on Metals), *Synopsis of Candidate's Thesis.* Moscow (1983).
248. Tovbin, Yu. K. *Kinetika i Kataliz*, **23**: 813-820 (1982).
249. Tovbin, Yu. K. *Kinetika i Kataliz*, **23**: 1231-1239 (1982).
250. Pindyurin, V. A., Surovtsev, S. Yu., and Tovbin, Yu. K. *Zh. Fiz. Khimii.*, **60**: 946-949 (1986).
251. Kiselev, A. V., Kovaleva, N. V., Korolev, A. Ya., and Shcherbakov, K. D. *Dokl. AN SSSR*, **124**: 617-620 (1959).
252. Lahrer, Y. *J. Chem. Soc., Faraday Trans. Part I*, **70**: 320 (1974).
253. Gerot, B. and Duval, K. *C. r. Acad. sci. Serie C*, **265**: 285-287 (1967).
254. Pryde, J. A. and Titcomb, C. G. *Trans. Faraday Soc.*, **65**: 2758-2765 (1969).
255. Veleckis, E. and Edwards, R. K. *J. Phys. Chem.*, **73**: 683-692 (1969).
256. Tovbin, Yu. K. *Zh. Fiz. Khimii*, **57**: 1016-1018 (1983).
257. Brennan, D., Hayward, D., and Trapnell, B. *Proc. Roy. Soc. London A*, **256**: 81-105 (1960).
258. Dobrovolsky, N. N., Ostrovsky, V. E., Rubashev, A. M., and Temkin, M. I. *Dokl. AN SSSR*, **183**, 4: 1120-1122 (1968).
259. Ostrovsky, V. E. and Dobrovolsky, N. N. *Tr. IV Mezhdunar. kongr. po katalizu* (Transactions of 4th Int. Congress on Catalysis), Preprint No. 46. Moscow (1968).
260. Beeck, O. In: W. Frankenburg, V. Komarewsky, *et al.* (eds.). *Advances in Catalysis and Related Subjects.* New York (1948-1952).
261. Bagg, J. and Tompkins, F. C. *Trans. Faraday Soc.*, **51**: 1071 (1955).
262. Hayward, D. O., Taylor, N., and Tompkins, F. S. *Discuss. Faraday Soc.*, 41: 75-86 (1966).
263. Tovbin, Yu. K. *Fiz. Tv. Tela*, **18**: 1404-1406 (1976).
264. Temkin, M. I. *Voprosy khimicheskoi kinetiki, kataliza i reaktsionnoi sposobnosti* (Questions of Chemical Kinetics, Catalysis, and Reactivity). Moscow: Acad. of Sci., USSR (1955).
265. Timashev, S. F. and Fedyanin, V. K. *Dokl. AN SSSR*, **191**, 6: 1333-1334 (1970).
266. Timashev, S. F. and Fedyanin, V. K. *Fiz. Tv. Tela*, **13**: 1196-1198 (1971).
267. Fedyanin, V. K. *Zh. Fiz. Khimii*, 45: 2867-2870 (1971).
268. Surovtsev, S. Yu., Pindyurin, V. A., and Tovbin, Yu. K., *Zh. Fiz. Khimii*, **60**: 1465-1470 (1986).
269. Pindyurin, V. A., Surovtsev, S. Yu., and Tovbin, Yu. K. *Zh. Fiz. Khimii*, **60**: 1471 (1986).
270. Brodowsky, H. and Poeschel, E. *Ztschr. phys. Chem.*, **44**: 143 (1965).
271. Harrison, W. *Uspekhi Fiz. Nauk*, **108**: 285-292 (1972).
272. Leonas, V. *Uspekhi Fiz. Nauk*, **107**: 29-56 (1972).

273. Kaplan, I. G. and Rodimova, O. B. *Uspekhi Fiz. Nauk*, **126**: 403-450 (1978).
274. Zubkov, V. A., Kolesov, B. I., and Gernshtein, T. M. *Uspekhi Khimii*, **52**: 1057-1085 (1983).
275. Meath, M. J. and Aziz, R. A. *Mol. Phys.*, **52**: 225-243 (1984).
276. Pople, J. A. *Discuss. Faraday Soc.*, **15**: 35-43 (1953).
277. Gray, C. G. and Lo, B. W. N. *Chem. Phys. Lett.*, **25**: 55-58 (1974).
278. Tsykalo, A. L. and Doroshenko, Zh. F. *Zh. Tekhn. Fiziki*, **47**: 1978-1981 (1977).
279. Kontsov, M. M. and Tsykalo, A. L. *Zh. Tekhn. Fiziki*, **53**: 1783-1786 (1983).
280. Tsykalo, A. L. *Zh. Tekhn. Fiziki*, **47**: 2601-2607 (1977).
281. Goldam, S. *J. Chem. Phys.*, **69**: 3775-3781 (1978).
282. Tolpygo, K. B. and Troitskaya, E. P. *Fiz. Tv. Tela*, **16**: 795-799 (1974).
283. Taggert, G. B. *Phys. Rev. B — Solid State*, **19**: 2895-2904 (1979).
284. Rac, Z. and Collins, M. F. *Phys. Rev. B — Solid State*, **20**: 229-237 (1980).
285. Binder, K. and Landau, D. P. *Surface Sci.*, **180**: 503-504 (1981).
286. Milchev, A. *J. Chem. Phys.*, **78**: 1994-1998 (1983).
287. Selka, W., Binder, K., and Kinzel, W. *Surface Sci.*, **125**: 74-93 (1983).
288. D'Elia, S. and Ceva, H. *Phys. Rev. B — Solid State*, **28**: 4001-4005 (1983).
289. Mariton, A., Stell, A., and Vandersande, C. *Phys. Rev. B — Solid State*, **29**: 519-521 (1984).
290. Penson, K. A. *Phys. Rev. B — Solid State*, **29**: 2404-2409 (1984).
291. Tovbin, Yu. K. *Zh. Fiz. Khimii*, **61**: 2711-2716 (1987).
292. Dienes, G. J. *Acta met.*, **6**: 278-282 (1958).
293. Wriedt, H. A. *Trans. AIME*, **221**: 377-383 (1961).
294. Sundquist, B. E. *Trans. AIME*, **236**: 1111-1122 (1966).
295. Tovbin, Yu. K. In: *Tez. dokl. IV Vsesoyuz. konf. po metodam polucheniya i analiza vysokochistykh veshchestv* (Abstracts of Reports to 4th All-Union Conference on Methods of Preparation and Analysis of Highly Pure Substances), p. 9. Gorki: Inst. of Chem., Acad. of Sci., USSR (1981).
296. Tovbin, Yu. K. *Kinetika i Kataliz*, **23**: 821-828 (1982).
297. Tovbin, Yu. K. and Surovtsev, S. Yu. *Kinetika i Kataliz*, **26**: 1181-1187 (1985).
298. Weissman, D. L., Shen, M. L., and Spicer, W. E. *Surface Sci.*, **92**: L59-L66 (1980).
299. Taylor, J. L. and Weinberg, W. H. *Surface Sci.*, **78**: 259-273 (1978).
300. Taylor, J. L., Ibbotson, D. E., and Weinberg, W. H. *Surface Sci.*, **79**: 349-384 (1979).
301. Kuppers, J. and Plagge, A. *J. Vac. Sci. and Technol.*, **13**: 259-263 (1976)
302. Fair, J. and Madix, R. J. *J. Chem. Phys.*, **73**: 3480-3485 (1980).
303. Surovtsev, S. Yu. and Tovbin, Yu. K. In: *Materialy II Vsesoyuz. konf. "Nestatsionarnye protsessy v katalize"* (Materials of All-Union Conference "Nonstationary Processes in Catalysis"), Part 1, pp. 10-13. Novosibirsk: IK SO AN SSSR (1983).
304. Tovbin, Yu. K. In: *Fizika mezhfaznykh yavlenii: Sb. nauch. st.* (Physics of Interphase Phenomena: Collected Scientific Articles), pp. 30-37. Nal'chik, KBGU (1981).
305. Tovbin, Yu. K. and Fedyanin, V. K. *Kinetika i Kataliz*, **19**: 989-996 (1978).
306. Fedyanin, V. K. *Kinetika i Kataliz*, **12**: 1353-1355 (1971).

307. Bonzel, H. P. and Ru, R. *Surface Sci.*, **33**: 91-106 (1972).
308. Kuchaev, V. L. In: *Materialy Vsesoyuz. seminara "Izotopnye metody izucheniya mekhanizma kataliza"* (Materials of All-Union seminar "Isotope Methods of Studying the Mechanism of Catalysis"), Preprint A2. Novosibirsk: IK SO AN SSSR (1981).
309. Tovbin, Yu. K. In: *Materialy III Vsesoyuz. konf. "Mekhanizmy kataliticheskikh reaktsii"* (Materials of 3rd All-Union Conference "Mechanisms of Catalytic Reactions"), Part 1, pp. 13-16. Novosibirsk: IK SO AN SSSR (1982).
310. Golchet, A. and White, J. M. *J. Catal.*, **53**: 266-279 (1978).
311. Hugo, P. *Ber. Bunsenges. phys. Chem.*, **74**: 121-127 (1970).
312. Wicke, E., Kummann, P., Keil, W., *et al. Ber. Bunsenges. phys. Chem.*, **84**: 315-323 (1980).
313. Barkowski, D., Haul, R., and Kretschmer, V. *Surface Sci.*, **107**: L329-L333 (1981).
314. Turner, J. E., Sales, B. C., and Maple, M. B. *Surface Sci.*, **103**: 54-74 (1981).
315. Turner, J. E., Sales, B. C., and Maple, M. B. *Surface Sci.*, **109**: 591-604 (1981).
316. Dagonnier, R. and Nuyts, J. *J. Chem. Phys.*, **65**: 2061-2065 (1976).
317. Dagonnier, R., Dumont, M., and Nuyts, J. *J. Catal.*, **66**: 130-146 (1980).
318. Lagos, R. E., Sales, B. C., and Suhl, H. *Surface Sci.*, **82**: 525 (1979).
319. Steinbruchel, C. and Schmidt, L. D. *Surface Sci.*, **40**: 693-702 (1973).
320. Suhl, H. *Surface Sci.*, **107**: 88-100 (1981).
321. Sales, B. C., and Turner, J. E., and Maple, M. B. *Surface Sci.*, **114**: 381-394 (1982).
322. Mukesh, D., Morton, W., Kenney, C. N., *et al. Surface Sci.*, **138**: 237-257 (1984).
323. Berman, A. D. and Krylov, O. V. *Probl. Kinetiki i Kataliza*, **18**: 102-115 (1978).
324. Borker, D. and Wicke, E. *Ber. Bunsenges. phys. Chem.*, **89**: 629-633 (1985).
325. Rickert, L. *Ber. Bunsenges. phys. Chem.*, **85**: 297-299 (1981).
326. Takondis, C. G., Schmidt, L. D., and Aris, R. *Surface Sci.*, **105**: 325-333 (1981).
327. Yablonsky, G. S., Bykov, V. I., and Elokhin, V. I. *Kinetika model'nykh reaktsii geterogennogo kataliza* (The Kinetics of Model Reactions of Heterogeneous Catalysis). Novosibirsk: Nauka (1984).
328. Norton, P. R., Bindner, P. E., Griffiths, K., *et al. J. Chem. Phys.*, **80**: 3859-3865 (1984).
329. Cox, M. P., Ertl, G., and Imbihl, R. *Phys. Rev. Lett.*, **54**: 1725-1728 (1985).
330. Imbihl, R., Cox, M. P., and Ertl, G. *J. Chem. Phys.*, **84**: 3519-3534 (1986).
331. Imbihl, R., Cox, M. P., Ertl, G., *et al. J. Chem. Phys.*, **83**: 1578-1587 (1985).
332. Hosler, W., Ritter, E., and Behm, R. J. *Ber. Bunsenges. phys. Chem.*, **90**: 205-208 (1986).
333. Yeates, R. C., Turner, J. E., and Gellman, A. J. *et al. Surface Sci.*, **149**: 175-190 (1985).
334. Schwartz, S. B. and Schmidt, L. D. *Surface Sci.*, **183**: L269-L278 (1987).
335. Moller, P., Wetzl, K., Eiswirth, M., *et al. J. Chem. Phys.*, **85**: 5328-5336 (1986).
336. Alkhazov, T. G. and Margolis, L. Ya. *Glubokoe kataliticheskoe okislenie organicheskikh veshchestv* (The Deep Catalytic Oxidation of Organic Substances). Moscow: Khimiya (1985).
337. Anderson, I. S. *Proc. Roy. Soc. London A*, **185**: 69-89 (1946).
338. Rees, A. L. G. *Trans. Faraday Soc.*, **50**: 335-342 (1954).
339. Atlas, L. M. *J. Phys. and Chem. Solids*, **29**: 91-100 (1968).

340. Tovbin, Yu. K. and Pindyurin, V. A. In: *Abstracts of Contributions Int. School on Physics of Ionic Solvation*, Lvov, 1983, p. 108. Kiev: ITP, Acad. of Sci., USSR (1983).
341. Tretyakov, Yu. D. *Fizika ferritov* (The Physics of Ferrites). Leningrad: Khimiya (1967).
342. Kofstad, P. and Hed, A. Z. *J. Electrochem. Soc.*, **115**: 102-104 (1968).
343. Lau, K. H. and King, S. C. *Phys. Rev. Lett.*, **44**: 1222-1225 (1980).
344. Kanomori, J. and Kawakami, K. *Solid State Commun.*, **39**: 733-735 (1981).
345. Ying, S. C. and Roelofs, L. D. *Surface Sci.*, **125**: 218-226 (1983).
346. Inaoka, T. and Yoshimori, A. *Surface Sci.*, **115**: 301-308 (1982).
347. Inaoka, T. and Yoshimori, A. *Surface Sci.*, **149**: 241-255 (1985).
348. Zhdanov, V. P. *Kinetika i Kataliz*, **27**: 597-600 (1986).
349. Ma, S.-K. *Modern Theory of Critical Phenomena.* Reading (1976).
350. Tovbin, Yu. K. and Fedyanin, V. K. *Zh. Fiz. Khimii*, **54**: 3127-3131 (1980).
351. Kondratyev, V. N. and Nikitin, E. E. *Khimicheskie protsessy v gazakh* (Chemical Processes in Gases). Moscow: Nauka (1981).
352. Dovesi, R., Pisani, C., Ricca, F. *et al. Surface Sci.*, **77**: 409-415 (1978).
353. Upton, T. H. and Goddard, W. A. *Phys. Rev. Lett.*, **42**: 472-476 (1979).
354. Hellsing, B. and Mallo, A. *Surface Sci.*, **144**: 336-346 (1984).
355. Tovbin, Yu. K. In: *Extended Abstracts 37th Meeting Int. Soc. of Electrochemistry*, Vol. 3, pp. 566-568. Vilnius (1986).
356. Fedyanin, V. K. *Kinetika i Kataliz*, **10**: 1391-1392 (1969).
357. Kikuchi, R. *J. Chem. and Phys. Solids*, **20**: 17-34 (1961).
358. Amandykov, K. K. and Gurov, K. P. *Fizika Metallov i Metallovedenie*, **38**: 903-911 (1974).
359. Barker, J. A. *J. Chem. Phys.*, **20**: 1526-1531 (1952).
360. Elenin, G. G. and Krylov, V. V. *Differents. Uravneniya*, **19**: 1223-1230 (1983).
361. Elenin, G. G., Krylov, V. V., and Tovbin, Yu. K. *Differents. Uravneniya*, **19**: 1230-1235 (1983).
362. Khazin, L. G. Private communication (1985).
363. Tovbin, Yu. K. and Fedyanin, V. K. *Teoret. i Mat. Fizika*, **24**: 129-133 (1975).
364. Yates, J. T., Jr. and Madey, T. E. *J. Chem. Phys.*, **54**: 4969-4978 (1971).
365. Glazov, V. M. and Pavlova, L. M. *Khimicheskaya termodinamika i fazovye ravnovesiya* (Chemical Thermodynamics and Phase Equilibria). Moscow: Metallurgiya (1981).
366. Langmuir, I. *J. Amer. Chem. Soc.*, **54**: 2798-2832 (1932).
367. Kobozev, N. I. and Geldfeld, Yu. M. *Zh. Fiz. Khimii*, **10**: 261-278 (1937).
368. Geldfeld, Yu. M. and Kobozev, N. I. *Zh. Fiz. Khimii*, **15**: 275-295 (1941).
369. Roberts, J. K. *Proc. Roy. Soc. London A*, **161**: 141-153 (1937).
370. Roberts, J. K. *Some Problems in Adsorption.* Cambridge University Press (1939).
371. Chang, T. S. *Proc. Cambridge Philos. Soc.*, **35**: 70-83 (1939).
372. Okamoto, G., Huriuti, J., and Hirota, K. *Sci. Pap. Inst. Phys. Chem. Res.*, **29**: 223-251 (1936).
373. Trapnell, B. M. W. *Proc. Roy. Soc. London A*, **206**: 39-50 (1951).
374. Couper, A. and Eley, D. D. *Proc. Roy. Soc. London A*, **211**: 536-543 (1952).

375. Laidler, K. I. *J. Phys. Chem.*, **53**: 318-319 (1953).
376. Temkin, M. I. *Kinetika i Kataliz*, **13**: 555-565 (1972).
377. Snagovsky, Yu. S., Khomenko, A. A., Ivanova, R. A., Shub, F. S., and Temkin, M. I. *Kinetika i Kataliz*, **18**: 942-949 (1978).
378. Peshev, O. *Uspekhi Khimii*, **35**: 1830-1852 (1966).
379. Volkenshtein, F. F. *Kinetika i Kataliz*, **12**: 176-184 (1971).
380. Toya, T. *J. Vac. Sci. and Technol.*, **9**: 890 (1972).
381. Goymour, C. G. and King, D. A. *J. Chem. Soc.*, *Faraday Trans.* Part I, **69**: 749-760 (1973).
382. Goymour, C. G. and King, D. A. *J. Chem. Soc.*, *Faraday Trans.* Part I, **69**: 736-748 (1973).
383. Alnot, M., Ehrhardt, J. J., Fusy, J., and Cassuto, A. *Surface Sci.*, **46**: 81-100 (1974).
384. Ducros, R., Alnot, M., Ehrhardt, J. J., Housley, M., Pignard, G., and Cassuto, A. *Surface Sci.*, **94**: 154-168 (1980).
385. Alnot, M., Fusy, J., and Cassuto, A. *Surface Sci.*, **72**: 467-484 (1978).
386. Wilf, M. and Dawson, P. T. *Surface Sci.*, **65**: 399-418 (1977).
387. Bowker, M. *Surface Sci.*, **100**: L472-L474 (1980).
388. Jones, R. C. and Perry, D. L. *Surface Sci.*, **71**: 59-74 (1978).
389. Jones, R. C. and Perry, D. L. *Surface Sci.*, **82**: 540-548 (1979).
390. Zhdanov, V. P. *Poverkhnost'. Fizika, Khimiya, Mekhanika*, **10**: 23-28 (1984).
391. Bridge, M. E. and Lambert, R. M. *J. Catal.*, **46**: 143-150 (1977).
392. Bridge, M. E. and Lambert, R. M. *Proc. Roy. Soc. London A*, **370**: 545-560 (1980).
393. Zhdanov, V. P. *Surface Sci.*, **133**: 469-483 (1983).
394. Ichi, S., Asada, H., and Ohno, Y. *Surface Sci.*, **179**: L441-L445 (1986).
395. Adams, D. L. *Surface Sci.*, **42**: 12-36 (1974).
396. Tovbin, Yu. K. Adsorbtsiya na ideal'noi regulyarnoi strukture pri vzaimodeistvii mezhdu adatomami (Adsorption on an Ideal Regular Structure with Interaction between the Adatoms). *Synopsis of Candidate's Thesis.* Moscow (1974).
397. Benziger, J. B. *J. Chem. Soc.*, *Faraday Trans.* Part I, **76**: 49-59 (1980).
398. Zhdanov, V. P. and Zamaraev, K. I. *Uspekhi Fiz. Nauk*, **149**: 635-670 (1986).
399. Tovbin, Yu. K. and Fedyanin, V. K. In: *Materialy III Vsesoyuz. konf. po kinetike geterogenno-katalit. reaktsii* (Materials of 3rd All-Union Conference on Kinetics of Heterogeneous Catalytic Reactions), Vol. 2, pp. 430-440. Kalinin (1980).
400. Benziger, J. B. and Madix, R. J. *Surface Sci.*, **79**: 394-412 (1979).
401. Zhdanov, V. P. *Kinetika i Kataliz*, **22**: 253-255 (1981).
402. Zhdanov, V. P. *Surface Sci.*, **102**: L35-L39 (1981).
403. Zhdanov, V. P. *Surface Sci.*, **111**: 63-79 (1981).
404. Zhdanov, V. P. *Surface Sci.*, **123**: 106-116 (1982).
405. Zhdanov, V. P. *Surface Sci.*, **137**: 515-526 (1984).
406. Stiles, M. and Metiu, H. *Chem. Phys. Lett.*, **128**: 337-342 (1986).
407. Silverberg, M. and Ben-Shul, A. *Chem. Phys. Lett.*, **134**: 491-496 (1987).
408. Tovbin, Yu. K. In: *Materialy II Vsesoyuz. konf. "Nestatsionarnye protsessy v katalize"* (Materials of All-Union Conference "Nonstationary Processes in Catalysis"), Part I, pp. 132-135 Novosibirsk: IK SO AN SSSR (1983).

409. Zhdanov, V. P. *Surface Sci.*, **148**: L691-L693 (1984).
410. Zhdanov, V. P. *Poverkhnost'. Fizika, Khimiya, Mekhanika*, **1**: 14-22 (1986).
411. Kaza, K. R. and Sunderesan, S. *Chem. Eng. Commun.*, **32**: 333-355 (1985).
412. Nagai, K. *Surface Sci.*, **176**: 193-218 (1986).
413. Zhdanov, V. P. *Surface Sci.*, **171**: L461-L463 (1986).
414. Zhdanov, V. P. *Surface Sci.*, **169**: L278-L280 (1986).
415. Hori, G. K. and Schmidt, L. D. *J. Catal.*, **38**: 335-350 (1975).
416. Zhdanov, V. P. *Surface Sci.*, **169**: 1-13 (1986).
417. Christmann, K., Ertl, G., and Pignet, T. *Surface Sci.*, **54**: 365-392 (1976).
418. Zhdanov, V. P., Sobolev, V. I., and Sobyanin, V. A. *Surface Sci.*, **175**: L747-L752 (1986).
419. Benziger, J. B. and Schoof, G. V. *J. Phys. Chem.*, **88**: 4439-4444 (1984).
420. Kustov, V. L. and Pekarchuk, S. B. *Zh. Fiz. Khimii*, **57**: 2962-2966 (1983).
421. Bowker, M. and King, D. A. *Surface Sci.*, **71**: 583-598 (1978).
422. Chumak, A. A. and Tarasenko, A. A. *Surface Sci.*, **91**: 694-706 (1980).
423. Reed, D. A. and Ehrlich, G. *Surface Sci.*, **102**: 588-609 (1981).
424. Tarasenko, A. A. and Chumak, A. A. *Poverkhnost'. Fizika, Khimiya, Mekhanika*, **2**: 35-43 (1983).
425. Johansson, P. K. *Chem. Phys. Lett.*, **65**: 366-379 (1979).
426. Leuthausser, U. *Ztschr. Phys. B*, **37**: 65-67 (1980).
427. Pfnur, H., Feulner, P., Engelhardt, H. A. *et al. Chem. Phys. Lett.*, **59**: 481-486 (1978).
428. Zhdanov, V. P. *Surface Sci.*, **III**: L662-L666 (1981).
429. Gijzeman, O. L. J., Zandvoort, M. M. J., von Labohm, F., *et al. J. Chem. Soc., Faraday Trans.* Part II, **80**: 771-782 (1984).
430. Christmann, K. *Ztschr. Naturforsch. A*, **34**: 22-31 (1979).
431. Oguz, E. *Surface Sci.*, **134**: 777-802 (1983).
432. Asada, H. *Surface Sci.*, **166**: 271-282 (1986).
433. Coulomb, J. P., Bienfalt, M., and Thore, P. *Phys. Rev. Lett.*, **42**: 733-735 (1979).
434. Tringides, M. and Gomer, R. *Surface Sci.*, **155**: 254-278 (1985).
435. Żhdanov, V. P. *Surface Sci.*, **177**: L896-L900 (1985).
436. Stechel, E. B. and Knotek, M. L. *Surface Sci.*, **167**: 297-312 (1986).
437. Zhdanov, V. P. *Surface Sci.*, **165**: L31-L34 (1986).
438. King, D. A. and Wells, M. G. *Proc. Roy. Soc. London A*, **339**: 245-269 (1974).
439. Singh-Bparai, S. P., Bowker, M., and King, D. A. *Surface Sci.*, **53**: 55-73 (1975).
440. Cassuto, A. and King, D. A. *Surface Sci.*, **102**: 388-404 (1981).
441. Nagai, K. *Phys. Rev. Lett.*, **54**: 2159-2162 (1985).
442. Nagai, K. and Kirushima, A. *Chem. Phys. Lett.*, **118**: 401-404 (1985).
443. Opila, R. and Gomer, R. *Surface Sci.*, **112**: 1-22 (1981).
444. Opila, R. and Gomer, R. *Surface Sci.*, **127**: 569-597 (1983).
445. Asada, H. *J. Res. Inst. Catal. Hokkaido Univ.*, **30**: 55-59 (1982).
446. Asada, H. *J. Res. Inst. Catal. Hokkaido Univ.*, **31**: 1-6 (1983).
447. Snagovsky, Yu. S., Vardomsky, E. Yu., and Zyskin, A. G. *Teoret. i Eksper. Khimiya*, **10**: 480-486 (1974).
448. Tovbin, Yu. K. *Teoret. i Eksper. Khimiya*, **17**: 28-34 (1981).
449. Bridge, M. E. and Lambert, R. M. *Surface Sci.*, **63**: 315-324 (1977).

450. Tarasenko, A. A. *Poverkhnost'. Fizika, Khimiya, Mekhanika*, 5: 29-35 (1985).
451. Tarasenko, A. A. and Chumak, A. A. *Fiz. Tv. Tela*, **22**: 2939-2944 (1980).
452. Tarasenko, A. A. and Chumak, A. A. *Fiz. Tv. Tela*, **24**: 2972-2977 (1982).
453. Asada, H. and Masuda, M. *Surface Sci.*, **99**: L429-L433 (1980).
454. Zwerger, W. *Ztschr. Phys. B*, **42**: 333-347 (1981).
455. Zhdanov, V. P. *Surface Sci.*, **133**: 469-483 (1983).
456. Zhdanov, V. P. *Surface Sci.*, **149**: L13-L17 (1985).
457. Zhdanov, V. P. *Poverkhnost'. Fizika, Khimiya, Mekhanika*, 3: 20-27 (1985).
458. Sadig, A. and Binder, K. *Surface Sci.*, **128**: 350-382 (1983).
459. Tringides, M. and Gomer, R. *Surface Sci.*, **145**: 121-144 (1984).
460. Tringides, M. and Gomer, R. *Surface Sci.*, **166**: 419-439 (1986).
461. Zhdanov, V. P. and Mordvintsev, Yu. N. *Poverkhnost'. Fizika, Khimiya, Mekhanika*, 9: 45-48 (1986).
462. Tovbin, Yu. K. *Poverkhnost'. Fizika, Khimiya, Mekhanika*, **10**: 45-50 (1982).
463. Sales, J. L. and Zgrablich, G. *Phys. Rev. B — Solid State*, **35**: 9520-9528 (1987).
464. Surda, A. and Karasova, I. *Surface Sci.*, **109**: 605-620 (1981).
465. Sunderesan, S. and Kaza, K. R. *Chem. Eng. Commun.*, **35**: 1-22 (1985).
466. Sunderesan, S. and Kaza, K. R. *Surface Sci.*, **160**: 103-121 (1985).
467. Geldart, D. J. W., Kzewzer, H. J., and Rys, F. S. *Surface Sci.*, **176**: 284-294 (1986).
468. Binder, K. (ed.). *Monte Carlo Methods in Statistical Physics.* Heidelberg: Springer (1979).
469. Fedoseev, D. V., Chuzhko, R. K., and Grivtsev, A. G. *Geterogennaya kristallizatsiya iz gazovoi fazy* (Heterogeneous Crystallization from the Gas Phase). Moscow: Nauka (1978).
470. Bowker, M. and King, D. A. *Surface Sci.*, **72**: 208-212 (1978).
471. Murch, E. *Philos. Mag. A*, **43**: 871-882 (1981).
472. Natori, A. and Ohtsubo, H. *Surface Sci.*, **171**: 13-44 (1986).
473. Chen, J.-R. and Gomer, R. *Surface Sci.*, **79**: 413-444 (1979).
474. Reed, D. A. and Ehrlich, G. *Surface Sci.*, **105**: 603-628 (1981).
475. Zeizi, Y., Redondo, A., and Goddard, W. A. *Surface Sci.*, **131**: 221-238 (1983).
476. Redondo, A., Zeizi, Y., and Goddard, W. A. *Surface Sci.*, **136**: 41-58 (1984).
477. Mintz, M. H., Atzmony, U., and Shamir, N. *Surface Sci.*, **185**: 413 (1987).
478. Jones, R. G. and Tong, A. W.-L. *Surface Sci.*, **188**: 87 (1987).
479. Gutleben, H. and Bechtold, E. *Surface Sci.*, **191**: 157 (1987).
480. Mak, C. H., Brand, J. L., Kochler, B. G., and George, S. M. *Surface Sci.*, **191**: 108 (1987).
481. Pak, H. and Evans, J. W. *Surface Sci.*, **186**: 550 (1987).
482. Luque, J. J. and Cordoba, A. *Surface Sci.*, **187**: 611 (1987).
483. Evans, J. W., Hoffman, D. R., and Pak, H. *Surface Sci.*, **192**: 475 (1987).
484. Paunov, M. and Bauer, E. *Surface Sci.*, **188**: 123 (1987).
485. Zhdanov, V. P. *Surface Sci.*, **179**: L57 (1987).
486. Zhdanov, V. P. *Poverkhnost'. Fizika, Khimiya, Mekhanika*, **8**: 5 (1987).
487. Sobyanin, V. A. and Zhdanov, V. P. *Surface Sci.*, **181**: 1163 (1987).
488. Joyce, K., Grout, P. J., and March, N. H. *Surface Sci.*, **181**: L141 (1987).
489. Beyeler, H. V., Brüesch, P., Pietroner, L., Schneider, W. R., Strassler, S., and

Zeller, H. R. In: M. B. Salamon (ed.). *Physics of Superionic Conductors.* Berlin: Springer (1979).
490. Tovbin, Yu. K. and Kolchin, A. M. *Zh. Fiz. Khimii*, **62**: 3306-3311 (1988).
491. Tovbin, Yu. K. and Kuznetsova, O. A. *Zh. Fiz. Khimii*, **64**: 426-433 (1990).
492. Tovbin, Yu. K. and Pak, V. M. *Zh. Fiz. Khimii*, **63**: 788-790 (1989).
493. Tovbin, Yu. K. and Pindyurin, V. A. *Zh. Fiz. Khimii*, **63**: 785-788 (1989).
494. Tovbin, Yu. K. and Surovtsev, S. Yu. *Zh. Fiz. Khimii*, **62**: 3354-3356 (1988).
495. Tovbin, Yu. K. and Chelnokova, O. V. *Zh. Fiz. Khimii*, **63**: 2556-2559 (1989).
496. Tovbin, Yu. K. and Chelnokova, O. V. *Izv. AN SSSR*, *Ser. Khimiya*, 9: 1943-1948 (1989).
497. Tovbin, Yu. K. *Progress in Surface Sci.*, v. 34, 1991.

INDEX

To the Reader

Mir Publishers would be grateful for your comments on the content, translation and design of this book. We would also be pleased to receive any other suggestions you may wish to make.

Our address is:

Mir Publishers
2 Pervy Rizhsky Pereulok
I-110, GSP, Moscow, 129820
USSR